Lecture Notes in Control and Information Sciences

Edited by M. Thoma and A. Wyner

Vol. 117: K.J. Hunt
Stochastic Optimal Control Theory
with Application in Self-Tuning Control
X, 308 pages, 1989.

Vol. 118: L. Dai
Singular Control Systems
IX, 332 pages, 1989

Vol. 119: T. Başar, P. Bernhard
Differential Games and Applications
VII, 201 pages, 1989

Vol. 120: L. Trave, A. Titli, A. M. Tarras
Large Scale Systems:
Decentralization, Structure Constraints
and Fixed Modes
XIV, 384 pages, 1989

Vol. 121: A. Blaquière (Editor)
Modeling and Control of Systems
in Engineering, Quantum Mechanics,
Economics and Biosciences
Proceedings of the Bellman Continuum
Workshop 1988, June 13–14, Sophia Antipolis, France
XXVI, 519 pages, 1989

Vol. 122: J. Descusse, M. Fliess, A. Isidori,
D. Leborgne (Eds.)
New Trends in Nonlinear Control Theory
Proceedings of an International
Conference on Nonlinear Systems,
Nantes, France, June 13–17, 1988
VIII, 528 pages, 1989

Vol. 123: C. W. de Silva, A. G. J. MacFarlane
Knowledge-Based Control with
Application to Robots
X, 196 pages, 1989

Vol. 124: A. A. Bahnasawi, M. S. Mahmoud
Control of Partially-Known
Dynamical Systems
XI, 228 pages, 1989

Vol. 125: J. Simon (Ed.)
Control of Boundaries and Stabilization
Proceedings of the IFIP WG 7.2 Conference
Clermont Ferrand, France, June 20–23, 1988
IX, 266 pages, 1989

Vol. 126: N. Christopeit, K. Helmes
M. Kohlmann (Eds.)
Stochastic Differential Systems
Proceedings of the 4th Bad Honnef Conference
June 20–24, 1988
IX, 342 pages, 1989

Vol. 127: C. Heij
Deterministic Identification
of Dynamical Systems
VI, 292 pages, 1989

Vol. 128: G. Einarsson, T. Ericson,
I. Ingemarsson, R. Johannesson,
K. Zigangirov, C.-E. Sundberg
Topics in Coding Theory
VII, 176 pages, 1989

Vol. 129: W. A. Porter, S. C. Kak (Eds.)
Advances in Communications and
Signal Processing
VI, 376 pages, 1989.

Vol. 130: W. A. Porter, S. C. Kak,
J. L. Aravena (Eds.)
Advances in Computing and Control
VI, 367 pages, 1989

Vol. 131: S. M. Joshi
Control of Large Flexible Space Structures
IX, 196 pages, 1989.

Vol. 132: W.-Y. Ng
Interactive Multi-Objective Programming
as a Framework for Computer-Aided Control
System Design
XV, 182 pages, 1989.

Vol. 133: R. P. Leland
Stochastic Models for Laser Propagation
in Atmospheric Turbulence
VII, 145 pages, 1989.

Vol. 134: X. J. Zhang
Auxiliary Signal Design in Fault
Detection and Diagnosis
XII, 213 pages, 1989

Vol. 135: H. Nijmeijer, J. M. Schumacher (Eds.)
Three Decades of Mathematical System Theory
A Collection of Surveys at the Occasion of the
50th Birthday of Jan C. Willems
VI, 562 pages, 1989

Vol. 136: J. Zabczyk (Ed.)
Stochastic Systems and Optimization
Proceedings of the 6th IFIP WG 7.1
Working Conference,
Warsaw, Poland, September 12–16, 1988
VI, 374 pages. 1989

For information about Vols. 1–116 please contact your bookseller or Springer-Verlag

Lecture Notes
in Control and Information Sciences 177

Editors: M. Thoma and W. Wyner

I. Karatzas, D. Ocone (Eds.)

Applied Stochastic Analysis

Proceedings of a US-French Workshop,
Rutgers University, New Brunswick, N.J.,
April 29 - May 2, 1991

Springer-Verlag
Berlin Heidelberg GmbH

Editors

Ioannis Karatzas
Department of Statistics
Box 10 Mathematics
Columbia University
New York, NY 10027
USA

Daniel Ocone
Department of Mathematics
Rutgers University
Hill Center
New Brunswick, N.J. 08903
USA

ISBN 978-3-540-55296-3 ISBN 978-3-540-47017-5 (eBook)
DOI 10.1007/978-3-540-47017-5

Originally published by Springer-Verlag Berlin Heidelberg New York in 1992

Typesetting: Camera ready by authors

60/3020 5 4 3 2 1 0 Printed on acid-free paper

PREFACE

This volume contains the Proceedings of a four-day *Workshop on Applied Stochastic Analysis* which took place at Rutgers University (Center for Continuing Education) from Monday, 29 April to Thursday, 2 May 1991. The Workshop covered the following topics:

(i) Singular Stochastic Control,

(ii) Queueing Networks,

(iii) The mathematical theory of *Stochastic Optimization and Filtering*, as well as associated numerical techniques,

(iv) Adaptive Control, and

(v) Estimation for Random Fields and its connections with simulated annealing, statistical mechanics, and combinatorial optimization.

The purpose of the Workshop was to help the interaction of specialists in these areas, by providing an up-to-date picture of current issues and outstanding problems. We hope that these Proceedings will convey this picture to a larger audience. Professor Robert Azencott was the plenary speaker and delivered two one-hour lectures on Simulated Annealing and Random Field Estimation. Other key speakers included Professors Alain Bensoussan, Wendell H. Fleming, J. Michael Harrison, Nikolai V. Krylov, P.R. Kumar, Pierre-Louis Lions, Etienne Pardoux and J. Michael Steele.

The effort was sponsored by the National Science Foundation, INRIA, and the CNRS under the auspices of the *US-France Collaborative Research Effort in Stochastic Control.* NSF grant NSF-INT-89-06965 provided the financial support for the US participants. We are grateful to Drs. A. Bensoussan and E. Bloch for their leadership role in promoting the scientific interaction between US and French researchers; special thanks also go to Drs. W.H. Fleming, C. Glenday, A. Manitius, and P.-L. Lions for their encouragement and support. Professor E. Pardoux was responsible for organizing the French participation in the Workshop. In organizing the Workshop, we benefited greatly from the expert help of Ms. Kathy Parker of the Mathematics Department at Rutgers, who provided valuable administrative assistance, and of Ms. M. McDonald, who helped us use the excellent facilities of the Rutgers University Center of Continuing education effectively and smoothly.

This Proceedings volume contains 21 papers that were presented at the Workshop, or submitted after invitation from participants. All contributions to the volume were refereed. We hope that, in this form, they will provide an authoritative account of current directions in Applied Stochastic Analysis.

Ioannis Karatzas

Daniel L. Ocone

List of Participants

V. ANANTHARAM, Dept. of Electrical Engineering, Cornell University, Ithaca, New York 14853, USA

R. AZENCOTT, Dépt. de Mathématiques, Batiment 425, Université Paris–Sud, 91405 Orsay, France

F. BACCELLI, INRIA, Sophia–Antipolis, 2004, route des Luciolles, 06565 Valbonne Cedex, France

V. BENEŠ, New Jersey Institute of Technology, Newark, New Jersey 07102, USA

A. BENSOUSSAN, INRIA, Domaine de Voluceau, Rocquencourt, BP 105, 78153 le Chesnay Cedex, France

P. BOUGEROL, Dépt. de Mathématiques, Université de Nancy I, BP 239, 54506 Vandoeuvre les Nancy, France

D. BRIDGE, Dept. of Mathematics, Carnegie Mellon University, Pittsburgh, Pennsylvania 15213, USA

F. CAMPILLO, INRIA, Sophia–Antipolis, 2004, route des Luciolles 06565, Valbonne Cedex, France

O. CATONI, Laboratoire de Mathématiques, Dépt. de Mathématiques et d'Informatique, Ecole Normale Superieure, 45, rue d'Ulm 75230, Paris Cedex 05, France

F. COMETS, Centre de Mathématiques Appliquées, Ecole Polytechnique, 91128 Palaiseau, France

R.W.R. DARLING, Dept. of Mathematics, University of Southern Florida, Tampa, Florida 33620–5700, USA

P. DUPUIS, Division of Applied Mathematics, Brown University, Providence, Rhode Island 02912, USA

T. DUNCAN, Dept. of Mathematics, University of Kansas, Lawrence, Kansas 66045, USA

O. ENCHEV, Dept. of Mathematics, Boston University, Boston, Massachusetts 02215, USA

W.H. FLEMING, Division of Applied Mathematics, Brown University, Box F, Providence, Rhode Island, 02912, USA

P. FLORCHINGER, Départment de Mathématiques et d'Informatique, Université de Metz, Ile de Saulcy, 57045 Metz Cedex, France

B. GIDAS, Division of Applied Mathematics, Brown University, Providence, Rhode Island, 02912, USA

J.M. HARRISON, Graduate School of Business, Stanford University, Stanford, California 94305, USA

U. HAUSSMANN, Department of Mathematics, University of British Columbia, Vancouver, British Columbia, V6T 1W5, Canada

A. HEINRICHER, Department of Mathematics, University of Kentucky, Lexington, Kentucky 40506, USA

K. HELMES, Dept. of Mathematics, University of Kentucky, Lexington, Kentucky 40506, USA

O. HIJAB, Dept. of Mathematics, Temple University, Philadelphia, PA 19122, USA

M. JAMES, Dept. of Mathematics, University of Kentucky, Lexington, Kentucky 40506, USA

I. KARATZAS, Dept. of Statistics, Columbia University, New York, New York 10027, USA

N. KRYLOV, Dept. of Mathematics, University of Minnesota, Minneapolis, Minnesota 55455, USA

P.R. KUMAR, Dept. of Electrical Eng'g & Coordinated Science Laboratory, University of Illinois, 1101 West Springfield Ave., Urbana, Illinois 61801, USA

T.G. KURTZ, Dept. of Mathematics, University of Wisconsin, 480 Lincoln Drive, Madison, Wisconsin 53706, USA

H. KUSHNER, Division of Applied Mathematics, Brown University, Box F, Providence, Rhode Island, 02912, USA

F. LE GLAND, INRIA, Sophia–Antipolis, 2004, route des Luciolles 06565, Valbonne Cedex, France

P.L. LIONS, Ceremade, UA CNRS #749, Universite Paris – Dauphine, Place du Marechal de Lattre de Tassigny, 05775 Paris Cedex 16, France

D. OCONE, Dept. of Mathematics, Rutgers University, Hill Center, New Brunswick, New Jersey, 08903, USA

E. PARDOUX, UER de Mathématiques, Université de Provence, 3 place Victor–Hugo, 13331 Marseille Cedex 3, France

B. PASIK–DUNCAN, Dept. of Mathematics, University of Kansas, Lawrence, Kansas 66045, USA

J. PICARD, Dépt. de Mathématiques Appliquées, Universite Blaise Pascal, 63177 Aubiere, France

M. PICQUÉ Dépt. de Mathématiques, Ecole Normale Superieure de Cachan, 61 Avenue de President Wilson, 94230 Cachan, France

J.P. QUADRAT, INRIA, Domaine de Voluceau, Rocquencourt, BP 105, 78153 le Chesnay Cedex, France

R. RISHEL, Dept. of Mathematics, University of Kentucky, Lexington, Kentucky 40506, USA

F. RUSSO, UER de Mathematiques, Universite de Provence, 3 place Victor–Hugo, 13331 Marseille Cedex 3, France

S. SHREVE, Dept. of Mathematics, Carnegie Mellon University, Pittsburgh, Pennsylvania 15213, USA

J.M. STEELE, Dept. of Statistics, Wharton School, University of Pennsylvania, 3000 Steinberg Hall – Dietrich Hall, Philadelphia, PA 19104–6302, USA

R. STOCKBRIDGE, Dept. of Statistics, University of Kentucky, Lexington, Kentucky 40506, USA

M. TAKSAR, Dept. of Applied Mathematics & Statistics, State University of New York, Stonybrook, New York 11794, USA

A. TROUVÉ, LMENS–DIAM, Ecole Normale Superieure, 45 rue d'Ulm, 75230 Paris Cedex 05, France

R. VANDERBEI, AT&T Bell Laboratories, 600 Mountain Ave., Murray Hill, New Jersey 07974, USA

R. WILLIAMS, Dept. of Mathematics, University of California–San Diego, La Jolla, California 92093, USA

Q. ZHANG, Faculty of Management, 246 Bloor Street West, University of Toronto, Toronto, Ontario, M5S 1V4, Canada

H. ZHU, Division of Applied Mathematics, Brown University, Box F, Providence, Rhode Island, 02912, USA

TABLE OF CONTENTS

Estimates of Cycle Times in Stochastic Petri Nets

François BACCELLI
and
Panagiotis KONSTANTOPOULOS
INRIA Sophia Antipolis
2004 route des Lucioles, 06565 Valbonne, France

Abstract

This paper focuses on the derivation of bounds and estimates for cycle times of strongly connected stochastic event graphs with i.i.d. holding times. We use association properties satisfied by partial sums of the holding times in order to prove that the firing epochs compare for stochastic ordering with the last birth in a multitype branching process, the structure of which is determined from the characteristics of the event graph using simple algebraic manipulations. Classical large deviation estimates are then used to compute the growth rate of this last birth epoch, following the method of Kingman and Biggins. The method allows one to derive a computable upper bound for the cycle time, and is exemplified on tandem queueing networks with communication blocking.

1 Introduction

This paper focuses on computational problems arising in the analysis of Stochastic Decision Free Petri Nets (SDFPN). The SDFPNs under consideration consist of a subclass of stochastic Petri nets, and are also called marked graphs or event graphs in the literature. A brief description of such networks together with the basic stochastic ordering concepts to be used in the paper are provided in §2 and §3.

The precise aim of this paper is the derivation of bounds for cycle times of strongly connected stochastic event graphs with i.i.d. holding times. Lower bounds based on convex ordering are already available (see [4]) under general statistical assumptions, and we will here focus on the derivation of computable upper bounds.

The derivation of our results is based on the evolution equations satisfied by firing times that were established in [1]. Section 4 summarizes related results on the association of the firing times that were obtained in [4], and that are of use in the estimates of the following sections

In §5, we first treat a simple example that allows one to introduce the techniques used in the paper. The general case is considered in §6. We use association properties satisfied by partial sums of the holding times in order to prove that the firing epochs are bounded from above for the stochastic ordering by the last birth in a multitype branching process, the structure of which is determined from the characteristics of the event graph. Classical large deviation estimates are then used to compute the growth rate of this last birth epoch, following the method of Kingman and Biggins. The method is exemplified on two simple examples of finite capacity tandem queueing networks with communication blocking.

This method also provides a way for analyzing the stability region of non strongly connected event graphs using the stability theorem based on the comparison of the cycle times of the strongly connected components of the graph proved in [1].

Large deviation results were already used in the literature for analyzing the growth of the longest branch in random graphs arising in computer science applications (see in particular [8]); Estimates of the type used in the uniform bounds of §6 were also shown to be useful for analyzing the stability region of certain models with infinitely many resources (see [2]). The contributions of the present paper are to point out that the method of Kingman ([12]) and Biggins ([7]) can be applied to the estimation of the cycle time (and hence to the stability region) of any finite stochastic event graph, and to provide the algebraic transformations that should be operated on the structure of the event graph in order to generate the relevant multitype branching process.

2 Notation and Definitions

2.1 Model Description

The basic model of this paper is a Stochastic Decision Free Petri Net with recycled transitions. We assume that tokens incur no sojourn times in places. The definition of this class of Petri net is sketched below (see [1] for more details on the matter).

- $\mathcal{T} = \{1, \ldots, I\}$: the set of transitions;
- $\pi(\cdot)$: the predecessor function ($\pi(j)$ is the set of transitions preceding j);
- $\sigma(\cdot)$: the successor function ($\sigma(j)$ the set of transitions that follow transition j);
- $\Gamma = (\mathcal{V}, \mathcal{E})$: the directed graph defined by the precedence relation π on the set $\mathcal{V} = \mathcal{T}$;
- $\mathcal{P}$: the set of places. Each place is preceded and followed by exactly one transition (this is the so called decision free property). It is assumed that there is at most one place between two transitions. There is a place between j and j' iff $(j, j') \in \mathcal{E}$; this place will be denoted (j, j');
- $\mu(j, j') \in \mathbb{N}$: the initial marking in place $(j, j') \in \mathcal{P}$.
- $M = \max_{(j,j') \in \mathcal{P}} \mu(j, j')$ is the maximum initial marking;
- $\alpha_j(k) \in \mathbb{R}^+$ is the holding time of the k-th firing of transition $j \in \mathcal{T}$, $k \geq 1$, namely the time it takes for transition j to fire when it is enabled for the k-th time.

All the transitions are assumed to be FIFO (see [1]). A simple condition for a transition to be FIFO is that it be *recycled* (vz., for all $1 \leq j \leq I$, $j \in \pi(j)$ and the place (j, j) has an initial marking $\mu(j, j) = 1$). A transition with constant holding times is FIFO too. The evolution of the SDFPN is characterized by the circulation of *tokens*, which stay in places, and are consumed and created by transitions. A transition j is enabled to *fire* when there is at least one token in each of the places (i, j), $i \in \pi(j)$. The firing consumes one token of each of these places and creates, after some holding time $\alpha_j(k)$, $k \geq 1$, one token into each of the places (j, j'), $j' \in \sigma(j)$. We assume that the firing of a transition, takes place as soon as it is enabled. In the literature, SDFPN's are also called *marked graphs* ([10]) or *event graphs* ([1,9]).

Without loss of generality, we can assume that the SDFPN is connected. Moreover, in order to guarantee the *liveness* of the SDFPN (i.e., each transition fires infinitely many times), we assume that for each cycle in the graph Γ, there is at least one place with a positive initial marking ([10]).

2.2 Statistical Assumptions and Definitions

Throughout this paper, we assume that the sequences $\{\alpha_j(k)\}_{k=1}^{+\infty}$, $j = 1, \ldots, I$, are mutually independent sequences of i.i.d. non-negative and integrable RV's (random variables) defined on a common probability space (Ω, F, P).

We will use the notion of association of random variables, and the notion of stochastic ordering between random variables, that will be denoted $\leq_{st}$ (see [6] for the relevant definitions and for the basic properties that will be used in the sequel). The use that will be made of these two notions is essentially that advocated in [5]. Equivalence in law will be denoted $=_{st}$.

3 Evolution Equations

In this subsection, we summarize results that were obtained in [1] In comparison to the results of this reference, we limit ourselves to the case of holding times in transitions. Let $X_j(k)$ denote the time when transition j starts firing for the k-th time. Whenever the initial condition is zero, these variables satisfy the evolution equation

$$X_j(k) \quad = \max_{\{i \in \pi(j)\}} (X_i(k - \mu(i,j)) + \alpha_i(k - \mu(i,j))), \quad k = 1, 2, \ldots, \tag{1}$$

where, by convention, the maximum over an empty set is $-\infty$. The way to define the initial condition of this equation is somewhat intricate, and we will skip this question for this formulation of the evolution equation.

Since the decision free net is live, the numbering of the transitions can be chosen in such a way that for all (j,k), $j = 1 \ldots, I$, $k \geq 1$, the variables $X_{j'}(k')$ that are found in the R.H.S of (1) are always such that either $k' < k$ or $k' = k$, $j' < j$. Therefore, the state variables $X_j(k)$ can be computed recursively in the order

$$X_1(1), X_2(1), \ldots, X_I(1), X_1(2), X_2(2), \ldots, X_I(2), \ldots, X_1(k), X_2(k), \ldots, X_I(k), \ldots. \tag{2}$$

Consider the semi-ring $(I\!R, \oplus, \otimes)$, where $\oplus$ is max and $\otimes$ is $+$. If the SDFPN under consideration is live, it is shown in [1] that one can rewrite this equation in matrix form in this semi-ring

$$X(k) = X(k - M) \otimes A(k - M, k) \oplus \cdots \oplus X(k-1) \otimes A(k-1, k), \; k = 1, 2, \ldots. \tag{3}$$

In this equation, the row vector $X(k) = (X_1(k), \ldots, X_J(k))$, $J \leq I$, is the restriction of $(X_1(k), \ldots, X_I(k))$ to the set of transitions followed by at least one place with at least one token in the initial marking.

The initial condition is $X(0)$. This initial condition can for instance be taken equal to $(0, \ldots, 0)$. Any other licit initial condition $X(0) \in \mathbb{R}^I$ will lead to the same cycle times. Hence, we will assume in the sequel without loss of generality that $X(0) = 0$.

In this equation, the matrix $A(k-l, k)$ is defined as follows: let $\mathcal{S}(j', j, l)$ be the set of paths in the graph Γ with at least two transitions, with initial vertex j' and final vertex j, and such that the first two transitions of the path are connected by a place with initial marking equal to l, while the other transitions are connected by places with zero initial marking. Then

$$A(k-l, k)_{j',j} = \bigoplus_{\{(j'=i_0, i_1 j_2, i_2 \ldots, i_{h-1}, i_h = j) \in \mathcal{S}(j', j, l)\}} \alpha_{j'}(k-l) \otimes \left(\bigotimes_{m=1}^{h-1} \alpha_{i_m}(k) \right), \qquad (4)$$

with the usual convention if the set $\mathcal{S}(j', j, l)$ is empty. The entry $A(k-l, k)_{j',j}$ is hence simply the length of the longest path in $\mathcal{S}(j', j, l)$.

4 Basic Model

In this paper, we will consider a live stochastic event graph with all its transitions recycled. We will assume this event graph to be strongly connected. We will also assume that the initial marking is $0-1$ valued. This last assumption is not a strong one as it was shown in [1] that for any stochastic event graph, one can always find an 'equivalent' event graph (with a possibly different topology, where the places and transitions that are added all have zero holding times), where the initial marking is $0-1$-valued. The matrix $A(k-1, k)$ will be denoted $A(k)$ for sake of simplicity. In this case the entries of $A(k)$ are found from equation (3):

$$X(k) = X(k-1) \otimes A(k), \qquad (5)$$

with initial condition $X(0)$, and formula (4) is written as

$$A(k)_{j',j} = \max_{\{(j'=i_0, i_1 j_2, i_2 \ldots, i_{h-1}, i_h = j) \in \mathcal{S}(j', j, 1)\}} \alpha_{j'}(k-1) + \alpha_{i_1}(k) + \ldots + \alpha_{i_{h-1}}(k). \qquad (6)$$

5 Statistical Properties

Lemma 5.1 *Under the foregoing statistical assumptions,*

$$\{A(k)_{i,j},\ X_j(k),\ i,j = 1,\ldots,J,\ k \geq 0\}$$

forms a set of associated random variables.

Proof. It is easy to prove that the i.i.d. assumption on the α's implies that the random variables $\{A(k-1,k)_{i,j},\ k \in \mathbb{N},\ i,j = 1,\ldots,J\}$ are associated in view of the fact that they are obtained as an increasing function of associated random variables (see (4)). The result for $X_j(k)$ follows immediately from (1). ∎

Lemma 5.2 *For all* $j_0, j_1, \ldots, j_n, \ldots \in \{1,\ldots,J\}$, *the random variables* $\{A(n)_{j_n,j_{n+1}},\ n \in \mathbb{N}\}$ *are mutually independent.*

Proof. The random variables $A(n)_{j_n,j_{n+1}}$ and $A(m)_{j_m,j_{m+1}}$ are clearly independent for $m > n+1$ in view of Equation (4) (for instance, $A(n)$ is measurable with respect to the σ-field endowed by the random variables $\alpha(n)$ and $\alpha(n-1)$, while $A(n+2)$ is measurable with respect to the one endowed by the random variables $\alpha(n+2)$ and $\alpha(n+1)$, and the independence follows from the independence assumptions on the α's).

We show that this independence property also holds for the random variables $A(n)_{j_n,j_{n+1}}$ and $A(n+1)_{j_{n+1},j_{n+2}}$. By looking at equation (6) we see that $A(n)_{j_n,j_{n+1}}$ is a function of $\alpha(n-1)$ and $\alpha(n)$, while $A(n+1)_{j_{n+1},j_{n+2}}$ is a function of $\alpha_{j_{n+1}}(n)$ and $\alpha(n+1)$. By assumption $\{\alpha(n)\}_{n=1}^{\infty}$ is an independent sequence, so the only possible source of dependency of $A(n+1)_{j_{n+1},j_{n+2}}$ and $A(n)_{j_n,j_{n+1}}$ is the existence of the term $\alpha_{j_{n+1}}(n)$ in the expression (6) for $A(n)_{j_n,j_{n+1}}$. Recalling the definition of $S(j_n, j_{n+1}, 1)$ we conclude that this is possible if and only if there exists a circuit from j_{n+1} to j_{n+1} with zero initial marking in all the places of the circuit, which contradicts the liveness assumption. ∎

6 Simple Bounds on Cycle Times

Using sub-additive ergodic theory, the following lemma was shown in [1]:

Lemma 6.1 *Under the foregoing assumptions, there exists a positive constant* γ *such that*

$$\lim_{k\to\infty} X^*(k)/k = \lim_{k\to\infty} E[X^*(k)]/k = \gamma \qquad a.s., \tag{7}$$

where $X^*(k) = \max_j X_j(k)$. *In addition, if the event graph under consideration is strongly connected,*

$$\lim_{k\to\infty} E[X_j(k)]/k = \lim_{k\to\infty} X_j(k)/k = \gamma \quad a.s., \quad \forall\, j = 1, \ldots, J. \tag{8}$$

In the strongly connected case, the constant γ is called the cycle time of the Petri net. The reason for this terminology comes from the deterministic case that was studied in [9].

The basic data of the present section is a strongly connected event graph that satisfies the statistical assumptions of §2.2, and the assumption $M = 1$. Let

- N be the maximal out-degree of the transitions that are followed by at least one place with a non-zero initial marking (the out-degree of a vertex is the number of edges starting from this vertex).
- b be a random variable that is a $\leq_{st}$ upper bound of each of the random variables $A(1)_{i,j}$, namely

  $$A(1)_{i,j} \leq_{st} b, \forall\, i,j = 1, \ldots, J. \tag{9}$$

  Let $b^*(\theta) = E[e^{\theta b}]$ (since $b \geq 0$, this function is defined at least for $Re(\theta) \geq 0$).
- Let $M(x)$ be the Cramér-Legendre transform of the distribution function of the random variable b, namely

  $$M(x) = \sup_{\theta \in I\!R} \{\theta x - \log(b^*(\theta))\}. \tag{10}$$

The present section is devoted to the proof of the following result:

Proposition 6.2 *Under the foregoing assumptions, the cycle time of the event graph admits the upper bound*

$$\gamma \leq \inf\{x > Eb \text{ such that } M(x) > \log(N)\}. \tag{11}$$

We start with two preliminary lemmas.

Lemma 6.3 *For all* $\epsilon > 0$, *and for all* $j = 1, \ldots, J$,

$$\lim_{k\to\infty} P[\frac{X_j(k)}{k} < \gamma + \epsilon] = 1, \tag{12}$$

and

$$\lim_{k\to\infty} P[\frac{X_j(k)}{k} < \gamma - \epsilon] = 0. \tag{13}$$

Proof The property follows immediately from the fact that a.s. convergence implies convergence in probability and from (8). ∎

Lemma 6.4 *If $\beta \in I\!R$ is such that*

$$\lim_{k\to\infty} P[X_j(k) - k\beta \le 0] = 1, \tag{14}$$

for some $j = 1, \ldots, J$, then $\beta \ge \gamma$.

Proof Under the assumption (14),

$$\lim_{k\to\infty} P[\frac{X_j(k)}{k} \le \beta] = 1, \tag{15}$$

so that we cannot have $\beta = \gamma - \epsilon$ for some $\epsilon > 0$, in view of Lemma 6.3. Therefore, $\beta \ge \gamma$. ∎

Proof of Proposition 6.2 From Equation (5) with initial condition $X(0) = 0$, it is easily checked by induction that

$$\begin{aligned} X_j(k) &= \bigoplus_{j_0,\ldots,j_{k-1}\in\{1,\ldots,J\}} \bigotimes_{n=1}^{k} A(n)_{j_n,j_{n+1}} \\ &= \max_{j_0,\ldots,j_k\in\{1,\ldots,J\}} \sum_{n=1}^{k} A(n)_{j_n,j_{n+1}}, \end{aligned} \tag{16}$$

where $j_k = j$. Therefore,

$$P[X_j(k) - \beta k \le 0] = P[\max_{j_0,\ldots,j_k\in\{1,\ldots,J\}} \sum_{n=1}^{k} \tilde{A}(n)_{j_n,j_{n+1}} \le 0], \tag{17}$$

where $\tilde{A}(k)_{i,j} = A(k)_{i,j} - \beta$.

For k fixed, Lemma 6.3 implies that the variables $\sum_{n=1}^{k} \tilde{A}(n)_{j_n,j_{n+1}}$, where $j_0, \ldots, j_{k-1}$ varies over the set $\{1, \ldots, J\}^k$, are associated. Therefore (see [6])

$$P[\max_{j_0,\ldots,j_{k-1}\in\{i,\ldots,J\}} \sum_{n=1}^{k} \tilde{A}(n)_{j_n,j_{n+1}} \le 0] \ge \prod_{j_0,\ldots,j_{k-1}\in\{1,\ldots,J\}} P[\sum_{n=1}^{k} \tilde{A}(n)_{j_n,j_{n+1}} \le 0]. \tag{18}$$

Since the random variables $A(n)_{j_n,j_{n+1}}$ are independent (see Lemma 5.2), we have then

$$P[\sum_{n=1}^{k} \tilde{A}(n)_{j_n,j_{n+1}} \le 0] \ge P[\sum_{n=1}^{k} (b(n) - \beta) \le 0], \tag{19}$$

where $\{b(n)\}$ is a sequence of i.i.d. random variables with the same distribution function as b. Observe first that if $\beta < Eb$ then the R.H.S. of (18) converges to zero. This is not what we want, so we may as well assume at this point that $\beta > Eb$. Now, Chernoff's Theorem implies

$$P[\sum_{n=1}^{k} b(n) > \beta k] = e^{-M(\beta)k+o(k)}, \tag{20}$$

so that

$$P[X^j(k) - \beta k \leq 0] \ \geq \ (1 - e^{-M(\beta)k+o(k)})^{C_j(k)}, \tag{21}$$

where $C_j(k)$ denotes the number of paths $j_0, \ldots, j_{k-1}$ such that $\sum_{n=1}^{k} \tilde{A}(n)_{j_n,j_{n+1}} \neq -\infty$. (It is interesting to observe at this point that $C_j(k)$ can also be defined in terms of the adjacency matrix of the event graph in the usual path-counting way.) Therefore, if

$$C_j(k)e^{-kM(\beta)} \to 0, \tag{22}$$

when k goes to ∞, then

$$\lim_{k\to\infty} P[X^j(k) - \beta k \leq 0] = 1. \tag{23}$$

Clearly, the bound $C_j(k) \leq N^k$ holds, so that a sufficient condition for (23) to hold is $M(\beta) > \log(N)$. In other words, for β such that $M(\beta) + \log N < 0$, (23) holds, so that $\beta \geq \gamma$ in view of Lemma 6.4. ∎

In fact , we proved the following and more general result

Proposition 6.5 *If* $\log(C_j(k)) = Ck + o(k)$, *then*

$$\gamma \leq \inf\{\beta > Eb \ \text{ such that } M(\beta) > C\}. \tag{24}$$

In fact C is then necessarily the Perron-Frobenius eigenvalue of the adjacency matrix.

6.1 Example: 1-Buffer Blocking Queues

Consider the example of Figure 1, which represents a line of processors with blocking before service. Let J denote the number of processors, each of which is represented by a transition. In Figure 1, J=4. The first processor (on the left of the figure) has an infinite supply of items to serve. Between two successive processors, the buffer is of capacity two,

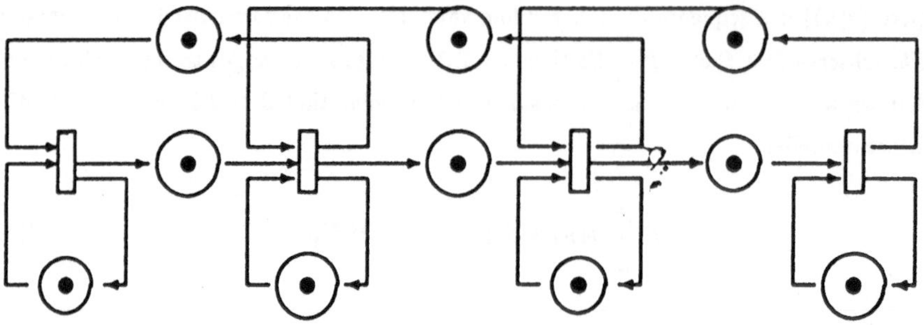

Figure 1: Communication Blocking: 4 Nodes, 1 Buffer

including the one in service (which is captured by the fact that there are two tokens in any of the upper circuits originating from a processor). The processors are single servers with a FIFO discipline (which is captured by the lower circuit associated with each transition) It is assumed that all transitions have exponentially distributed holding times with parameter 1. In this example, we have $N = 3$, $b^*(\theta) = (1-\theta)^{-1}$ and

$$\frac{1}{k}\log C_j(k) = 1 + 2\cos(\frac{\pi}{J+1}) + o(k) \leq 3 \tag{25}$$

(see the Appendix for the last formula). The Cramér-Legendre transform is given by

$$M(x) = \sup \theta \in I\!R \theta x + \log(1-\theta).$$

The derivative of the function $\theta x + \log(1-\theta)$ w.r.t. θ vanishes for $\theta = 1 - x^{-1}$ and this point is a minimum. Therefore

$$M(x) = x - \log(x) - 1.$$

As a direct application of Proposition 6.2, we get

$$\gamma \leq \inf\{x \mid x - \log(x) - 1 > \log(3)\}$$

which provides the following uniform bound in J:

$$\gamma \leq 3.33.$$

In other words, the throughput of the systems is always larger that .3, regardless of the number of processors. If we apply Proposition 6.5 using the more precise estimate of $C_j(k)$ given in (25), we get

$$\gamma \leq \inf\{x \mid 1 - x + \log(x) < \log(1 + 2\cos(\frac{\pi}{5}))\} \simeq 3.09$$

If the service times are Erlang-3 with mean 1, namely if $b^*(\theta) = (3/(3-\theta))^3$, we get in the same way from Proposition 6.2 that

$$\gamma \leq \inf\{x \mid 3(x - 1 - \log(x)) > \log(3)\} \simeq 2.11,$$

that is a throughput of .48.

7 General Case

We start with Equation (3) with $M = 1$, which reads

$$X_j(k) = \max_{i \in \pi^*(j)} \{X_i(k-1) + A(k)_{i,j}\}, \tag{26}$$

where $\pi^*(j)$ denotes the set of vertices $i \in \{1, \ldots, J\}$, such that $A(1)_{i,j} \neq -\infty$.

An age dependent branching process with J types (see [7]) is characterized by random variables $Z_{j,i}(t) \in \mathbb{R}$ counting the number of individuals of type i born in the first generation and by time $t \in \mathbb{R}^+$, from an individual of type j, where the offspring processes $Z_j(t) = (Z_{j,1}(t), \ldots, Z_{j,J}(t))$ associated with the individuals are assumed to be mutually independent in j.

Consider now the following specific age-dependent branching process associated with the stochastic event graph under consideration:

- there are as many types as transitions in the graph (more precisely as transitions followed by at least one place with a nonzero initial marking);
- the variables $Z_{j,i}(t)$ are defined as follows
  - for all j, the J stochastic processes $\{Z_{j,i}(t),\ t \in \mathbb{R}^+\}$, $i \in \{1, \ldots, J\}$ are mutually independent;
  - the law of $Z_{j,i}(t)$ is defined by $Z_{j,i}(t) =_{st} 1_{i \in \pi^*(j)} 1_{A(1)_{i,j} \leq t}$, for all $i, j \in \{1, \ldots, J\}$.

Finally, let $\hat{X}_j(k)$ be the time of the latest birth of the k-th generation in this branching process when initiated at time 0 by an individual of type j. Taking $\hat{X}_j(0) = 0$, for all j, it is easily checked that these random variables satisfy the recurrence relations

$$\hat{X}_j(k) = \max_{i \in \pi^*(j)} \hat{X}_i(k-1) + \overline{A}(k)_{i,j}. \tag{27}$$

These relations hold only because each individual can have *at most one* child of each type. For any random vector $B \in \mathbb{R}^K$, $\overline{B}$ denotes the 'product form version' of B, namely the vector defined in law by the relations $\overline{B}_j =_{st} B_j$, and such that its coordinates are mutually independent. For instance, the matrices $\overline{A}(k)$ are such that $\overline{A}(k)_{i,j} =_{st} A(k)_{i,j}$ and the random variables $\overline{A}(k)_{i,j}, k \geq 1, i, j = 1, \ldots, J$ are mutually independent.

Lemma 7.1 *Under the foregoing statistical assumptions, for all $j \in \{1, \ldots, J\}$ and $k \geq 1$,*

$$X_j(k) \leq_{st} \hat{X}_j(k). \tag{28}$$

Proof The proof is by induction on k. For $k = 1$, we have

$$\begin{aligned} X_j(1) &= \max_{i \in \pi^*(j)} A(1)_{i,j}, \\ &\leq_{st} \max_{i \in \pi^*(j)} \overline{A}(1)_{i,j}, \\ &=_{st} \hat{X}_j(1), \end{aligned} \tag{29}$$

where the first inequality comes from the association of the entries of $A(1)$ (Lemma 5.1) and from the basic theorem on the maximum of associated random variables (see [6]). Assume now that the property holds up to rank $k - 1$. Then, using the association property of lemma 5.1, we get

$$\begin{aligned} X_j(k) &= \max_{i \in \pi^*(j)} X_i(k-1) + A(k)_{i,j}, \\ &\leq_{st} \max_{i \in \pi^*(j)} \overline{X_i(k-1) + A(k)_{i,j}}, \\ &=_{st} \max_{i \in \pi^*(j)} \overline{X}_i(k-1) + \overline{A}(k)_{i,j}, \end{aligned} \tag{30}$$

where the last equivalence comes from the independence between $X_i(k-1)$ and $A(k)_{i,j}$ (see Lemma 5.2 and (16)). Now, from the independence assumption, $X_i(k-1) \leq_{st} \hat{X}_i(k-1)$ for all $i \in \{1, \ldots, J\}$. Therefore, the monotonicity of the map $x \in I\!R^J \to \max_{i \in \pi^*(j)}\{x_i + \overline{A}(k)_{i,j}\}$ implies

$$\begin{aligned} X_j(k) &\leq_{st} \max_{i \in \pi^*(j)} \hat{X}_i(k-1) + \overline{A}(k)_{i,j}, \\ &=_{st} \hat{X}_j(k). \end{aligned} \tag{31}$$

∎

let $F_{j,i}(t)$ be the monotone function defined by the relation

$$F_{j,i}(t) = E[Z_{j,i}(t)] = P[A_{i,j}(1) \leq t], \; t \in I\!R^+, \tag{32}$$

(that is, $F_{j,i}(t)$ is the distribution of the first birth time) and $\Phi(\theta)$ be the $J \times J$ matrix with entries

$$\Phi_{j,i}(\theta) = \int_0^\infty e^{\theta t} F_{j,i}(dt), \tag{33}$$

if $i \in \pi^*(j)$ and zero otherwise. where θ is some real valued parameter. Whenever the integrals defining the entries converge, the matrix $\Phi(\theta)$ is positive, and we denote by $\phi(\theta)$ its Perron-Frobenius eigenvalue. It should be observed at this point that $\Phi(0)$ is simply the 0-1 adjacency matrix of the event graph. Its Perron-Frobenius eigenvalue, $\phi(0)$, is therefore strictly larger than 1 (except in trivial cases). Let $M(x)$ be the Cramér-Legendre transform of $\phi(\theta)$:

$$M(x) = \sup \theta > 0(\theta x - \log \phi(\theta)), \tag{34}$$

It is well known that $M(x)$ is increasing for $x \geq 0$. Let Γ be defined by

$$\Gamma = \inf\{x \mid M(x) > 0\}. \tag{35}$$

We are now in a position to state the main theorem, which generalizes Proposition 6.2:

Theorem 7.2 *Under the foregoing statistical assumptions, the cycle time γ of the event graph satisfies the bound*

$$\gamma \leq \Gamma. \tag{36}$$

Proof From Lemma 7.1, for all bounded and nondecreasing function f,

$$E[f(\frac{X_j(k)}{k})] \leq E[f(\frac{\hat{X}_j(k)}{k})], \tag{37}$$

for all $j = 1, \ldots, J$.

We first prove that

$$\limsup_k \frac{[\hat{X}_j(k)]}{k} \leq \Gamma, \quad a.s., \tag{38}$$

following closely the method proposed by Biggins [7] for analyzing this type of branching processes (in fact Biggins proves that $\lim_k [\hat{X}_j(k)]/k = \Gamma$, and we only summarize here the relevant part of his paper allowing one to establish (38)).

Denoting by $Z_{j,i}^k(t) \in \mathbb{N}$ the total number of individuals of type i of the k-th generation born by time t from an initial individual of type j, we get the vectorial relation

$$E[\int_0^\infty e^{\theta t} dZ_j^{k+1}(t) | F_k dt] = (\int_0^\infty e^{\theta t} dZ_j^k(t) dt) \Phi(\theta),$$

where F_k denotes the σ-field of the events up to the k-th generation. Taking expectations in the last expression, we get the relation

$$E[\int_0^\infty e^{\theta t} dZ_{j,i}^k(t)] = \Phi_{j,i}^k(\theta), \tag{39}$$

where Φ^k denotes the k-th power of Φ. Let $v(\theta)$ be the right eigenvector associated with the maximal eigenvalue $\phi(\theta)$. We get from (39) that

$$< E[\int_0^\infty e^{\theta t} dZ_j^k(t)dt], v(\theta) > \quad = \phi^k(\theta) v_j(\theta), \tag{40}$$

so that

$$< E[\int_0^\infty e^{\theta t} dZ_j^k(t)dt], 1 > \quad \leq \phi^k(\theta) v_j(\theta) u(\theta), \tag{41}$$

where $u(\theta) = (\min_i v_i(\theta))^{-1}$ ($v(\theta)$ is strictly positive due to Perron-Frobenius). Now, since $\hat{X}_j(k) = \sup\{t \mid \exists i = 1, \ldots, J, Z_{j,i}^k(t) = 0\}$ we have

$$E[e^{\theta \hat{X}_j(k)}] \leq \quad < E[\int_0^\infty e^{\theta t} Z_j^k(dt)], 1 >, \tag{42}$$

In addition, for $\theta \geq 0$, we have the obvious bound

$$P[\frac{\hat{X}_j(k)}{k} \geq \beta] \quad \leq \quad E[e^{\theta(\frac{\hat{X}_j(k)}{k} - \beta)}]. \tag{43}$$

This plus (41) and (42) in turn imply

$$\lim_k \frac{1}{k} \log P[\hat{X}_j(k) \geq k\beta] \quad \leq \quad \sup_{\theta > 0}(\theta\beta - \log(\phi(\theta))) = -M(\beta). \tag{44}$$

Therefore, for all β such that $M(\beta) > 0$, $\sum_{k \geq 1} P[\hat{X}_j(k) \geq k\beta] < \infty$, so that the Borel-Cantelli Lemma immediately implies (38).

In view of (8), we get from Lebesgue dominated convergence theorem that for f bounded and continuous

$$\lim_k E[f(\frac{\hat{X}_j(k)}{k})] = f(\gamma). \tag{45}$$

For f continuous, monotone nondecreasing and bounded, we also have

$$\limsup_k E[f(\frac{\hat{X}_j(k)}{k})] \leq E[\limsup_k f(\frac{\hat{X}_j(k)}{k})] = E[f(\limsup_k \frac{\hat{X}_j(k)}{k})] \leq f(\Gamma), \tag{46}$$

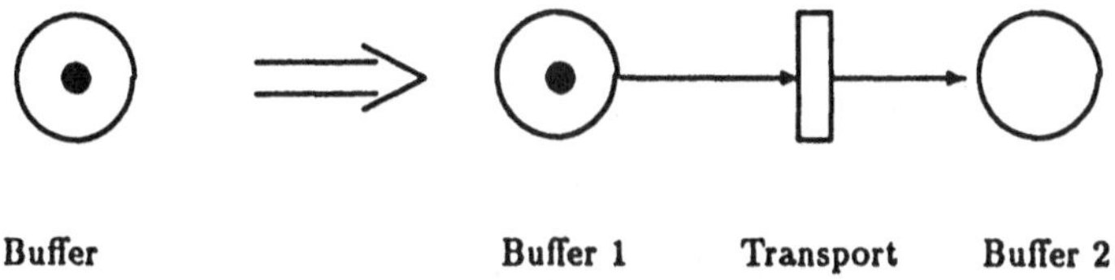

Figure 2: Blocking with Transportation Times

where we successively used Fatou's lemma, the monotonicity and continuity of f and (38). Therefore $f(\gamma) \leq f(\Gamma)$ for all nondecreasing continuous and bounded f, which immediately implies (36). ∎

Observe that in the particular case where all non $-\infty$ entries of $A(1)$ have the same distribution characterized by the function $b^*(\theta)$, the eigenvalue of interest is precisely

$$\phi(\theta) = b^*(\theta)C, \tag{47}$$

where C is the Perron-Frobenius eigenvalue of the adjacency matrix associated with the matrix A, namely the maximal eigenvalue of the matrix $\Phi(0)$.

7.1 Example: Blocking Queues with Transportation Times

The example is that of the line of processors described previously, but with a deterministic transportation time between processors. The associated event graph is obtained from that of Figure 1 by replacing each buffer by two buffers connected by a transportation transition with deterministic holding times δ as shown in Figure 2. The statistical assumptions concerning the holding times of transitions associated with processors are those of the previous example. In this example, we have

$$\Phi(\theta) = \begin{pmatrix} \frac{1}{1-\theta} & \frac{1}{1-\theta} & 0 & 0 \\ \frac{e^{\delta\theta}}{1-\theta} & \frac{1}{1-\theta} & \frac{1}{1-\theta} & 0 \\ 0 & \frac{e^{\delta\theta}}{1-\theta} & \frac{1}{1-\theta} & \frac{1}{1-\theta} \\ 0 & 0 & \frac{e^{\delta\theta}}{1-\theta} & \frac{1}{1-\theta} \end{pmatrix}. \tag{48}$$

It is proved in the Appendix that the Perron-Frobenius eigenvalue of this matrix is

$$\phi(\theta) = \frac{1}{1-\theta}\left(1 + 2e^{\frac{\delta\theta}{2}}\cos(\frac{\pi}{J+1})\right). \tag{49}$$

The technique is then the same as above for deriving the upper bound Γ. The lower bound g given in the following arrays are those obtained by convex ordering following the method indicated in [4]. For $J = 4$, one gets

δ	0	1	2	3
Γ	3.1	3.3	3.7	4.2
g	1	1.5	2	2.5

For J large, the lower bound is unchanged, and we get the following upper bound

δ	0	1	2	3
Γ	3.3	3.6	4.0	4.4

8 Appendix: 1-Buffer Blocking Queues

We give here a probabilistic solution to the Path Counting Argument of interest here. Let Q denote the adjacency matrix of the event graph, namely the $J \times J$ matrix such that $Q_{i,j} = 1$ if $i \in \pi(j)$, and zero otherwise. For $J = 4$, we get for instance

$$Q = \begin{pmatrix} 1 & 1 & 0 & 0 \\ 1 & 1 & 1 & 0 \\ 0 & 1 & 1 & 1 \\ 0 & 0 & 1 & 1 \end{pmatrix}. \tag{50}$$

It is easily checked by induction that $Q^k_{i,j}$ counts the number of paths of length k from i to j. Let M be the substochastic matrix defined by $M = Q/3$, and let λ_J denote the Perron-Frobenius eigenvalue associated with M. From the irreducibility of M, we get

$$C_j(k) = \sum_{i=1}^{J} Q^k_{i,j} = \Theta(3\lambda_J)^k.$$

In order to evaluate λ_J, we introduce the Markov chain Z_k with substochastic transition matrix M and uniform initial measure. We have then

$$P[Z_k = j] = \Theta(\lambda_J)^k$$

We can now evaluate $P[Z_k = j]$ using the recurrence relations

$$\begin{aligned} P[Z_{k+1} = j] &= P[Z_k = j-1]/3 + P[Z_k = j]/3 + P[Z_k = j+1]/3, \quad 1 < j < J, \\ P[Z_{k+1} = 1] &= P[Z_k = 1]/3 + P[Z_k = 2]/3, \\ P[Z_{k+1} = J] &= P[Z_k = J-1]/3 + P[Z_k = J]/3. \end{aligned} \tag{51}$$

Let

$$P(x,y) = \sum_{k=0}^{\infty}\sum_{j=1}^{J} x^k y^j P[Z_k = j].$$

From (51), we get

$$P(x,y) = \frac{G(y) - F(x)(1+y^{J+1})}{3 - x(y+1+y^{-1})}, \tag{52}$$

where $F(x)$ is the function

$$F(x) = x\sum_{k=0}^{\infty} P[Z_k = 1]x^k, \tag{53}$$

and

$$G(y) = 1 + y + y^2 + \ldots + y^J. \tag{54}$$

The denominator of (52) vanishes for $x = x(y) = 3/(y+1+y^{-1}) \leq 1$. Therefore, necessarily

$$F(x) = \frac{G(y)}{1+y^{J+1}}, \tag{55}$$

for $x = x(y)$. The poles of $F(x(y))$ are for $y^{J+1} = -1$, namely for

$$y(l) = e^{\frac{i\pi+2il\pi}{J+1}}, \quad l = 0,\ldots,J.$$

We have

$$x(y(l)) = \frac{3}{1+2\cos\frac{\pi(1+2l)}{J+1}}, \tag{56}$$

the smallest of which is for $l = 0$. Therefore, from classical theorems on generating functions,

$$P[Z_k = 1] = \Theta(x(y(0))^{-k},$$

or equivalently

$$\lambda_J = \frac{1+2\cos\frac{\pi}{J+1}}{3} \tag{57}$$

which in turn implies

$$C_1(k) = 1 + 2\cos\frac{\pi}{J+1}. \tag{58}$$

9 Appendix: Blocking Queues with Transportation Delay

The technique is the same as in the previous section. Let M be the substochastic matrix defined by $M = \Phi(\theta)/K$, where $K = 2/(1-\theta) + 1/(1-\theta)^2$. In order to compute the

Perron-Frobenius eigenvalue of M, we introduce the dissipative Markov chain Z_k with transition matrix M. This matrix has the form

$$M = \begin{pmatrix} \alpha & \alpha & 0 & 0 \\ \beta & \alpha & \alpha & 0 \\ 0 & \beta & \alpha & \alpha \\ 0 & 0 & \beta & \alpha \end{pmatrix}, \tag{59}$$

where $\alpha = 1/K(1-\theta)$ and $\beta = 1 - 2\alpha$. The associated functional equation with initial condition $Z_0 = 1$ reads

$$P(x,y) = \frac{y - x(\alpha F(x) + \beta y^{J+1} G(x))}{1 - x(y\beta + \alpha + y^{-1}\alpha)}, \tag{60}$$

where $F(x)$ and $G(x)$ are the functions

$$\begin{aligned} F(x) &= x \sum_{k=0}^{\infty} P[Z_k = 1] x^k, \\ G(x) &= x \sum_{k=0}^{\infty} P[Z_k = J] x^k. \end{aligned} \tag{61}$$

The denominator of (60) vanishes for $y = y(x)$, where $y(x)$ is the algebraic function defined by the equation

$$y^2 x\beta + y(x\alpha - 1) + x\alpha = 0. \tag{62}$$

This algebraic function has two real valued branch points and is analytic when cut between the two branch points. The cut is the interval of the real axis where the discriminant

$$\Delta(x) = (x\alpha - 1)^2 - 4x^2\alpha\beta \tag{63}$$

is negative. The algebraic function $y(x)$ is two-valued with

$$\begin{aligned} y^+(x) &= \frac{1 - x\alpha + \sqrt{\Delta}(x)}{2x\beta}, \\ y^-(x) &= \frac{1 - x\alpha - \sqrt{\Delta}(x)}{2x\beta}. \end{aligned} \tag{64}$$

On the cut, let $y^+(x) = \rho(x)e^{i\theta(x)}$ and $y^-(x) = \rho(x)e^{-i\theta(x)}$. We get immediately that $\rho^2(x) = \alpha/\beta$ and that

$$\cos(\theta(x)) = \frac{1 - x\alpha}{2x\sqrt{\beta}\sqrt{\alpha}}. \tag{65}$$

From (60) we get that for x on the cut

$$\begin{aligned} y^+(x) &= x\alpha F(x) + \beta(y^+(x))^{J+1} G(x), \\ y^-(x) &= x\alpha F(x) + \beta(y^-(x))^{J+1} G(x), \end{aligned} \tag{66}$$

which implies that

$$\beta G(x) = \frac{y^+(x) - y^-(x)}{(y^+(x))^{J+1} - (y^-(x))^{J+1}}, \tag{67}$$

that is

$$\beta G(x) = \frac{\rho(x)\sin\theta(x)}{\rho(x)^{J+1}\sin((J+1)\theta(x))}. \tag{68}$$

This function is then continued analytically from the region of the cut. The poles of this function are located for x such that $(J+1)\theta(x) = l\pi$. In view of (65), this implies

$$\cos\frac{l\pi}{J+1} = \frac{1 - x\alpha}{2x\sqrt{\beta}\sqrt{\alpha}}, \tag{69}$$

that is

$$x = \frac{1}{\alpha + 2\cos(\frac{l\pi}{J+1})\sqrt{\beta}\sqrt{\alpha}}. \tag{70}$$

The smallest of these poles is for $l = 1$. Hence

$$P[Z_k = 1] = \Theta(\lambda_J^k) \tag{71}$$

where

$$\lambda_J = \alpha + 2\cos(\frac{\pi}{J+1})\sqrt{\beta}\sqrt{\alpha}. \tag{72}$$

Therefore,

$$\phi(\theta) = K(\alpha + 2\cos(\frac{\pi}{J+1})\sqrt{\beta}\sqrt{\alpha}). \tag{73}$$

References

[1] F. Baccelli, "Ergodic Theory of Stochastic Petri Networks," Rapport INRIA No. 1037, 1989, To appear in *Annals of Probability*, Jan. 92.

[2] F. Baccelli, T. Konstantopoulos, V. Malyshev, "On the Stability of a Model with Infinitely Many Resources" In preparation.

[3] F. Baccelli, Z. Liu, "On a Class of Stochastic Recursive Equations Arising in Queueing Systems," INRIA Research Report, No. 984, March, 1989. To appear in *Annals of Probability*, Jan. 1992.

[4] F. Baccelli, Z. Liu, "Comparison Properties of Stochastic Decision Free Petri Nets", INRIA Research Report No. 1433, May 1991; also submitted to IEEE Tr. Aut. Cont.

[5] F. Baccelli, A. M. Makowski, "Queueing Models for Systems with Synchronization Constraints," *Proceedings of the IEEE*, Special Issue on Dynamics of Discrete Event Systems, Vol. 77, pp. 138-161, Jan. 1989.

[6] R. Barlow, F. Proschan, *Statistical Theory of Reliability and Life Testing.* Holt, Rinehart and Winston, 1975.

[7] J. D. Biggins, *The First and Last Birth Problem for a Multitype Age-Dependent Branching Process* Adv. Appl. Prob. 8, pp. 446-459, 1976.

[8] C. S. Chang, R. Nelson, "Bounds on the Speedup and Efficiency of Partial Synchronization in Parallel Processing Systems" IBM RC 16474, T. J. Watson, 1991.

[9] G. Cohen, D. Dubois, J. P. Quadrat, M. Viot, "A Linear-System-Theoretic View of Discrete-Event Processes and Its Use for Performance Evaluation in Manufacturing," *IEEE Trans. on Automatic Control*, Vol. AC-30, pp. 210-220, 1985.

[10] F. Commoner, A. W. Holt, S. Even, A. Pnuelly. "Marked Directed Graphs," *J. Computer and System Science*, Vol. 5, pp. 511-523, 1971.

[11] J. F. C. Kingman, "Subadditive Ergodic Theory," *Ann. Probab.*, Vol. 1, pp. 883-909, 1973.

[12] J. F. C. Kingman, "The First Birth Problem for Age-Dependent Branching Processes," *Ann. Probab.*, Vol. 3, pp. 790-801, 1975.

[13] J. F. C. Kingman, *Subadditive Processes*, in Ecole d'Eté de Probabilité de Saint-Flour, P.-L. Hennequin Editor, Lecture Notes in Mathematics, 539. Springer-Verlag, Berlin, pp. 165-223, 1976.

On Bellman Equations of Ergodic Control in R^n

A. Bensoussan* J. Frehse†

Introduction

The problem of ergodic control is the following. Consider a stochastic differential equation

$$\begin{aligned} dy &= g(y,v)dt + \sigma(y)dw \\ y(0) &= x \end{aligned}$$

where $w(t)$ is a Wiener process and $v(t)$ stands for the control. Consider the payoff

$$u_\alpha(x) = \inf_{v(.)} \int_0^\infty e^{-\alpha t} \ell(y(t), v(t))dt.$$

One is interested in the behaviour of $u_\alpha(x)$ as α tends to 0.

One can give a purely analytic version of this problem, using the Bellman equation. This is the approach of this article. The main difficulty stems from the fact that the equation is posed in the whole space R^n and not in a bounded domain. In the case of a bounded domain with periodic or Neumann boundary conditions the problem has been well studied by many authors (see A. Bensoussan, J. Frehse [1], P. Gimbert [3], J.M. Lasry [4], P.L. Lions [5]).

In the whole space there is a lack of compactness, which introduces specific difficulties. Local arguments have to be used.

1 Motivation

1.1 General comments

The problem is formulated as follows

$$Lu + H(x, Du) + \lambda = q(x), \quad x \in R^n$$

where $L = -D_i a_{ik} D_k$ is a 2nd order differential operator, and H is a nonlinear function of Du, called the Hamiltonian.

The *Unknown* are the pair (u, λ) where λ is a constant.

We are interested in studying the *Existence* and *Uniqueness* of a solution.

An other way of looking at the problem, and a constructive way to prove existence is to study the limit of the equation

$$Lu_\alpha + H(x, Du_\alpha) + \alpha u_\alpha = q$$

as α tends to 0.

*University Paris Dauphine and INRIA
†Institüt für Angewandte Mathematik der Universität Bonn

1.2 Explicit calculations

This is an advantage of R^n. Consider, to begin with, the case

$$-\Delta u_\alpha + |Du_\alpha|^2 + \alpha u_\alpha = |x|^2.$$

There is an explicit solution given by

$$u_\alpha(x) = \frac{-\alpha + \sqrt{\alpha^2 + 16}}{8}\left(|x|^2 + \frac{2n}{\alpha}\right).$$

It is the *unique positive solution.* We can state immediately that

$$\begin{aligned} \alpha u_\alpha(x) &\to n = \lambda \\ u_\alpha(x) - u_\alpha(0) &\to \frac{1}{2}|x|^2 = u(x) \end{aligned}$$

and the pair $u, \lambda = n$ satisfy the equation

$$-\Delta u + |Du|^2 + n = |x|^2.$$

We now turn to the *General Quadratic case.* The equation is written as follows

$$\begin{aligned} &-\frac{1}{2}tr(Bx + b)^* D^2 u_\alpha (Bx + b) - Du_\alpha.(Fx + f) \\ &+ \frac{1}{4}N^{-1}(G^* Du_\alpha + 2n).(G^* Du_\alpha + 2n) \\ &+ \alpha u_\alpha = Mx.x + 2m.x \end{aligned}$$

in which

$$B \in L(R^n\ ;\ L(R^n\ ;\ R^n)), \quad b \in L(R^n\ ;\ R^n).$$

The function u_α is the value function of the following stochastic control problem. The state equation is given by

$$\begin{aligned} dy &= (Fy + Gv + f)dt + (By + b)dw \\ y(0) &= x \end{aligned}$$

and the payoff by

$$J_x(v(.)) = E\int_0^\infty e^{-\alpha t}[My.y + 2m.y + Nv.v + 2n.v]dt.$$

Then we have

$$u_\alpha(x) = \inf_{v(.)} J_x(v(.)).$$

We assume $M > 0$. There exists an explicit solution given by

$$u_\alpha(x) = P_\alpha x.x + 2p_\alpha x + r_\alpha.$$

The matrix P_α is a solution of

$$B^* P_\alpha B + F^* P_\alpha + P_\alpha F + M - P_\alpha G N^{-1} G^* P_\alpha = \alpha P_\alpha. \tag{1.1}$$

To make precise the notation, note that $L(R^n\ ;\ R^n)$ is equipped with the scalar product $M.\tilde{M} = tr\ M\tilde{M}^*$. Now if $Bx = \sum_i B_i x_i$ with $B_i \in L(R^n\ ;\ R^n)$ the matrix $B^* P_\alpha B$ is given explicitly by

$$(B^* P_\alpha B)_{ij} = tr B_i^* P_\alpha B_j.$$

Once P_α is defined, we find p_α and r_α as follows

$$(F^* - P_\alpha GN^{-1}G^*)p_\alpha + B^*P_\alpha b + P_\alpha(f - GN^{-1}n) + m = \alpha p_\alpha \tag{1.2}$$

$$tr\, b^*P_\alpha b + 2p_\alpha.f - N^{-1}(G^*p_\alpha + m).(G^*p_\alpha + n) = \alpha r_\alpha. \tag{1.3}$$

We look for a solution of (1.2) of the form $p_\alpha = P_\alpha q_\alpha$. Since $(F^* - P_\alpha GN^{-1}G^* - \alpha)P_\alpha$ is bijective, this situation defines q_α in a unique way.

We can now let α tend to 0. As $\alpha \to 0$ we have $P_\alpha \to P$, $p_\alpha \to p$ which are the solutions of

$$\begin{aligned} B^*PB + F^*P + PF + M - PGN^{-1}G^*P &= 0 \\ (F^* - PGN^{-1}G^*)p + B^*Pb + P(f - GN^{-1}n) + m &= 0. \end{aligned}$$

Moreover

$$r_\alpha \to +\infty$$

but

$$u_\alpha(x) - u_\alpha(0) \Rightarrow Px.x + 2p.x = u$$

$$\alpha u_\alpha \to tr\, b^*Pb + 2p.f - N^{-1}(G^*p + n)(G^*p + n) = \lambda$$

and the pair (u, λ) is solution of

$$\begin{aligned} &-\frac{1}{2}tr(Bx + b)^*D^2u(Bx + b) - \nabla u.(Fx + b) \\ &+\frac{1}{4}N^{-1}(G\nabla u + 2n).(G^*\nabla u + 2n) + \lambda = Mx.x + 2m.x. \end{aligned}$$

2 Existence

2.1 Approximation

We consider the equation

$$Lu_\alpha + H(x, Du_\alpha) + \alpha u_\alpha = q(x) \tag{*}$$

We make the following assumptions

$$\begin{gathered} L = -\sum D_i a_{ik} D_k \\ a_{ik} \in L^\infty_{loc}(R^n) \\ \sum_{i,k} a_{ik}(x)\xi_i\xi_k \geq c(x)|\xi|^2, \quad x, \xi \in R^n \\ c > 0 \quad \frac{1}{c} \in L^\infty_{loc}(R^n) \end{gathered} \tag{2.1}$$

(locally uniform ellipticity).

$$q \in L^\infty_{loc}(R^n) \tag{2.2}$$

$$k(x)|\eta|^2 - \hat{k}(x) \leq H(x, \eta) \leq K(x)|\eta|^2 + \hat{K}(x) \tag{2.3}$$

$$k > 0, \quad \frac{1}{k} \in L^\infty_{loc}(R^n), \quad \hat{k} > 0$$

$$\inf_{|x|=r}(q(x) - \hat{K}(x)) \to \infty \ as \ r \to \infty \tag{2.4}$$

$$\sup_{|x|>1} K(x)/c(x) < \infty. \tag{2.5}$$

Then we can state the

Theorem 1 *Under the assumptions (2.1) to (2.5), there exists a solution* $u_\alpha \in H^1_{loc} \cap L^\infty_{loc}$ *of* $(*)$ *such that*

$$\begin{array}{lcl} \int_{B_r} |Du_\alpha|^2 dx & \leq K_r & r \geq 1 \\ \alpha \|u_\alpha\|_{L^\infty(B_r)} & \leq & K_r \\ \operatorname{osc}_{B_r} u_\alpha & \leq & K_r \end{array}$$

uniformly as $\alpha \to 0$.
K_r *is a generic constant possibly depending on* r.

2.2 Limit behaviour as $\alpha \to 0$

We can now state the following

Theorem 2 *Under the assumptions of Theorem 1, there exists a subsequence*

$$\begin{array}{lcl} u_\beta & = & u_\beta - u_\beta(0) \to v \\ \beta u_\beta & \to & \lambda \text{ constant} \end{array}$$

strongly in $L^\infty_{loc} \cap H^1_{loc}(R^n)$. *The limit pair* (v, λ) *satisfies*

$$Lv + H(x, \nabla v) + \lambda = q.$$

3 Uniqueness

We state now a uniqueness theorem for the equation

$$Lu + H(x, Du) + \lambda = q \tag{**}$$

$$L = -D_i a_{ij} D_j + b.D,$$

where we make the following assumptions

$$\begin{array}{lcl} a_{ij} & \in & L^\infty_{loc}, \text{ locally elliptic} \\ b & \in & L^\infty_{loc} \end{array} \tag{3.1}$$

$$H(x,p) \text{ is Caratheodory, locally Lipschitz in } p \tag{3.2}$$

$$H(x,p) \geq c \sum a_{ij} p_j p_i, \ c > 0 \tag{3.3}$$

$$H \text{ satisfies } H(x, \beta p) \geq \beta^2 H(x,p), \quad \beta \geq 1 \tag{3.4}$$

$$q \in L^\infty_{loc}, \quad q(\infty) = \infty. \tag{3.5}$$

We can state the

Theorem 3 *Under the assumptions (3.1) to (3.5), there exists a unique pair* (u, λ) *(u up to a constant,* λ *constant), such that* $u \in W^{1,\infty}_{loc}$, $u(\infty) = \infty$, *and u is a solution of space* $(**)$.

4 A brief sketch of the proof of existence

4.1 Proof of Theorem 1

Since the problem is posed in the whole space, we use an approximation by a Neumann problem as follows

$$Lu + H(x, Du) + \alpha u = q \text{ on } B_R$$

with Neumann boundary conditions. In variational form, this is equivalent to

$$\sum_{i,k} \int a_{ik} D_k u D_i \varphi dx + \int [H(x, Du)\varphi + \alpha u \varphi] dx = \int q \varphi dx$$

$$\forall \varphi \in L^\infty \cap H^1(B_R).$$

There are solutions of this equation. We denote by $u_{\alpha,R}$ a solution.

One first gets estimates. These amount to

$$\alpha u_{\alpha,R} \geq -K_0 \tag{4.1}$$

Take $r \geq 1$, $R \geq 2r$, then one has the following estimates

$$\alpha \|u_{\alpha R}\|_{L^\infty(B_r)} \leq C_r \tag{4.2}$$

$$|Du_{\alpha,R}|_{L^2(B_r)} \leq C_r \tag{4.3}$$

$$osc\{u_{\alpha,R} \mid x \in B_r\} \leq C_r \tag{4.4}$$

where

$$osc\{u_{\alpha,R} \mid x \in B_r\} = \sup_{x,x_2 \in B_r} |u_{\alpha,R}(x_1) - u_{\alpha,R}(x_2)|$$

To prove (4.4), consider a number $\gamma_r = \gamma_{r,\alpha,R}$ such that

$$\begin{aligned} \mu\{x \in B_r \mid u_{\alpha,R}(x) - \gamma_r \geq 0\} &\geq \frac{1}{2}\mu(B_r) \\ \mu\{x \in B_r \mid u_{\alpha,R}(x) - \gamma_r \leq 0\} &\leq \frac{1}{2}\mu(B_r) \\ |\gamma_r| &\leq \|u_{\alpha,R}\|_{L^\infty(B_r)}, \end{aligned}$$

where μ is the Lebesgue measure, then

$$u_{\alpha,R} - \gamma_r \leq C_r \text{ on } B_r \tag{4.5}$$

$$|u_{\alpha,R} - \gamma_r|_- \leq C_r \tag{4.6}$$

Proof of (4.1). Let

$$\begin{aligned} [\xi]_+ &= \xi \quad if \quad \xi \geq 0, \quad 0 \text{ otherwise} \\ [\xi]_- &= \xi \quad if \quad \xi \leq 0, \quad 0 \text{ otherwise} \\ |\xi|_- &= -[\xi]_- \end{aligned}$$

Take $\varphi = 1 - E$, $E = \exp[\lambda |u + L|_-]$, λ, L to be chosen (note that $\varphi \leq 0$).

Set χ_- = characteristic function of $u + L \le 0$. Then we get

$$\begin{aligned}
\lambda \int_{B_R} a_{ik} D_k u D_i u E \chi_- dx &+ \alpha \int_{B_R} u(1-E)dx \\
&= \int_{B_R} (q-H)(1-E)dx \\
&\le \int_{B_R} (q-\hat{K})(1-E)dx + \int_{B_R} K|Du|^2(E-1)dx \\
&\le \int_{B_R} (q-\hat{K})(1-E)dx + \int_{B_R} K|Du|^2 E\chi_- dx
\end{aligned}$$

Choose λ large enough (depending on R), we obtain

$$\alpha \int_{B_R} u(1-E)dx \le \int_{B_R} (q-\hat{K})(1-E)dx.$$

But $q - \hat{K} \ge 0$ on $R^n - B_{r_0}$ and $|q - \hat{K}| \le C_{r_0}$ on B_{r_0}. Therefore

$$\int_{B_R} \alpha L|1-E|dx \le C_{r_0} \int_{B_R} |1-E|dx.$$

It is then enough to choose $L = 2\frac{C_{r_0}}{\alpha}$ which implies $1 - E = 0$.

Proof of (4.3). We test by $\varphi = \tau^2$ with $\tau = 1$ on B_r, 0 outside B_{2r}, and $|D\tau| \le K/r$. We deduce easily (4.3) and

$$\alpha |u_{\alpha,R}|_{L^1(B_r)} \le C_r.$$

Proof of (4.2). We use the regularized Green function. Let $z \in B_r$ and, define G_ρ^z by

$$\sum_{i,k} \int_{B_{2r}} a_{ik} D_k \varphi D_i G_\rho dx = \fint_{B_\rho(z)} \varphi dx \quad \varphi \in L^\infty \cap H_0^1(B_{2r})$$

where $\fint$ stands for the average.

We test by $\varphi = \alpha\tau^2 G_\rho$, with $\tau \in H_0^{1,\infty}(B_{2r})$

$$\tau = 1 \text{ on } B_{3r/2}, \quad |D\tau| \le K/r.$$

Note that $\|G_\rho\|_{H^{1,\mu}(B_{2r})} \le C_r$ for $\mu < \frac{n}{n-1}$.

The Green function creates a term of the form $\alpha \fint_{B_\rho(z)} u dx$, which converges to $\alpha u(z)$ as $\rho \to 0$. This method permits one to estimate $\alpha u(z)$ for any z in B_r.

Proof of (4.5). We prove $|(u-\gamma_r)^+|_{L^\infty} \le K_r$. For that we first prove the following estimate

$$|(u-\gamma_r)^+|_{L^p(B_{\sigma r})} \le K_r$$

$\forall p > \frac{n}{2}$, and $\sigma = \sigma(p) \in]1,2[$.

One begins with $p = \frac{2n}{n-2}$.

We next test with $\varphi = (u-\gamma_r)^+ G_\rho \tau^2$ and the result obtains.

Proof of (4.6). One first proves the estimate

$$\int_{B_R} |Du_{\alpha,R}|^2 \exp[\lambda\, |u_{\alpha,R} - \gamma_r|_-]dx \le K_r$$

for any λ fixed, λ sufficiently large, the right hand side constant depending on λ.

This is obtained by testing with

$$\varphi = 1 - e^{\lambda|u-\gamma_r|_-}$$

and chosing

$$\lambda > 2 \sup_x K(x)/c(x).$$

It should be noticed that Neumann approximation is important here. Then test by

$$\varphi = f_0((u-\gamma_r)_-)G_\rho \tau^2$$

with

$$f_0(\xi) = 1 - e^{-\lambda\xi}$$

$\tau = 1$ on B_{2r}, 0 outside B_{3r}, $|D\tau| \le K/r$, $G_\rho \in L^\infty \cap H_0^1(B_{3r})$, with singularity $z \in B_r$.

Letting $R \to \infty$, and using uniform estimates, the proof of Theorem 1 is completed.

4.2 Proof of Theorem 2

The function $v_\alpha = u_\alpha - u_\alpha(0)$ satisfies

$$\|v_\alpha\|_{L^\infty(B_r)} + |Dv_\alpha|_{L^2(B_r)} \le K_r.$$

This and the equation imply that there exists a subsequence $v_\alpha \to v$ in $H^1_{loc}(R^n)$ (strongly) and $\alpha \fint_{B_1} u_\alpha dx \to \lambda$.

Now

$$\alpha u_\alpha = \alpha(u_\alpha - \fint_{B_1} u_\alpha dx) + \alpha \fint_{B_1} u_\alpha dx \to \lambda$$

$H(x, Dv_\alpha) \to H(x, Dv)$ in $L^1_{loc}(R^n)$; this is a consequence of Vitali's theorem.

Passing to the limit, the pair v, λ is a solution of

$$Lv + H(x, Dv) + \lambda = q.$$

5 Uniqueness

5.1 Behaviour at infinity

This is a preliminary step to uniqueness. We begin with

Proposition 1 *The solution v which has been constructed in the proof of Theorem 2 has the property*

$$v \ge -K_0.$$

Proof. This follows from an estimate

$$u_\alpha(x) - \fint_{B_r} \alpha dx \ge -K_r \quad \forall r, \forall x$$

obtained by proving

$$u_{\alpha,R}(x) - \fint_{B_r} u_{\alpha,R} dx \ge -K_r \quad \forall x \in B_R.$$

This inequality is obtained by testing with

□ $$\varphi = 1 - e^{\lambda|u_{\alpha,R} - \bar{u}_{\alpha,R,r} + K_r|_-}$$

Proposition 2 *Assume $a_{ij} \in W^{1,\infty}$, $K(x)$ bounded and $q - \hat{K} \geq c_0 > 0$ for $|x| > \rho_0$ then a solution of*

$$Lv + H(x, Dv) + \lambda = q$$

satisfying

$$v \in W^{2,p}_{loc}, \quad v \geq -K$$

satisfies $\inf_{|x|=\rho} v(x) \to +\infty$, *as* $\rho \to \infty$.

Proof. Let x_0 with $|x_0| = \rho$. Let ρ sufficiently large and $Q(x) = \beta\rho\left(1 - 4\frac{|x - x_0|^2}{\rho^2}\right)$ with β an appropriate constant. Set $z = v - Q$ then

$$z \mid_{\partial B_{\rho/2}(x_0)} = v \mid_{\partial B_{\rho/2}(x_0)} \geq -K.$$

One checks that z satisfies the maximum principle on $B_{\rho/2}(x_0)$, hence $z(x_0) \geq -K$ and $\inf_{|x|=\rho} v(x) \geq \beta\rho - K$.

5.2 Proof of Theorem 3

Let $(u, \lambda_1), (v, \lambda_2)$ be two solutions with $\lambda_1 \geq \lambda_2$. We pick R large enough so that

$$q(x) \geq \lambda_1 + c_0 \qquad c_0 > 1, \qquad |x| \leq R.$$

Then the main point is to prove that

$$\max_{x \in B_R}(u - v) = \max_{x \in R^n}(u - v),$$

and then let $x^* \in B_R$ a point of maximum of $u - v$.

Set $\psi = u(x^*) - v(x^*) - u(x) + v(x)$, then $\psi \geq 0$, $\psi(x^*) = 0$ and on any domain Ω, ψ satisfies

$$L\psi + K_\Omega . D\psi = \lambda_1 - \lambda_2 \geq 0.$$

We may assume $x^* \in \Omega$, then from Harnack's inequality $\psi = 0$ on a ball $B_\rho(x^*)$. Continuing the procedure we prove that $\psi = 0$, hence $u - v =$ constant, and $\lambda_1 = \lambda_2$.

Remark. The uniqueness theorem holds for H quadratic, but also for cases like

$$H(x, p) = c\sum a_{ij}p_j p_i + \frac{c|p|^4}{|p|^2 + 1},$$

or for the super-quadratic Hamiltonian

$$H(x, p) = c\sum a_{ij}p_j p_i + c|p|^4.$$

6 Eigenvalue problem

We consider here the particular case

$$-D_i a_{ij} D_j u + b.Du + a_{ij} D_j u . D_i u + \lambda = q$$

with

$$a_{ij} \in W^{1,\infty} \quad a_{ij}\xi_j\xi_i \geq \alpha|\xi|^2, \ \forall \xi \in R^n, \ \alpha > 0,$$

$$q \in L^{\infty}_{loc}, \quad q(\infty) = \infty.$$

Then there exists one and only one solution u, λ with $u \in W^{1,\infty}_{loc}$, λ = constant, $u(\infty) = \infty$.

By the exponential transformation $\varphi = e^{-u}$ we get the existence and uniqueness (up to a multiplicative constant) of a solution

$$\varphi > 0, \quad \varphi(\infty) = 0, \quad \varphi \in W^{1,\infty}_{loc}$$

$$-D_i a_{ij} D_j \varphi + b.D\varphi + q\varphi = \lambda\varphi.$$

It can be approximated by

$$-D_i a_{ij} D_j \varphi_\alpha + b.D\varphi_\alpha + q\varphi_\alpha + \alpha\varphi_\alpha \log \varphi_\alpha = 0.$$

For direct developments in this direction, we refer to A. Bensoussan, H. Nagaï [2].

References

[1] A. Bensoussan, J. Frehse : On Bellman equation of ergodic type with quadratic growth Hamiltonian. Contributions to Modern Calculus of Variations, Pitman Res. Notes Math. Series, ed. L. Cesari, Longman, Vol. 148(1987), 13-26.

[2] A. Bensoussan, H. Nagai : An ergodic control arising from the principal eigenfunction of an elliptic operator, J. Math. Soc. Japan, Vol. 43, n° 1, 1991.

[3] P. Gimbert : Problèmes de Neumann quasi-linéaires, J. Funct. Analysis 68 (1985), 65-72.

[4] J.M. Lasry : Contrôle stochastique ergodique, Thèse d'Etat, Université Paris Dauphine, 1974.

[5] P.L. Lions : Quelques remarques sur les problèmes elliptiques quasi-linéaires du 2ème ordre, J. Anal. Math. 45 (1985), 234-254.

Some results on the filtering Riccati equation with random parameters

Philippe BOUGEROL
Université Pierre et Marie Curie
Laboratoire de Probabilités
4, Place Jussieu
75252 Paris Cedex 05 France

Abstract: We consider the Kalman filter in a random stationary environment. The associated Riccati equation has random parameters. We first describe some recent results we have obtained on the asymptotic behavior of this equation under a.s. stabilizability and detectability assumptions. They depend on contraction properties of Hamiltonian matrices. Then we give a simple self-contained proof under a stronger detectability condition.

1. Introduction.

Let $\{(A_n, B_n, C_n),\ n \in \mathbf{Z}\}$ be a stationary ergodic sequence, where A_n, B_n and C_n are matrices with size $d \times d$, $d \times p$ and $q \times d$ respectively, defined on a probability space $(\Omega, \mathcal{F}, \mathbf{P})$. We consider the Riccati equation

(1)
$$
\begin{aligned}
P_{n+1} &= B_n B_n^* + A_n P_n A_n^* - K_n C_n P_n A_n^* \\
K_n &= A_n P_n C_n^* (I + C_n P_n C_n^*)^{-1}
\end{aligned}
$$

where P_n, $n \geq 0$, are nonnegative symmetric $d \times d$ matrices (K_n is called the gain matrix associated with P_n). We will describe the behaviour of P_n as $n \to +\infty$.

These equations arise in the following situation: let us consider the linear system

(2)
$$
\begin{aligned}
X_{n+1} &= A_n X_n + B_n \epsilon_{n+1} \\
Y_{n+1} &= C_n X_n + \eta_{n+1}
\end{aligned}
$$

where $X_n \in \mathbf{R}^d$, $Y_n \in \mathbf{R}^q$. We suppose that there is a sub-sigma algebra $\mathcal{F}_0$ of $\mathcal{F}$ such that, for $n \geq 0$,

(a) A_n, B_n, C_n are $\mathcal{F}_n$-measurable, where $\mathcal{F}_n = \sigma(\mathcal{F}_0, Y_1, \ldots, Y_n)$.

(b) $(\varepsilon_{n+1}, \eta_{n+1})$ is a gaussian $\mathbf{R}^p \times \mathbf{R}^q$-valued random vector, with mean 0 and covariance matrix equal to the identity, independent of $\sigma(\mathcal{F}_n, X_n)$.

(c) Conditionally on $\mathcal{F}_0$, the random vector X_0 has a gaussian law.

Then, in this so-called conditionally gaussian set-up, the Kalman recursive equations for

$$\widehat{X}_n = \mathbf{E}(X_n/\mathcal{F}_n), \qquad P_n = \mathbf{E}\left((X_n - \widehat{X}_n)(X_n - \widehat{X}_n)^*/\mathcal{F}_n\right)$$

are given by (1) and by

$$\widehat{X}_{n+1} = A_n\widehat{X}_n + K_n(Y_{n+1} - C_n\widehat{X}_n)$$

(see for instance Whittle (1982, p. 260)). Some real situations which are modelized by these equations are described in Bougerol (1991a). Let us just give one example: X_n is the position of a plane, Y_n is an observation of X_n and C_n is a random matrix which describes the state of the sky at time n. Either the sky is clear in which case C_n is the identity matrix or the sky is cloudy and C_n is equal to zero. The sequence $\{(A_n, B_n, C_n), n \in \mathbf{Z}\}$ can be, for instance, almost periodic, or i.i.d., or a function of an ergodic Markov chain.

We shall see that under weak conditions, the matrices P_n converge a.s. exponentially fast to a stationary ergodic process. In particular, they converge in law, there is no explosion, and the filtering process is successful. Two kinds of conditions arise naturally. The first one (called, in the sequel, ***weak controllability/observability***) says that in some sense, at least with some positive probability, the system is controllable/observable from time to time. (Notice that it is not required as usual that the system is uniformly controllable/observable; such strong condition would not hold for our plane in a cloudy sky). The second condition is a notion of ***almost sure stabilizability/detectability***. These results are described in the sections 2 and 3 below. They are proved in Bougerol (1991a,b). In the last section we provide a simple new demonstration under a strengthened detectability assumption.

2. A contraction property of the filtering Riccati equation.

We shall need the following condition :

Hypothesis (H). *The sequence $\{(A_n, B_n, C_n), n \in \mathbf{Z}\}$ is strictly stationary and ergodic;the matrices A_n are invertible, and $\log^+ \|A_n\|, \log^+ \|B_n\|, \log^+ \|C_n\|$ are integrable.*

This condition does not depend on the chosen norms. We associate with the system (2) the Hamiltonian matrices

$$M_n = \begin{pmatrix} A_n + S_n A_n^{*-1} R_n & S_n A_n^{*-1} \\ A_n^{*-1} R_n & A_n^{*-1} \end{pmatrix}, \tag{3}$$

where $R_n = C_n^* C_n$ and $S_n = B_n B_n^*$. (Note that the system (2) is slightly different from the one considered in Bougerol (1991a), thus the Hamiltonian matrices are not exactly the same. The only reason of our choice here is that this situation is more symmetric).

These Hamiltonian matrices are in the symplectic group $\mathrm{Sp}(d,\mathbf{R})$. Let $\mathcal{P}$ (resp. $\mathcal{P}_0$) be the set of $d \times d$ symmetric nonnegative (resp. positive) matrices, let

$$\mathcal{H} = \left\{ \begin{pmatrix} A & B \\ C & D \end{pmatrix} \in \mathrm{Sp}(d,\mathbf{R})\,;\ D \text{ is invertible, } CD^* \in \mathcal{P},\ D^*B \in \mathcal{P} \right\}$$

and

$$\mathcal{H}_0 = \left\{ \begin{pmatrix} A & B \\ C & D \end{pmatrix} \in \mathrm{Sp}(d,\mathbf{R})\,;\ CD^* \in \mathcal{P}_0,\ D^*B \in \mathcal{P}_0 \right\}.$$

Then $\mathcal{H}$ is a multiplicative semigroup and $\mathcal{H}_0$ is an ideal of $\mathcal{H}$. In some sense, $\mathcal{H}_0$ is an analogue of the semigroup of matrices with positive entries. The matrices $M = \begin{pmatrix} A & B \\ C & D \end{pmatrix}$ in $\mathcal{H}$ act on $\mathcal{P}$ and on $\mathcal{P}_0$ by the formula

$$M \cdot T = (AT + B)(CT + D)^{-1}, \qquad T \in \mathcal{P}. \tag{4}$$

Let δ be the Riemannian metric on $\mathcal{P}_0$ defined by

$$\delta(T_1, T_2) = \left\{ \sum_{i=1}^{d} \log^2 \lambda_i(T_1, T_2) \right\}^{1/2}, \quad T_1, T_2 \in \mathcal{P}_0,$$

where $\lambda_i(T_1, T_2)$, $i = 1, 2, \ldots, d$, are the eigenvalues of $T_1 T_2^{-1}$. For any M in $\mathcal{H}$, let

$$c(M) = \sup \left\{ \frac{\delta(M \cdot T_1, M \cdot T_2)}{\delta(T_1, T_2)}\,;\ T_1, T_2 \in \mathcal{P}_0 \right\}.$$

We have shown in Bougerol (1991a) that

Theorem 2.1. *If M is in $\mathcal{H}$, then $c(M) \leq 1$, and if M is in $\mathcal{H}_0$ then $c(M) < 1$.*

The Hamiltonian matrices M_n of (3) are in $\mathcal{H}$, and the equation (1) can be written as

$$P_{n+1} = M_n \cdot P_n \tag{5}$$

(see e.g. Whittle (1982, Thm 5.7.1)). Moreover $M_n \ldots M_1$ is a.s. in $\mathcal{H}_0$ for n large enough if and only if $\{(A_n, B_n, C_n),\ n \in \mathbf{Z}\}$ is weakly observable and controllable in the following sense.

Definition 2.2. *We say that the system $\{(A_n, B_n, C_n),\ n \in \mathbf{Z}\}$ is weakly controllable if for some $n \geq 1$, $\sum_{k=1}^{n} A_n \ldots A_{k+1} B_k B_k^* A_{k+1}^* \ldots A_n^*$ is nondegenerate with positive probability. This system is called weakly observable when its dual $\{(A_{-n}^*, C_{-n}^*, B_{-n}^*), n \in \mathbf{Z}\}$ is weakly controllable.*

Under these conditions, the matrices M_n act on $\mathcal{P}_0$ by contractions, and at least for some products these contractions are strict. This property is the main point of the proof of the following theorem, given in Bougerol (1991a).

Theorem 2.3. *We suppose that (H) holds and that $\{(A_n, B_n, C_n),\ n \in \mathbf{Z}\}$ is weakly observable and controllable. Then there exists a unique stationary ergodic process $\{\overline{P}_n, n \in \mathbf{Z}\}$, with values in $\mathcal{P}_0$, which is a solution of (1).*

It will be generalized in Theorem 3.5 below. This approach to the study of the asymptotic behaviour of the Kalman filter, using contractions, is new. It can be useful in other situations.

As shown in Bougerol (1990) this theorem is linked to the fact (proved in another context by Wojtkowski (1985)) that no Lyapunov exponent of the Hamiltonian matrices vanishes.

3. Almost sure stabilizability and detectability.

We first define a convenient notion of exponential stability.

Definition 3.1. *We say that a sequence $\{A_n, n \in \mathbf{Z}\}$ of $d \times d$ matrices is exponentially stable if there exists $\gamma > 0$ with the following property: For any $\varepsilon > 0$, there is $C > 0$ such that*

$$\|A_n A_{n-1} \dots A_{k+1}\| \leq C e^{(k-n)\gamma} e^{(|n|+|k|)\varepsilon},$$

for all $k, n \in \mathbf{Z}$ such that $k < n$.

When $d = 1$, if for some $\tau < 0$,

$$\lim_{n \to +\infty} \frac{1}{n} \log \|A_n \dots A_1\| = \lim_{n \to +\infty} \frac{1}{n} \log \|A_{-1} \dots A_{-n}\| = \tau,$$

then $\{A_n, n \in \mathbf{Z}\}$ is exponentially stable, but this implication does not hold in general when $d > 1$. Nevertheless, we have

Proposition 3.2. *Let $\{A_n, n \in \mathbf{Z}\}$ be a stationary ergodic sequence of $d \times d$ matrices such that $\log^+ \|A_n\|$ is integrable. Then this sequence is a.s. exponentially stable if and only if the upper Lyapunov exponent $\tau := \inf_{n>0} \frac{1}{n} \mathrm{E}(\log \|A_n \dots A_1\|)$ is negative.*

The following definition is a natural generalization of the classical one.

Definition 3.3. *The linear system* (2) *associated with (A_n, B_n, C_n) is called almost surely stabilizable if there exists a sequence $\{F_n, n \in \mathbf{Z}\}$ of random $p \times d$ matrices such that, almost surely,* a) $\lim_{|n| \to +\infty} \frac{1}{n} \log^+ \|F_n\| = 0$, b) $\{A_n + B_n F_n, n \in \mathbf{Z}\}$ *is exponentially stable. It is is called a.s. detectable if its dual, associated with $(A^*_{-n}, C^*_{-n}, B^*_{-n})$, is a.s. stabilizable.*

Using Theorem 2.3, we have shown in Bougerol (1991b) that:

Theorem 3.4. *We suppose that (H) holds. Then*

(a) Weak controllability implies a.s. stabilizability and weak observability implies a.s. detectability.

(b) If the system (2) *is a.s. stabilizable and detectable, there exists a unique stationary ergodic process $\{\overline{P}_n, n \in \mathbf{Z}\}$ with values in $\mathcal{P}$, which is a solution of the Riccati equation* (1). *The sequence $\{A_n - \overline{K}_n C_n, n \in \mathbf{Z}\}$ is a.s. exponentially stable (where $\overline{K}_n$ is the gain matrix associated with $\overline{P}_n$). Moreover, for any $P \in \mathcal{P}$, let $\{P_n, n \geq 0\}$ be the solution of* (1) *such that $P_0 = P$, and K_n the associated gain matrices. Then $\{A_n - K_n C_n, n \in \mathbf{N}\}$ is a.s. exponentially stable and, a.s.,*

$$\overline{\lim}_{n \to +\infty} \frac{1}{n} \log \|P_n - \overline{P}_n\| < 0. \tag{6}$$

In general $\log^+ \|A_n - \overline{K}_n C_n\|$ is not integrable and the sequence $\{A_n - \overline{K}_n C_n, n \in \mathbf{Z}\}$ has no Lyapunov exponent. The occurrence of such a.s. exponentially stable sequences without moment is a real technical difficulty in the analysis of these Riccati equations. In the deterministic case, related results have been obtained by Anderson and Moore (1981) under uniform conditions.

4. A simple proof.

We now consider a strong detectability condition:

Condition (D). *There is a sequence $\{F_{-n}, n \in \mathbb{N}\}$ of random $d \times q$ matrices such that for some $\beta > 0$, $\sup_{n>0} \mathrm{E}\,\|F_{-n}\|^\beta$ is finite and such that*

$$\overline{\lim}_{n\to+\infty}\Big\{\mathrm{E}\,\|(A_{-1}-F_{-1}C_{-1})(A_{-2}-F_{-2}C_{-2})\dots(A_{-n}-F_{-n}C_{-n})\|^\beta\Big\}^{1/n} < 1.$$

For instance, this condition holds when the matrices A_n are i.i.d., of integrable norm and with a negative Lyapunov exponent. Let us give a self contained proof of the following proposition.

Proposition 4.1. *We suppose that for some $\alpha > 0$, $\|A_n\|^\alpha, \|B_n\|^\alpha, \|C_n\|^\alpha$ are integrable, that the system $\{(A_n, B_n, C_n), n \in \mathbb{Z}\}$ is a.s. stabilizable and that condition (D) holds. Then there exists a unique stationary ergodic solution $\overline{P}_n$ of (1) and $\mathrm{E}\,\|\overline{P}_n\|^r$ is finite when $r = \min(\alpha,\beta)/4$. The upper Lyapunov exponent of the sequence $\{A_n - \overline{K}_n C_n, n \in \mathbb{Z}\}$ exists and is negative.*

Proof: We use the notation introduced in (4). It is well known (see e.g. Whittle (1982, p. 62)) that if $P_{n+1} = M_n \cdot P_n$, i.e. if P_n is solution of the Riccati equation (1), then

$$P_{n+1} = \min_K (A_n - KC_n)P_n(A_n - KC_n)^* + KK^* + B_nB_n^*, \tag{7}$$

where K is an arbitrary $d \times q$ matrix. The minimum is obtained when K is the gain matrix K_n. Let, for $k > 0$,

$$P_{0,k} = (M_{-1}M_{-2}\dots M_{-k})\cdot O,$$

where O is the matrix whose coefficients are all identically equal to 0. Using (7), we see that the sequence $\{P_{0,k}, k > 0\}$ is increasing in $\mathcal{P}$. Let $\{F_{-n}, n \geq 0\}$ be the matrices given by condition (D), let $G_k = (A_{-1} - F_{-1}C_{-1})\dots(A_{-k} - F_{-k}C_{-k})$ and $J_k = F_{-k}F_{-k}^* + B_{-k}^*B_{-k}$. It follows from (7) that

$$P_{0,k} \leq J_1 + G_1J_2G_1^* + \dots + G_{k-1}J_kG_{k-1}^*.$$

It is easily seen that the right hand side is bounded in $L^r(\Omega, \mathcal{F}, \mathrm{P})$. Since $\{P_{0,k}, k \in \mathbb{N}\}$ is an increasing sequence in $\mathcal{P}$, it converges, a.s., to a limit $\overline{P}_0$ in L^r. Similarly, for each $n \in \mathbb{Z}$, $(M_{n-1}\dots M_{n-k})\cdot O$ converges almost surely to a random matrix $\overline{P}_n$ as k tends to $+\infty$. Then it is clear that the sequence $\{\overline{P}_n, n \in \mathbb{Z}\}$ is a solution of (1). It is stationary and ergodic because it is a function of the stationary ergodic process (M_n). Since $\overline{P}_n$ is in L^r, it follows from the Borel Cantelli lemma that, a.s., $\lim_{n\to+\infty} \frac{1}{n}\log^+\|\overline{P}_n\| = 0$. Moreover

$$\overline{P}_{n+1} = (A_n - \overline{K}_nC_n)\overline{P}_n(A_n - \overline{K}_nC_n)^* + \overline{K}_n\overline{K}_n^* + B_nB_n^*,$$

where $\overline{K}_n$ is the associated gain matrix. Note that $\log^+\|A_n - \overline{K}_nC_n\|$ is integrable since $\overline{P}_n \in L^r$. Thus we can apply the next lemma to $M_n = A_n - \overline{K}_nC_n$, $N_n = (\overline{K}_n, B_n)$, and $R_n = \overline{P}_n$: we see that the upper Lyapunov exponent of $\{A_n - \overline{K}_n C_n, n \in \mathbf{Z}\}$ is negative. Now, suppose that (P'_n) is another stationary sequence of nonnegative symmetric matrices which is solution of (1). Then

$$P'_n = (M_{n-1} \dots M_{n-k}) \cdot P'_{n-k-1} \geq (M_{n-1} \dots M_{n-k}) \cdot O,$$

so that $P'_n \geq \overline{P}_n$. Since, for each $n \in \mathbf{N}$,

$$P'_{n+1} - \overline{P}_{n+1} \leq (A_n - \overline{K}_n C_n)(P'_n - \overline{P}_n)(A_n - \overline{K}_n C_n)^*,$$

we see that $P'_n - P_n$ converges a.s. to O, as $n \to +\infty$. By stationarity, this implies that P'_n has the same law as P_n. But $P'_n \geq \overline{P}_n$, so that $P'_n = P_n$ for all $n \in \mathbf{Z}$. The stationary solution $(\overline{P}_n)$ is thus unique.

Lemma 4.2. *Let $\{(M_n, N_n), n \in \mathbf{Z}\}$ be a stationary ergodic a.s. stabilizable sequence, where M_n (resp. N_n) are $d \times d$ (resp. $d \times p$) matrices. We suppose that $\log^+ \|M_n\|$ is integrable and that there exists a sequence $R_n, n \in \mathbf{Z}$, of symmetric nonnegative matrices such that $R_{n+1} = M_n R_n M_n^* + N_n N_n^*$ and such that $\lim_{n\to+\infty} \frac{1}{n} \log^+ \|R_n\| = 0$, a.s. Then the upper Lyapunov exponent of the sequence $\{M_n, n \in \mathbf{Z}\}$ is negative.*

Proof: We will first suppose for simplicity that M_n is invertible. For any $n \in \mathbf{Z}$, let $T_n = N_n N_n^*$ and $M_{n,k} = M_n M_{n-1} \dots M_{n-k+1}$. For each $m \geq 1$,

$$R_{n+1} = \sum_{k=0}^{m-1} M_{n,k} T_{n-k} M_{n,k}^* + M_{n,m} R_{n-m+1} M_{n,m}^*.$$

Therefore, the increasing sequence $\{\sum_{k=0}^{m-1} M_{n,k} T_{n-k} M_{n,k}^*, m \geq 1\}$ is dominated by R_{n+1}. Let Q_{n+1} be its limit as $m \to +\infty$. Then (M_n, N_n, Q_n) is a stationary ergodic process, Q_n is nonnegative and

$$Q_{n+1} = M_n Q_n M_n^* + T_n. \tag{8}$$

Since M_n is supposed to be invertible, we see that the rank of Q_n is increasing. By ergodicity this implies that this rank is a.s. some constant q. Let now $\{\Gamma_n, n \in \mathbf{Z}\}$ be a stationary sequence of orthogonal $d \times d$ matrices such that $\overline{Q}_n := \Gamma_n^* Q_n \Gamma_n$ is a diagonal matrix in which the last $d-q$ diagonal terms are vanishing. Let $\overline{M}_n = \Gamma_{n+1}^{-1} M_n \Gamma_n$ and $\overline{N}_n = \Gamma_{n+1}^{-1} N_n$. It is readily seen using (8) that we can write

$$\overline{M}_n = \begin{pmatrix} M_n^1 & M_n^2 \\ 0 & A_n^3 \end{pmatrix}, \quad \overline{N}_n = \begin{pmatrix} N_n^1 \\ 0 \end{pmatrix}, \quad \overline{Q}_n = \begin{pmatrix} D_n & 0 \\ 0 & 0 \end{pmatrix},$$

where the size of $M_n^1, M_n^2, M_n^3, N_n^1, D_n$ is $q \times q, q \times (d-q), (d-q) \times (d-q), q \times p$ and $q \times q$ respectively. The sequence (M_n, N_n) is a.s. stabilizable. This implies immediately that, a.s.,

$$\overline{\lim}_{n\to+\infty} \frac{1}{n} \log \|M_n^3 \dots M_1^3\| < 0,$$

so that the upper Lyapunov exponent τ_3 of the sequence (M_n^3) is negative. On the other hand since

$$\sum_{k=0}^{r-1} M_n^1 \dots M_{n-k+1}^1 N_{n-k}^1 (M_n^1 \dots M_{n-k+1}^1 N_{n-k}^1)^*$$

increases to the almost surely nondegenerate matrix D_n, we can choose an integer $r \geq 1$ such that this expression is nondegenerate with positive probability. Without loss of generality we can suppose that $r = 1$ to simplify the notations (when $r > 1$ one make use of the sequence (R_{nr}) and prove that the upper exponent of the sequence $(M_{(n+1)r} \dots M_{nr+1})$ is negative; this exponent is r times the upper exponent of the sequence (M_n)). We remark that

$$D_{n+1} = M_n^1 D_n M_n^{1*} + N_n^1 N_n^{1*}$$

so that, if $H_n = D_{n+1}^{-1/2} M_n^1 D_n^{1/2}$ and $S_n = D_{n+1}^{-1/2} N_n^1 N_n^{1*} D_{n+1}^{-1/2}$, then $I = H_n H_n^* + S_n$. This implies that $\|H_n\| \leq 1$ and that $\mathbf{P}(\|H_n\| = 1) \leq \mathbf{P}(\det S_n > 0)$. But $\mathbf{P}(\det S_n > 0) < 1$ since we have supposed that $r = 0$. Thus it follows from Birkhoff's ergodic theorem that, a.s.,

$$\overline{\lim}_{n \to +\infty} \frac{1}{n} \log \|D_{n+1}^{-1/2} M_n^1 \dots M_1^1 D_1^{1/2}\| \leq \lim_{n \to +\infty} \frac{1}{n} \sum_{k=1}^{n} \log \|H_k\| < 0.$$

On the other hand,

$$\|M_n^1 \dots M_1^1\| \leq \|D_{n+1}^{1/2}\| \|D_{n+1}^{-1/2} M_n^1 \dots M_1^1 D_1^{1/2}\| \|D_1^{-1/2}\|.$$

Thus we can use the fact that $\lim_{n \to +\infty} \frac{1}{n} \log^+ \|R_n\| = 0$ to conclude that

$$\overline{\lim}_{n \to +\infty} \frac{1}{n} \log \|M_n^1 \dots M_1^1\| < 0,$$

so that the upper Lyapunov exponent τ_1 of the sequence (M_n^1) is negative. Finally, we know that the upper exponent of the sequence $\overline{M}_n$ is $\max(\tau_1, \tau_3)$ (see Furstenberg and Kifer (1983)). Thus this exponent is negative. It is obviously equal to the upper Lyapunov exponent of the sequence (M_n). This proves the lemma when M_n is a.s. invertible. Actually, we have only needed the fact that the matrices Q_n have a constant rank, a.s. In the general case one uses the induced systems on the sets $\Omega_r = \{\omega \in \Omega \, ; \, \mathrm{rank}(Q_0(\omega)) = r\}$; the proof is standard and not difficult and we don't give the details.

Remark: Following for instance the proof given in Anderson and Moore (1979, p. 81), it is easy (at least when P_0 is in $\mathcal{P}_0$) to deduce directly from Proposition 4.1 that (6) holds.

Bibliography

Anderson B.D.O. and Moore J.B. (1979), "Optimal filtering", Prentice Hall, New Jersey.

Anderson B.D.O. and Moore J.B. (1981), "Detectability and stabilizability of time-varying discrete time linear systems", *Siam J. Control and Optimization*, **19**, 20–32.

Bougerol Ph. (1990), "Filtre de Kalman Bucy et exposants de Lyapounov". To appear in the Proceedings of the Oberwolfach Conference on Lyapunov Exponents, Springer Lecture Notes.

Bougerol Ph. (1991a), " Kalman filtering with random coefficients and contractions", Preprint.

Bougerol Ph. (1991b), " Almost sure stabilizability and Riccati's equation for linear systems with random parameters", Preprint.

Furstenberg H. and Kifer Y. (1983), "Random matrix products and measures on projective spaces". *Israel J. Math.*, **46**, 12–32.

Kingman J.F.C. (1973), "Subadditive ergodic theory", *Ann. of Probab.*, **1**, p 883–909.

Whittle P. (1982),"Optimization over time", Vol. 1, Wiley.

Wojtkowski M. (1985),"Invariant families of cones and Lyapunov exponents", *Ergodic Th. Dyn. Systems*, **5**, 145–161.

MULTI-DIMENSIONAL FINITE–FUEL SINGULAR STOCHASTIC CONTROL

D.S. Bridge and S.E. Shreve
Department of Mathematics
Carnegie Mellon University
Pittsburgh, PA 15213

1. Introduction.

In their seminal paper, Beneš, Shepp & Witsenhausen (1980) considered the problem of singular control of a one–dimensional Brownian motion so as to minimize the infinite horizon, discounted, integrated second moment of the state process, subject to a "fuel constraint" on the total amount of control which could be exerted. Let $X(t)$ be the state of the controlled Brownian motion at time t, and let $Y(t)$ be the amount of remaining fuel. Beneš, Shepp and Witsenhausen constructed a function g such that the optimal control process is the one which forces $|X(t)| \leq g(Y(t))$ to hold for all times after the initial time, until fuel is exhausted. They had apparently considered the same problem for n–dimensional Brownian motion, and conjectured that a similar result would hold with $\|X(t)\|$ replacing $|X(t)|$ in the above inequality, and that g' is a Bessel function if n is even, and an elementary function if n is odd.

This paper provides a detailed solution to the n–dimensional problem. We find the form of the optimal control process claimed by Beneš, Shepp & Witsenhausen, except that our function g' is apparently neither a Bessel function nor an elementary function.

The solution constructed in this paper is another in a long line of solutions to singular stochastic control problems in which the value function is twice continuously differentiable, even across the "free boundary" where the optimal control process acts. This so–called "principle of smooth fit" has played a fundamental role in the solution of one–dimensional problems (e.g., Harrison, Sellke & Taylor (1983), Harrison & Taksar (1983), Harrison & Taylor (1978), Jacka (1983), Karatzas (1983), Karatzas (1985), Karatzas & Shreve (1986), Lehoczky & Shreve (1986), Shreve, Lehoczky & Gaver (1984), Sun (1987), Taksar (1985)), and has been observed in the few multi–dimensional problems which have been explicitly solved. The Beneš, Shepp and Witsenhausen finite–fuel problem is such an example, even when the controlled Brownian motion is one–dimensional, because the fuel process adds a second dimension to the state. It is known that under fairly general assumptions, including convexity of the cost function, the value function for the problem of singular stochastic control of a two–dimensional Brownian motion is twice continuously differentiable (Soner & Shreve (1989)), but general results concerning the smoothness of the value function for the control of higher–dimensional Brownian motion are unknown. Smoothness is known to hold in higher dimensions if control can be exerted in only one direction (Soner & Shreve (1991)). In this paper, we obtain smoothness for a particular problem of control of an n–dimensional Brownian motion in which the direction of control is not specified in advance.

2. Formulation of the control problem.

Let $W = \{W(t), \mathcal{F}(t); t \geq 0\}$ be an n–dimensional Brownian motion $(n \geq 2)$ on some probability space $(\Omega,\mathcal{F},P)$. We assume that the filtration $\{\mathcal{F}(t)\}_{t\geq 0}$ satisfies the "usual conditions" of right-continuity and augmentation by the P-null sets of $\mathcal{F}$.

2.1 *Definition.* An $\mathbb{R}^n$-valued, *RCLL function on* $[0-,\infty)$ is a mapping $\xi : \mathbb{R} \to \mathbb{R}^n$ which is right-continuous with left-hand limits at every point in $\mathbb{R}$ and which is constant on $(-\infty,0)$, so that $\xi(0-) = \xi(t)$ for all $t < 0$. We denote by $\check{\xi}(t)$ the total variation (in the Euclidean norm) of ξ on $(-\infty,t]$. We call $\check{\xi}(t)$ the *total variation of* ξ *on* $[0-,t]$.

2.2 *Definition.* Let $y \geq 0$ be given. An *admissible control process for initial fuel level* y is an $\{\mathcal{F}(t)\}$-adapted process whose every path is an $\mathbb{R}^n$-valued RCLL function ξ on $[0-,\infty)$, which satisfies $\xi(0-) = 0$, and whose total variation $\check{\xi}(\infty) \triangleq \lim_{t\to\infty} \check{\xi}(t)$ does not exceed y. We denote by $\mathcal{A}(y)$ the set of all such processes.

Given an *initial state* $x \in \mathbb{R}^n$, an *initial fuel level* $y \geq 0$, and a control process $\xi \in \mathcal{A}(y)$, we define the associated *fuel process*

$$Y(t) \triangleq y - \check{\xi}(t), \quad t \geq 0. \tag{2.1}$$

The associated *state process* is given by $X(0-) = x$ and

$$X(t) = x + W(t) - \xi(t), \quad t \geq 0. \tag{2.2}$$

The expected cost of using the control process ξ is

$$J(x,y;\xi) \triangleq E\int_0^\infty e^{-\alpha t}\,\|X(t)\|^2 dt, \tag{2.3}$$

where α is a positive discount factor. The optimal expected cost is

$$u(x,y) \triangleq \inf\{J(x,y;\xi);\ \xi\in\mathcal{A}(y)\}. \tag{2.4}$$

Our goal is to determine the *value function* u and to construct a control process which attains the infimum in (2.4).

3. The weak verification theorem.

The problem posed in §2 is radially symmetric in the state variable, so we expect the value function u to be of the form

$$u(x,y) = v(\|x\|^2,y), \quad \forall\ x \in \mathbb{R}^n,\ y \geq 0, \tag{3.1}$$

for some function v to be determined. We shall find v by solving the appropriate

Hamilton–Jacobi–Bellman variational inequality. In this section, we show that under certain weak conditions, any solution v to this variational inequality satisfies

$$u(x,y) \geq v(\|x\|^2,y), \quad \forall\ x \in \mathbb{R}^n,\ y \geq 0. \tag{3.2}$$

In §4 we show that under slightly stronger conditions, for each $y \geq 0$, there is a control process ζ^* satisfying $J(x,y;\ \zeta^*) = v(\|x\|^2,y)$. This fact forces equality in (3.2). Section 5 is devoted to the construction of a function v satisfying the conditions set forth in §§3 and 4. In §6, we show that this v has the required degree of smoothness, and in §7 we prove that v solves the Hamilton–Jacobi–Bellman variational inequality.

Actually, rather than constructing the function v directly, we construct the function v^* related to v by

$$v^*(r,y) = v(r^2,y), \quad r \geq 0, \quad y \geq 0. \tag{3.3}$$

When no control is exerted, the state process behaves like a Bessel process, whose infinitesimal generator is $\frac{n-1}{2r}\frac{\partial}{\partial r} + \frac{1}{2}\frac{\partial^2}{\partial r^2}$. This fact will be seen to correspond to the inequality

$$\alpha v^*(r,y) - \frac{n-1}{2r} v_r^*(r,y) - \frac{1}{2} v_{rr}^*(r,y) \leq r^2, \tag{3.4}$$

with equality holding whenever $(\|x_t\|,y_t)$ is in the region in whose interior no control should be exerted. (Here and elsewhere, partial derivatives are denoted by subscripts.) When control is exerted, it should displace the state directly toward the origin, reducing fuel by the amount of the displacement. This fact will be seen to correspond to the inequality

$$v_r^*(r,y) + v_y^*(r,y) \leq 0, \tag{3.5}$$

with equality holding whenever $(\|x_t\|,y_t)$ is in a region where control should be exerted.

For the function v related to v^* by (3.3), or equivalently

$$v(\rho,y) = v^*(\sqrt{\rho},y), \quad \rho \geq 0, \quad y \geq 0, \tag{3.6}$$

inequalities (3.4) and (3.5) take the form

$$\alpha v(\rho,y) - n v_\rho(\rho,y) - 2\rho v_{\rho\rho}(\rho,y) \leq \rho, \tag{3.7}$$

$$2\sqrt{\rho}\, v_\rho(\rho,y) + v_y(\rho,y) \leq 0. \tag{3.8}$$

The variables (r,y) are convenient because (3.5) is considerably simpler than its analogue (3.8). However, (3.4) has a singularity at $r = 0$ which is more easily handled in (3.7). Thus, we find it convenient to switch between the functions v and v^*.

The fact that we expect for each $(\rho,y) \in [0,\infty)^2$ to have equality in either (3.7) or (3.8) is captured by the *Hamilton–Jacobi–Bellman variational inequality*

$$\max\{\alpha v - n v_\rho - 2\rho v_{\rho\rho} - \rho,\ 2\sqrt{\rho}\, v_\rho + v_y\} = 0 \tag{3.9}$$

on $[0,\infty)^2$. We are prepared to state our first theorem.

3.1 ***Weak Verification Theorem.*** Suppose $v : [0,\infty)^2 \to [0,\infty)$ is continuous and the first and second partial derivatives of v are defined and continuous on $(0,\infty)^2$ with continuous extensions to all of $[0,\infty)^2$. Assume that the function $(x,y) \to v(\|x\|^2,y)$ is convex. Assume further that for some constant K,

$$0 \le v_\rho(\rho,y) \le K,\quad 0 \le v(\rho,y) \le K(1+\rho),\quad \forall\ \rho \ge 0,\ y \ge 0, \tag{3.10}$$

and v satisfies the variational inequality (3.9) on $[0,\infty)^2$. Then (3.2) holds.

3.2 ***Remark.*** Being the value function for a control problem in which the state depends linearly on the control and the cost depends quadratically on the state, the function u is fully convex. Therefore, if a function v is to satisfy (3.1) then it will also satisfy the convexity hypothesis of Theorem 3.1.

In order to prove Theorem 3.1, we need a lemma about real analysis.

3.3 ***Lemma.*** Let ξ and λ be $\mathbb{R}^n$–valued, RCLL functions on $[0-,\infty)$, and assume that $\check{\xi}(t) < \infty$ for every $t \ge 0$. Then

$$\left|\int_{0-}^{t} \lambda(s-)\cdot d\xi(s)\right| \le \int_{0-}^{t} \|\lambda(s-)\|\, d\check{\xi}(s),\quad \forall\ t \ge 0. \tag{3.11}$$

Proof: Let $t > 0$ and $\varepsilon > 0$ be given. The RCLL property of λ allows us to construct a piecewise constant, $\mathbb{R}^n$–valued, RCLL function γ on $[0-\infty)$ such that $\sum_{i=1}^{n} |\gamma_i(s) - \lambda_i(s)| \le \varepsilon$ for all $s \in \mathbb{R}$. In particular, there is a partition $0 = t_0 < t_1 < \ldots < t_m = t$ such that γ is constant on each $[t_{k-1},t_k)$. We have from the Cauchy–Schwarz inequality that

$$\left|\int_{0-}^{t}\gamma(s-)\cdot d\xi(s)\right| \le |\gamma(0-)\cdot(\xi(0)-\xi(0-)) + \sum_{k=1}^{n}\gamma(t_{k-1})\cdot(\xi(t_k)-\xi(t_{k-1}))|$$

$$\le \|\gamma(0-)\|\,\|\xi(0)-\xi(0-)\| + \sum_{k=1}^{m}\|\gamma(t_{k-1})\|\,\|\xi(t_k)-\xi(t_{k-1})\| \tag{3.12}$$

$$\le \int_{0-}^{t}\|\gamma(s-)\|d\check{\xi}(s).$$

Denoting by $\check{\xi}_j(t)$ the total-variation on $[0-,t]$ of the real-valued function ξ_j, we have

$$\left|\int_{0-}^{t}\lambda(s-)\cdot d\xi(s) - \int_{0-}^{t}\gamma(s-)\cdot d\xi(s)\right| \le \sum_{i=1}^{n}\int_{0-}^{t}|\lambda_i(s-)-\gamma_i(s-)|\,d\check{\xi}_i(s)$$

$$\le \varepsilon\sum_{i=1}^{n}\check{\xi}_i(t) = n\,\varepsilon\,\check{\xi}(t), \tag{3.13}$$

and

$$\left|\int_{0-}^{t}\|\lambda(s-)\|d\check{\xi}(s) - \int_{0-}^{t}\|\gamma(s-)\|d\check{\xi}(s)\right| \le \int_{0-}^{t}\|\lambda(s-)-\gamma(s-)\|d\check{\xi}(s) \le \varepsilon\check{\xi}(t). \tag{3.14}$$

Putting (3.12) – (3.14) together and letting $\varepsilon \downarrow 0$, we obtain (3.11).□

Proof of Theorem 3.1: Let $x \in \mathbb{R}^n$, $y \ge 0$ be given. If $u(x,y) = \infty$, then (3.2) holds. We assume therefore that $u(x,y) < \infty$. Let $\xi \in \mathscr{A}(y)$ be a control process for which $J(x,y;\xi) < \infty$. It follows from (2.2) that

$$d(\|X(t)\|^2) = 2X(t)\cdot dX(t) + n\,dt. \tag{3.15}$$

By assumption,

$$E\int_0^\infty e^{-\alpha t}\|X(t)\|^2 dt < \infty \text{ a.s.,} \tag{3.16}$$

so we can define a sequence of almost surely finite stopping times

$$\tau_n \triangleq \inf\{t \ge n;\ e^{-\alpha t}\|X(t)\|^2 \le \tfrac{1}{n}\}. \tag{3.17}$$

Then $\lim_{n\to\infty}\tau_n = \infty$, and

$$\lim_{n\to\infty} Ee^{-\alpha\tau_n}\|X(\tau_n)\|^2 = 0. \tag{3.18}$$

According to the Doléans–Dade/Meyer generalization of Itô's formula for RCLL process (e.g., Elliott (1982), Theorem 12.13), we have

$$E\, e^{-\alpha\tau_n} v(\|X(\tau_n)\|^2, Y(\tau_n))$$

$$= v(\|x\|^2,y) + E\int_0^{\tau_n} e^{-\alpha s}[-\alpha v + n v_\rho + 2\|X(s-)\|^2 v_{\rho\rho}]ds + 2E\int_0^{\tau_n} e^{-\alpha s} v_\rho\, X(s-)\cdot dW(s)$$

$$- E\int_{0-}^{\tau_n} e^{-\alpha s}[2v_\rho\, X(s-)\cdot d\xi(s) + v_y d\check{\xi}(s)] + E\sum_{0\le s\le \tau_n} e^{-\alpha s}[v(\|X(s)\|^2, Y(s)) \tag{3.19}$$

$$- v(\|X(s-)\|^2, Y(s-)) + 2v_\rho X(s-)\cdot(\xi(s) - \xi(s-)) + v_y(\check{\xi}(s) - \check{\xi}(s-))].$$

Here and below, v and all its derivatives are evaluated at $(\|X(s-)\|^2, Y(s-))$ unless otherwise indicated. Because v_ρ is bounded and (3.16) holds, the second integral on the right–hand side of (3.19) has expectation zero. According to the variational inequality (3.9), the first integral is bounded below by $-E\int_0^{\tau_n} e^{-\alpha s}\|X(s-)\|^2 ds = -E\int_0^{\tau_n} e^{-\alpha s}\|X(s)\|^2 ds$. From Lemma 3.3, (3.11), (3.10), and (3.9), we also have

$$- E\int_{0-}^{\tau_n} e^{-\alpha s}[2v_\rho\, X(s-)\cdot d\xi(s) + v_y d\check{\xi}(s)] \ge - E\int_{0-}^{\tau_n} e^{-\alpha s}[2v_\rho\|X(s-)\| + v_y]d\check{\xi}(s) \ge 0. \tag{3.20}$$

Finally, suppose that $s \ge 0$ is a number for which $\check{\xi}(s) - \check{\xi}(s-) \ne 0$. The convexity of v implies

$$v(\|X(s)\|^2, Y(s)) - v(\|X(s-)\|^2, Y(s-)) + 2v_\rho X(s-)\cdot(\xi(s) - \xi(s-) + v_y(\check{\xi}(s) - \check{\xi}(s-)) \ge 0. \tag{3.21}$$

Putting all these facts into (3.19), we obtain

$$v(\|x\|^2,y) \le E\, e^{-\alpha\tau_n} v(\|X(\tau_n)\|^2, Y(\tau_n)) + E\int_0^{\tau_n} e^{-\alpha s}\|X(s)\|^2 ds. \tag{3.22}$$

Letting $n \to \infty$ and using (3.10) and (3.18), we conclude that

$$v(\|x\|^2,y) \le J(x,y;\xi). \tag{3.23}$$

Since this is true for every control process ξ, (3.2) holds. □

4. The strong verification theorem.

In this section we begin with a function v satisfying the hypotheses of Theorem 3.1. We further assume that v divides $(0,\infty)^2$ into two connected regions, according to whether

$$\alpha v(\rho,y) - n\, v_\rho(\rho,y) - 2\rho\, v_{\rho\rho}(\rho,y) - \rho = 0 \tag{4.1}$$

or

$$2\sqrt{\rho}\, v_\rho(\rho,y) + v_y(\rho,y) = 0, \tag{4.2}$$

and these regions are separated by a "free boundary". We also assume the existence of a control process which causes reflection of $(\|X(t)\|,Y(t))$ at this free boundary. Under all these assumptions, we will prove that v determines the value function u via (3.1), and the just mentioned control process is optimal. The existence of the optimal control process then follows from Theorem 4.1 of Lions & Sznitman (1984).

4.1 ***Strong Verification Theorem.*** Suppose $v : [0,\infty)^2 \to [0,\infty)$ satisfies the hypotheses of Theorem 3.1, and assume the existence of a strictly decreasing function $g : [0,\infty) \to (0,\infty)$ such that:

(i) (4.1) holds whenever $y = 0$ and whenever $y > 0$, $0 \le \sqrt{\rho} \le g(y)$;

(ii) (4.2) holds whenever $y > 0$, $\sqrt{\rho} \ge g(y)$.

Assume also that for each $x \in \mathbb{R}^n$, $y \ge 0$, there is a control process ξ^* whose every path is continuous, except possibly at time zero, and which causes the associated state and fuel processes X^* and Y^* starting at $X^*(0-) = x$, $Y^*(0-) = y$ to satisfy

(iii) $d\xi^*(t) = \dfrac{X^*(t-)}{\|X^*(t-)\|}\, d\check{\xi}^*(t), \quad t \ge 0,$

(iv) if $\|x\| > g(y)$, then $\|X^*(0)\| = g(Y^*(0))$ or else $Y^*(0) = 0$ and $\|X^*(0)\| > g(0)$;

(v) if $\|x\| \le g(y)$, then $X^*(0) = x$;

(vi) $\|X^*(t)\| \le g(Y^*(t)), \quad 0 < t \le \tau^*$;

(vii) $\int_0^{\tau^*} 1_{\{X^*(t) \neq g(Y^*(t))\}}\, d\check{\xi}^*(t) = 0,$

where $\tau^* \triangleq \inf\{t \ge 0\,;\, Y^*(t) = 0\}$ is the time of fuel depletion. Then (3.1) holds and ξ^* is optimal, i.e.,

$$u(x,y) = v(\|x\|^2,y) = J(x,y;\xi^*). \tag{4.3}$$

4.2 ***Remark.*** The process ξ^* in Theorem 4.1 is continuous, except that it can cause the state to jump at the initial time. According to (iii)–(v), an initial jump occurs if and only if $\|x\| > g(y)$, and then it causes state displacement of X^* in the $-x$ direction so as to bring $(X^*(0),Y^*(0))$ to the free boundary, or, if that is not possible, to bring $Y^*(0)$ to 0. If $Y^*(0) > 0$, control is exerted thereafter only at the boundary (cf. (vii)) so as to ensure (vi), and when control is exerted, it always pushes the state directly toward the origin (cf. (iii)). After Y^* falls to zero, no further control is exerted.

4.3 ***Remark.*** Theorem 4.1 of Lions & Sznitman guarantees that the process ξ^* in our Theorem 4.1 can be constructed, provided that the function g is twice continuously differentiable. We construct this g in §5, and discover that it is actually of class C^∞.

Proof of Theorem 4.1: We return to the proof of Theorem 3.1 for the purpose of obtaining equality in (3.23) when ξ^* is substituted for ξ. We shall then have, from Theorem 3.1 and the definition of u, that $u(x,y) \le J(x,y;\xi^*) = v(\|x\|^2,y) \le u(x,y)$, and (4.3) will follow.

Let us reexamine the terms in (3.19). From conditions (i) and (vi), we see that for $t > 0$, $(X^*(t), Y^*(t))$ is in the region where (4.1) holds, so

$$E\int_0^{\tau_n} e^{-\alpha s}[-\alpha v + nv_\rho + 2\|X^*(s-)\|v_{\rho\rho}]ds = -E\int_0^{\tau_n} e^{-\alpha s}\|X^*(s)\|^2 ds.$$

From (ii), (iii) and (vii), we see that

$$-E\int_0^{\tau_n} e^{-\alpha s}[2v_\rho X^*(s-)\cdot d\xi^*(s) + v_y\, d\check{\xi}^*(s)] = -E\int_0^{\tau_n} e^{-\alpha s}[2\|X^*(s-)\|\, v_\rho + v_y]\, d\check{\xi}^*(s) = 0.$$

Finally, the summation appearing on the right–hand side of (3.19) is trivially zero unless $\|x\| > g(y)$. Assume $\|x\| > g(y)$; then $\frac{\xi(0)}{\|\xi(0)\|} = \frac{x}{\|x\|}$, and the summation contains the single term

$$v(\|X^*(0)\|^2, Y^*(0)) - v(\|x\|^2,y) + [2\|x\|v_\rho(\|x\|^2,y) + v_y(\|x\|^2,y)]\check{\xi}(0). \tag{4.4}$$

Let us define $\theta(\eta) = v((\|x\|-\eta)^2, y-\eta)$ for $0 \le \eta \le \|\xi(0)\|$. For $\eta = \|\xi(0)\|$, we have

$$\|x\|-\eta = \|x\|\left(1 - \frac{\|\xi(0)\|}{\|x\|}\right) = \left\|x - \frac{x}{\|x\|}\|\xi(0)\|\right\| = \|x-\xi(0)\| = \|X^*(0)\|, \tag{4.5}$$

$$y - \eta = y - \|\xi(0)\| = y - \check{\xi}(0) = Y^*(0). \tag{4.6}$$

Because $\|x\|-\eta - g(y-\eta)$ is strictly decreasing in η, there is at most one value of η which makes this expression zero. By (iv), (4.5) and (4.6), this expression is positive for $0 \le \eta < \|\xi(0)\|$, and (ii) implies that

$$\theta'(\eta) = -2(\|x\|-\eta)v_\rho((\|x\|-\eta)^2,y-\eta) - v_y((\|x\|-\eta)^2,y-\eta) = 0, \quad \forall\, \eta \in (0,\|\xi(0)\|).$$

Therefore, $\theta(\|\xi(0)\|) - \theta(0) = 0$ and $\theta'(0) = 0$, which shows that the expression in (4.4) is zero. From these considerations, we see that instead of (3.22) we have

$$v(\|x\|^2,y) = E\, e^{-\alpha\tau_n^*} v(\|X^*(\tau_n)\|^2, Y^*(\tau_n)) + E\int_0^{\tau_n^*} e^{-\alpha s}\|X^*(s)\|^2 ds. \tag{4.7}$$

The sequence of stopping times $\{\tau_n^*\}_{n=1}^\infty$ is constructed as in the proof of Theorem 3.1; this is possible because for $t \ge 0$, $\|X^*(t)\|^2 \le 2\|x+W(t)\|^2 + 2\|\xi_t^*\|^2 \le 2\|x+W(t)\|^2 + 2y^2$, so $\int_0^\infty e^{-\alpha t}\|X^*(t)\|^2 dt < \infty$ almost surely. Letting $n \to \infty$ in (4.7), we obtain $v(\|x\|^2,y) = J(x,y;\xi)$.□

5. Construction of the value function

As noted earlier, rather than constructing the function v of Theorem 4.1, we shall construct the function v^* related to v by (3.3). We will construct a positive number r_0 and a C^∞, strictly decreasing function $g:[0,\infty)\xrightarrow{onto}(0,r_0]$. Using g, we define three open regions

$$R_\ell^* \triangleq \{(r,y);\ y>0,\ 0<r<g(y)\},$$
$$R_m^* \triangleq \{(r,y);\ y>0,\ g(y)<r<r_0+y\},$$
$$R_r^* \triangleq \{(r,y);\ y>0,\ r>r_0+y\},$$

and their boundaries

$$B_\ell^* \triangleq \{(r,y);\ y>0,\ r=0\},\quad B_m^* \triangleq \{(r,y);\ y>0,\ r=g(y)\},$$

$$B_r^* \triangleq \{(r,y);\ y>0,\ r=r_0+y\},\quad B_0^* \triangleq \{(r,y);\ y=0,\ r>0\}.$$

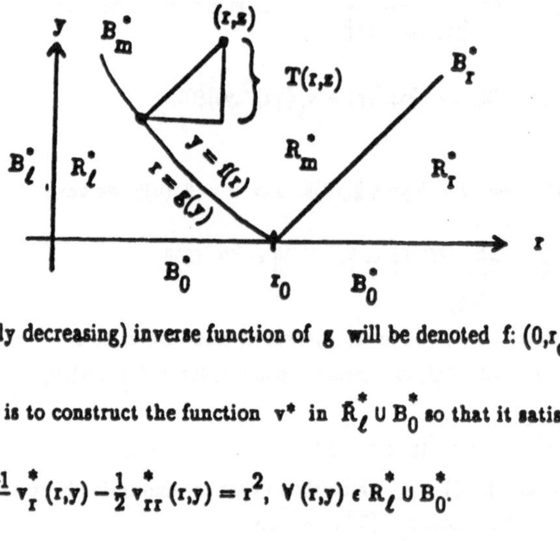

The (C^∞, strictly decreasing) inverse function of g will be denoted $f:(0,r_0]\xrightarrow{onto}[0,\infty)$.

The idea is to construct the function v^* in $\bar{R}_\ell^* \cup B_0^*$ so that it satisfies (cf. (3.4))

$$\alpha v^*(r,y) - \frac{n-1}{2r} v_r^*(r,y) - \frac{1}{2} v_{rr}^*(r,y) = r^2,\ \forall (r,y) \in R_\ell^* \cup B_0^*. \tag{5.1}$$

This is actually a second–order, *ordinary* differential equation for $v^*(\cdot,y)$ and so has the general solution

$$v^*(r,y) = \frac{r^2}{\alpha} + \frac{n}{\alpha^2} + A_1(y)\,\varphi_1(r) + A_2(y)\,\varphi_2(r), \tag{5.2}$$

where φ_1 and φ_2 are solutions to the homogeneous equation

$$\alpha\varphi(r) - \frac{n-1}{2r}\varphi'(r) - \frac{1}{2}\varphi''(r) = 0. \tag{5.3}$$

From the fact that

$$u(x,0) = E\int_0^\infty e^{-\alpha t}\|x+W(t)\|^2 dt = \frac{\|x\|^2}{\alpha} + \frac{n}{\alpha^2},$$

we see that $v^*(r,0)$ should be $\frac{r^2}{\alpha} + \frac{n}{\alpha^2}$, and thus $A_1(0) = A_2(0) = 0$. Because $u(\cdot,y)$ has a minimum at $x = 0$, we should have $v_r^*(0,y) = 0$, and thus $A_1(y)\varphi_1'(0) + A_2(y)\varphi_2'(0) = 0$ for every $y \in [0,\infty)$. This requires that either $\varphi_1'(0) = \varphi_2'(0) = 0$ and consequently $\varphi_1 = c\,\varphi_2$ for some constant c, or else $A_1 = -\frac{\varphi_2'(0)}{\varphi_1'(0)}A_2$. In either case, we can write $A_1(y)\varphi_1(r) + A_2(y)\varphi_2(r) = A(y)\varphi(r)$, where φ solves (5.3), and

$$A(0) = 0,\ \varphi'(0) = 0. \tag{5.4}$$

We may normalize φ by setting $\varphi(0) = 1$, absorbing the necessary constant into A. We can then compute

$$\varphi(r) = 1 + \sum_{k=1}^{\infty} \frac{\alpha^k r^{2k}}{k!n(n+2)\ldots(n+2k-2)},\ \forall\, r \geq 0. \tag{5.5}$$

With this function φ, we have

$$v^*(r,y) = \frac{r^2}{\alpha} + \frac{n}{\alpha^2} + A(y)\varphi(r),\ \forall\,(r,y) \in \bar{R}_\ell^* \cup B_0^*, \tag{5.6}$$

where $A(\cdot)$ is still to be defined.

Note from (5.5) that φ and all its derivatives are nonnegative and increasing, and

$$\varphi(r) = 1 + \frac{\alpha r^2}{n} + 0(r^4),\ \varphi'(r) = \frac{2\alpha r}{n} + 0(r^3),\ \varphi''(r) = \frac{2\alpha}{n} + 0(r^2) \geq \frac{2\alpha}{n}. \tag{5.7}$$

The function $r\varphi'(r) - \varphi(r)$ has derivative $r\varphi''(r)$, which is bounded below by $\frac{2\alpha r}{n}$. Because $(r\varphi'(r) - \varphi(r))|_{r=0} = -1$, there is a unique number r_0 satisfying

$$r\varphi'(r) - \varphi(r) < 0,\ \forall\, r \in [0,r_0),\quad r_0\varphi'(r_0) - \varphi(r_0) = 0,\ r_0 > \sqrt{\frac{n-1}{2\alpha}}. \tag{5.8}$$

The bound on r_0 is obtained by setting $r = \sqrt{\frac{n-1}{2\alpha}}$ in (5.3) and rewriting the resulting equation as

$$\sqrt{\frac{n-1}{2a}}\,\varphi'\Big(\sqrt{\frac{n-1}{2a}}\Big)-\varphi\Big(\sqrt{\frac{n-1}{2a}}\Big)=-\frac{1}{2a}\,\varphi''\Big(\sqrt{\frac{n-1}{2a}}\Big)<0.$$

Note also that $\frac{d}{dr}[r\varphi''(r)-\varphi'(r)]=r\varphi'''(r)>0$ for all $r>0$, and $[r\varphi''(r)-\varphi'(r)]\big|_{r=0}=0$, so

$$r\varphi''(r)-\varphi'(r)>0,\quad \forall\, r>0. \tag{5.9}$$

We want v^* to be C^2, and to satisfy (cf. (3.5))

$$v^*_r(r,y)+v^*_y(r,y)=0,\quad \forall\,(r,y)\in \bar{R}^*_m\cup\bar{R}^*_r. \tag{5.10}$$

In particular, (5.10) should hold along the free boundary B^*_m, from which we conclude that

$$\frac{2r}{a}+A(f(r))\varphi'(r)+A'(f(r))\varphi(r)=0,\quad \forall\, r\in(0,r_0]. \tag{5.11}$$

Differentiating (5.10) with respect to r, we obtain

$$v^*_{rr}(r,y)+v^*_{yr}(r,y)=0,\quad \forall\,(r,y)\in\bar{R}^*_m\cup\bar{R}^*_r, \tag{5.12}$$

and evaluation of this equation along the free boundary leads to

$$\frac{2}{a}+A(f(r))\varphi''(r)+A'(f(r))\varphi'(r)=0,\quad \forall\, r\in(0,r_0]. \tag{5.13}$$

We may solve (5.11) and (5.13) for

$$A(f(r))=\psi_1(r),\quad A'(f(r))=\psi_2(r),\quad \forall\, r\in(0,r_0], \tag{5.14}$$

where

$$\psi_1(r)\triangleq-\frac{2}{a}\left[\frac{r\varphi'(r)-\varphi(r)}{\Phi(r)}\right],\quad \psi_2(r)\triangleq-\frac{2}{a}\left[\frac{-r\varphi''(r)+\varphi'(r)}{\Phi(r)}\right],\quad \forall\, r\geq 0, \tag{5.15}$$

$$\Phi(r)\triangleq(\varphi'(r))^2-\varphi(r)\varphi''(r),\quad \forall\, r\geq 0. \tag{5.16}$$

Using (5.3), we obtain

$$\Phi'(r)=-\frac{(n-1)}{r}(\varphi'(r))^2+\frac{2a(n-1)}{r}(\varphi(r))^2-\frac{n(n-1)}{r^2}\varphi(r)\varphi'(r),\quad \forall\, r>0 \tag{5.17}$$

so that

$$\Phi(0)=\frac{-2a}{n},\quad \Phi'(0)\triangleq\lim_{r\downarrow 0}\Phi'(r)=0. \tag{5.18}$$

Direct computation relying on (5.3) also shows that

$$\frac{d}{dr}[r^{n-1}\Phi(r)] = -(n-1)r^{n-3}\varphi(r)\varphi'(r) < 0, \ \forall\, r > 0, \tag{5.19}$$

$$\frac{d}{dr}[r^{n+1}\Phi'(r)] = -2(n-1)r^{n-1}(\varphi'(r))^2 < 0, \ \forall\, r > 0. \tag{5.20}$$

It follows upon integration of (5.19) and (5.20) that both Φ and Φ' are strictly negative on $(0,\infty)$, and ψ_1 and ψ_2 are well-defined. From (5.8) and (5.9), we see that

$$\psi_1(r) < 0, \ \psi_2(r) < 0, \quad \forall\, r \in [0,r_0). \tag{5.21}$$

We differentiate the first equation in (5.14) with respect to r and divide by the second equation to obtain the formula

$$f'(r) = \frac{\psi_1'(r)}{\psi_2(r)} = -\frac{r\varphi''(r)\Phi(r) - [r\varphi'(r) - \varphi(r)]\Phi'(r)}{\Phi(r)[r\varphi''(r) - \varphi'(r)]}, \ \forall\, r \in (0,r_0]. \tag{5.22}$$

To have $f(r_0) = 0$, we now define

$$f(r) \triangleq -\int_r^{r_0} \frac{\psi_1'(s)}{\psi_2(s)}\, ds, \quad \forall\, r \in (0,r_0]. \tag{5.23}$$

In light of (5.8), (5.9), the negativity of Φ and Φ', and the positivity of φ'', we see from (5.22) that f is strictly decreasing.

5.1 ***Lemma.*** The strictly decreasing, C^∞ function f defined by (5.23) maps $(0,r_0] \xrightarrow{\text{onto}} [0,\infty)$.

Proof: It remains to show that $\lim_{r \downarrow 0} f(r) = \infty$. We have

$$\begin{aligned} f(r) &\geq \int_r^{r_0} \frac{s\varphi''(s)}{s\varphi''(s) - \varphi'(s)}\, ds \geq \frac{\varphi''(0)}{\varphi'''(r_0)} \int_r^{r_0} \frac{s\varphi'''(s)}{s\varphi''(s) - \varphi'(s)}\, ds \\ &= \frac{\varphi''(0)}{\varphi'''(r_0)} \ln\left[\frac{r_0\varphi''(r_0) - \varphi'(r_0)}{r\varphi''(r) - \varphi'(r)}\right], \end{aligned}$$

which has limit ∞ as $r \downarrow 0$. □

Let the mapping $g:[0,\infty) \xrightarrow{\text{onto}} (0,r_0]$ be the (C^∞, strictly decreasing) inverse of f. Using g, we may now convert the first part of (5.14) into a definition for the function A:

$$A(y) \triangleq \psi_1(g(y)), \quad \forall\, y \geq 0. \tag{5.24}$$

Note that $A(0) = \psi_1(r_0) = 0$ because of (5.8), so (5.4) holds. Furthermore, (5.22) implies that

$$A'(y) = \psi_1'(g(y))g'(y) = \frac{\psi_1'(g(y))}{f'(g(y))} = \psi_2(g(y)), \quad \forall\, y \geq 0,$$

so A satisfies the second equation in (5.14). Equations (5.14) imply (5.11) and (5.13), so A satisfies these equations as well.

With A given by (5.24) and φ given by (5.5), we define v^* on $\bar{R}_\ell^* \cup B_0^*$ by (5.6) and note that (5.1) holds. We also define $T : \bar{R}_m^* \to [0,\infty)$ by

$$T(r,y) \triangleq \inf\{t \geq 0;\ y-t = f(r-t)\}, \quad \forall\,(r,y) \in \bar{R}_m^*, \tag{5.25}$$

and we set

$$v^*(r,y) = \begin{cases} v^*(r-T(r,y), y-T(r,y)), & \forall\,(r,y) \in \bar{R}_m^*, \\ v^*(r-y,0), & \forall\,(r,y) \in \bar{R}_r^*, \end{cases} \tag{5.26}$$

so that v^* is constant along 45° lines in $\bar{R}_m^* \cup R_r^*$, as suggested by (5.10). This completes the construction of v^*.

6. Regularity of the value function

It is clear from the definition (5.6) that v^* is C^∞ on R_ℓ^*, and the derivatives of v^* have continuous extensions to $\bar{R}_\ell^*$. In this section we show that the first and second partial derivatives of v^* have continuous extensions to $[0,\infty)^2$. From this we shall conclude that v (related to v^* by (3.3) or (3.6)) satisfies the smoothness hypotheses of Theorem 3.1.

6.1 *Lemma*. The limits of the derivatives of v^* from inside R_ℓ^* on B_m^* satisfy $v_r^* + v_y^* = 0$, $v_{rr}^* + v_{ry}^* = 0$, and $v_{ry}^* + v_{yy}^* = 0$.

Proof: The first two equations follow from (5.11) and (5.13). The last equation is a result of the second equation, the strict negativity of f', and

$$\frac{d}{dr}\left[v_r^*(r,f(r)) + v_y^*(r,f(r))\right] = 0, \quad \forall\, r \in (0,r_0]. \qquad \square$$

6.2 ***Remark.*** Recalling (5.6), we may rewrite the equation $v^*_{ry}(r,f(r)) + v^*_{yy}(r,f(r)) = 0$ as

$$A'(f(r))\varphi'(r) + A''(f(r))\varphi(r) = 0, \quad \forall\, r \in (0,r_0]. \tag{6.1}$$

6.3 ***Lemma.*** For $(r,y) \in R^*_m$, the derivatives v_r, v_y, v_{rr}, v_{ry} and v_{yy} evaluated at (r,y) are equal to these same derivatives at $(r-T(r,y),\, y - T(r,y))$, the latter being defined as limits from within R^*_ℓ.

Proof: According to the Implicit Function Theorem, the mapping T of (5.25) is C^1. Indeed, since $y - T(r,y) = f(r-T(r,y))$, we have

$$T_r(r,y) = \frac{-f'(r-T(r,y))}{1-f'(r-T(r,y))} > 0, \quad T_y(r,y) = \frac{1}{1-f'(r-T(r,y))} > 0, \quad \forall\, (r,y) \in R^*_m. \tag{6.2}$$

From (5.26) and Lemma 6.1, we have

$$v^*_r(r,y) = (1-T_r)v^*_r(r-T,y-T) - T_r v^*_y(r-T,y-T) = v^*_r(r-T,y-T),$$

$$v^*_y(r,y) = -T_y v^*_r(r-T,y-T) + (1-T_y)v^*_y(r-T,y-T) = v^*_y(r-T,y-T),$$

$$v^*_{rr}(r,y) = (1-T_r)v^*_{rr}(r-T,y-T) - T_r v^*_{ry}(r-T,y-T) = v^*_{rr}(r-T,y-T),$$

$$v^*_{ry}(r,y) = -T_y v^*_{rr}(r-T,y-T) + (1-T_y)v^*_{ry}(r-T,y-T) = v^*_{ry}(r-T,y-T),$$

$$v^*_{yy}(r,y) = -T_y v^*_{ry}(r-T,y-T) + (1-T_y)v^*_{yy}(r-T,y-T) = v^*_{yy}(r-T,y-T).$$

□

6.4 ***Corollary.*** v^* is of class C^2 on $R^*_\ell \cup B^*_m \cup R^*_m$.

On R^*_r we have from (5.6), (5.26) that $v^*(r,y) = \frac{(r-y)^2}{\alpha} + \frac{n}{\alpha^2}$. This results in the formulas

$$v^*_r(r,y) = \frac{2}{\alpha}(r-y),\ v^*_y = -\frac{2}{\alpha}(r-y),\ v^*_{rr}(r,y) = \frac{2}{\alpha},\quad v^*_{ry}(r,y) = -\frac{2}{\alpha},\ v^*_{yy}(r,y) = \frac{2}{\alpha} \tag{6.3}$$

for the derivatives of v^*. As (r,y) approaches B^*_r, these expressions converge to those obtained by replacement of (r,y) by $(r_0,0)$. On R^*_m, we have the formulas in the proof of Lemma 6.3, and as (r,y) approaches B^*_r, these converge to the values computed from (5.6):

$$\begin{aligned} &v^*_r(r_0,0) = \frac{2r_0}{\alpha} + A(0)\varphi'(r_0), \quad v^*_y(r_0,0) = A'(0)\varphi(r_0), \\ &v^*_{rr}(r_0,0) = \frac{2}{\alpha} + A(0)\varphi''(r_0), \quad v^*_{ry}(r_0,0) = A'(0)\varphi'(r_0), \quad v^*_{yy}(r_0,0) = A''(0)\varphi(r_0). \end{aligned} \tag{6.4}$$

On the other hand,

$$A(0) = 0,\ A'(0) = -\frac{2r_0}{\alpha\varphi(r_0)} = -\frac{2}{\alpha\varphi'(r_0)},\ A''(0) = \frac{2}{\alpha\varphi(r_0)}. \tag{6.5}$$

The first of these equalities appears in (5.4); the second follows from the first, (5.11) and (5.8); the third is obtained from the second and (6.1). Substitution of (6.5) into (6.4) gives the same result as letting $(r,y) \to (r_0,0)$ in (6.3), and so the first and second partial derivatives of v^* are continuous across B_r^*.

Finally, observe that as $y \downarrow 0$, the formulas for v_r^* and v_{rr}^* in (6.3) agree with those obtained from (5.6) on B_0^*. We have proved the following theorem.

6.5 *Theorem.* The function v^* defined by (5.6), (5.26) is of class C^2 on $(0,\infty)^2$, and the first and second partial derivatives of v^* have continuous extensions to $[0,\infty)^2$.

6.6 *Corollary.* The function v related to v^* by (3.3), or equivalently, (3.6), is of class C^2 on $(0,\infty)^2$, and the first and second partial derivatives of v have continuous extensions to $[0,\infty)^2$.

Proof: The only item which needs to be checked is the existence of limits as $\rho \downarrow 0$ for

$$v_\rho(\rho,y) = \frac{1}{2\sqrt{\rho}} v_r^*(\sqrt{\rho}, y) = \frac{1}{\alpha} + \frac{1}{2\sqrt{\rho}} A(y)\varphi'(\sqrt{\rho}), \tag{6.6}$$

$$v_{\rho\rho}(\rho,y) = -\frac{1}{4\rho^{3/2}} v_r^*(\sqrt{\rho}, y) + \frac{1}{4\rho} v_{rr}^*(\sqrt{\rho}, y) = A(y)\left[-\frac{1}{4\rho^{3/2}}\varphi'(\sqrt{\rho}) + \frac{1}{4\rho}\varphi''(\sqrt{\rho})\right], \tag{6.7}$$

$$v_{\rho y}(\rho,y) = \frac{1}{2\sqrt{\rho}} v_{ry}^*(\sqrt{\rho}, y) = \frac{1}{2\sqrt{\rho}} A'(y)\varphi'(\sqrt{\rho}), \tag{6.8}$$

where we have taken advantage of the fact that as $\rho \downarrow 0$, the pair $(\sqrt{\rho}, y)$ is in R_ℓ^* so that $v^*(\sqrt{\rho}, y)$ is given by (5.6). The limits in (6.6)–(6.8) exist as $\rho \downarrow 0$ because of (5.7). □

7. Satisfaction of the Hamilton–Jacobi–Bellman conditions.

In the previous two sections, we constructed a C^2 function v on $[0,\infty)^2$ (see, in particular, (5.6), (5.26) and Corollary 6.6). In this section, we show that v satisfies the conditions of the Weak Verification Theorem 3.1 and conditions (i) and (ii) of the Strong Verification Theorem 4.1. In light of Theorem 4.1 and Remark 4.3, this will prove that the value function u for the finite–fuel problem is given by (4.3).

7.1 *Lemma.* The function A defined by (5.24) satisfies

$$-\frac{n}{\alpha^2} < A(y) < 0,\quad A'(y) < 0,\quad A''(y) > 0,\quad \forall\, y > 0. \tag{7.1}$$

Proof. The negativity of A follows from (5.21). The negativity of A′ is a consequence of (5.21) and the second equation in (5.14). The positivity of A′′ comes from (6.1) and the positivity of φ and φ'.

Because A is strictly decreasing, for every $y \in [0,\infty)$ we have from (5.14), (5.15), (5.18) and (5.7) that

$$A(y) > \lim_{z\to\infty} A(z) = \lim_{r\downarrow 0} A(f(r)) = \lim_{r\downarrow 0} -\frac{2}{\alpha\varphi(r)}[r\varphi'(r)-\varphi(r)] = -\frac{n}{\alpha^2}. \qquad \square$$

7.2 *Lemma.* We have $0 < v_r^*(r,y) \le \frac{2}{\alpha} r$ for all $r > 0,\ y \ge 0$.

Proof: For $(r,y) \in R_\ell^*$ we have $v_r^*(r,y) = \frac{2}{\alpha} r + A(y)\varphi'(r) \le \frac{2}{\alpha} r$ and because of (5.11),

$$v_r^*(r,y) = \frac{2}{\alpha} r + A(y)\varphi'(r) \ge \frac{2}{\alpha} r + A(f(r))\varphi'(r) = -A'(f(r))\varphi(r) > 0.$$

We now use Lemma 6.3 to conclude that for $(r,y) \in R_m^*$,

$$0 < v_r^*(r - T(r,y),\ y - T(r,y)) = v_r^*(r,y) \le \frac{2}{\alpha}(r - T(r,y)) \le \frac{2}{\alpha} r.$$

For $(r,y) \in \bar{R}_r$, we have from (5.6), (5.26) that

$$v_r^*(r,y) = v_r^*(r-y,\ 0) = \frac{2}{\alpha}(r-y) \in (0, \frac{2}{\alpha} r]. \qquad \square$$

7.3 *Corollary.* The function $v(\rho,y) = v^*(\sqrt{\rho}, y)$ satisfies (3.10).

Proof: Lemma 7.1 and equation (5.6) imply that $v^*(0,y) = \frac{n}{\alpha^2} + A(y) \in (0, \frac{n}{\alpha^2}]$. According to Lemma 7.2, $v^*(r,y) = v^*(0,y) + \int_0^r v_r^*(s,y)ds \in (0, \frac{n}{\alpha^2} + \frac{r^2}{\alpha}]$, so $v(\rho,y) \in (0, \frac{n}{\alpha^2} + \frac{\rho}{\alpha}]$. Lemma 7.2 and (3.6) give us that $v_\rho(\sqrt{\rho},y) \in [0, \frac{1}{\alpha})$, for every $(\rho,y) \in (0,\infty)^2$. $\square$

7.4 *Lemma.* For $(r,y) \in R_\ell^* \cup B_\ell^*$ we have

$$(v_r^* + v_y^*)_r > 0, \quad (v_r^* + v_y^*)_y > 0, \quad v_r^* + v_y^* < 0.$$

Proof: From Lemmas 7.1 and 6.1, we have for $(r,y) \in R_\ell^* \cup B_\ell^*$,

$$(v_r^*+v_y^*)_r = \frac{\partial}{\partial r}\left[\frac{2r}{\alpha} + A(y)\varphi'(r) + A'(y)\varphi(r)\right]$$
$$= \frac{2}{\alpha} + A(y)\varphi''(r) + A'(y)\varphi'(r)$$
$$> \frac{2}{\alpha} + A(y)\varphi''(g(y)) + A'(y)\varphi'(g(y))$$
$$= v_{rr}^*(y,g(y)) + v_{ry}^*(y,g(y)) = 0.$$

Now define $q(r) = \frac{\varphi'(r)}{\varphi(r)}$ and use (5.3) to compute $q'(r) = -\frac{\Phi(r)}{\varphi^2(r)} > 0$ for all $r \geq 0$. Because q is increasing, we have from (6.1) that for $(r,y) \in R_\ell^* \cup B_\ell^*$,

$$(v_r^*+v_y^*)_y = A'(y)\varphi'(r) + A''(y)\varphi(r)$$
$$= A'(y)\varphi'(r) - A'(y)q(g(y))\varphi(r)$$
$$= A'(y)\varphi(r)[q(r) - q(g(y))] > 0.$$

According to Lemma 6.1, $v_r^* + v_y^* = 0$ on B_m^*, and because of the derivatives just computed, we must have $v_r^* + v_y^* < 0$ in $R_\ell^* \cup B_\ell^*$. □

7.5 *Remark.* From Lemmas 6.1, 6.3 and 7.2 we conclude that $v_r^* + v_y^* \leq 0$ on $\overline{R_\ell^* \cup R_m^*}$ with equality holding on $\bar{R}_m^*$. The formulas (6.3) show that equality also holds on $\bar{R}_r^*$. Because $2\sqrt{\rho}\, v_\rho(\rho,y) + v_y(\rho,y) = v_r^*(\sqrt{\rho}, y) + v_y^*(\rho,y)$, condition (ii) of Theorem 4.1 holds. To obtain condition (i) of Theorem 4.1, we observe that for $\rho > 0$, $y \geq 0$,

$$\alpha v(\rho,y) - n v_\rho(\rho,y) - 2\rho v_{\rho\rho}(\rho,y) - \rho = \alpha v^*(\sqrt{\rho},y) - \frac{n-1}{2\sqrt{\rho}} v_r^*(\sqrt{\rho},y) - \frac{1}{2} v_{rr}^*(\sqrt{\rho},y) - \rho,$$

and so condition (i) is equivalent to the assertion that

$$\alpha v^*(r,y) - \frac{n-1}{2r} v_r^*(r,y) - \frac{1}{2} v_{rr}^*(r,y) = r^2, \ \forall\, (r,y) \in R_\ell^* \cup B_m^* \cup B_0^*. \tag{7.2}$$

Equality (7.2) follows immediately from (5.6) and (5.3). □

The following lemma completes the proof that v satisfies the variational inequality (3.9).

7.6 *Lemma.* We have

$$\alpha v^*(r,y) - \frac{n-1}{2r} v_r^*(r,y) - \frac{1}{2} v_{rr}^*(r,y) < r^2, \ \forall\, (r,y) \in R_m^* \cup B_r^* \cup R_r^*. \tag{7.3}$$

Proof: Let $(r,y) \in R_m^* \cup B_r^* \cup R_r^*$ be given, and set

$$z_0 = \begin{cases} f(r) & \text{if } 0 < r < r_0, \\ 0 & \text{if } r \geq r_0. \end{cases}$$

We define a function $k : [z_0,y] \to \mathbb{R}$ by

$$k(z) \triangleq \alpha v^*(r,z) - \frac{n-1}{2r} v_r^*(r,z) - \frac{1}{2} v_{rr}^*(r,z) - r^2, \quad \forall z \in [z_0,y].$$

Then k is continuous and $k(z_0) = 0$ because $(r,z_0) \in \bar{R}_\ell^* \cup B_0^*$, (7.2) holds, and v^* is C^2. We prove the lemma by showing that k is strictly decreasing, so $k(y) < 0$.

Let $z \in (z_0,y)$ be given. There are three possibilities: $(r,z) \in R_r^*$, $(r,z) \in R_m^*$ and $(r,z) \in B_r^*$. We show in the first two cases that $k'(z)$ is defined and is strictly negative. The last case occurs only for the single value $z = r - r_0$, and so the conclusions of the first two cases are sufficient to guarantee that k is strictly decreasing.

If $(r,z) \in R_r^*$, which occurs if and only if $r_0 < r - z$, we have $v^*(r,z) = v^*(r-z,0) = \frac{(r-z)^2}{\alpha} + \frac{n}{\alpha^2}$, and $k(z) = (r-z)^2 + \frac{(n-1)z}{\alpha r} - r^2$. Therefore

$$k'(z) = -2(r-z) + \frac{n-1}{\alpha r} < -2r_0 + \frac{n-1}{\alpha r_0},$$

which is negative because of the bound $r_0 > \sqrt{\frac{n-1}{2\alpha}}$ in (5.8).

Finally, we consider $(r,z) \in R_m^*$. Then $v^*(r,z) = v^*(r-T(r,z), z-T(r,z))$. Recall Lemma 6.3 and recall also that v^* is C^∞ on R_ℓ^* and its derivatives have continuous extensions to B_m^*, where $(r-T(r,z), z-T(r,z))$ resides. We can thus compute

$$\begin{aligned} k'(z) &= \frac{\partial}{\partial z}\left[\alpha v^*(r,z) - \frac{n-1}{2r} v_r^*(r,z) - \frac{1}{2} v_{rr}^*(r,z) - r^2\right] \\ &= \alpha v_y^*(r,z) - \frac{n-1}{2r} v_{ry}^*(r,z) - \frac{1}{2}\frac{\partial}{\partial z} v_{rr}^*(r-T(r,z), z-T(r,z)) \\ &= \alpha v_y^*(r-T,z-T) - \frac{n-1}{2r} v_{ry}^*(r-T,z-T) - \frac{1}{2} v_{rry}^*(r-T,z-T) \\ &\quad + \frac{1}{2}\left[v_{rrr}^*(r-T,z-T) + v_{rry}^*(r-T,z-T)\right] T_y \end{aligned}$$

$$= A'(z-T)\Big[a\varphi(r-T) - \frac{n-1}{2(r-T)}\varphi'(r-T) - \frac{1}{2}\varphi''(r-T)\Big]$$

$$+ A'(z-T)\left(\frac{n-1}{2}\right)\left(\frac{1}{r-T} - \frac{1}{r}\right)$$

$$+ \frac{1}{2}\left[v^*_{rrr}(r-T,z-T) + v^*_{rry}(r-T,z-T)\right]Ty,$$

where T and T_y are evaluated at (r,z). The first term in the last expression vanishes because φ satisfies (5.3). The second term $A'(z-T)\left(\frac{n-1}{2}\right)\left(\frac{1}{r-T} - \frac{1}{r}\right)$ is negative (Lemma 7.1). The last term equals $[A(z-T)\varphi'''(r-T) + A'(z-T)\varphi''(z-T)]\, T_y(r,z)$, which is negative because of (6.2) and Lemma 7.1. □

In order to show that v satisfies the conditions of the Weak Verification Theorm 3.1 and conditions (i) and (ii) of the Strong Verification Theorem 4.1, it remains only to show that the mapping

$$(x,y) \to v(\|x\|^2,y) = v^*(\|x\|,y) \tag{7.4}$$

is convex in $(x,y) \in \mathbb{R}^n \times [0,y]$.

7.7 Lemma. The mapping (7.4) is convex.

Proof: We first show that the mapping

$$(r,y) \to v^*(r,y) \tag{7.5}$$

is convex in $(r,y) \in [0,\infty)^2$. For this, it suffices to prove positive semidefiniteness of $\nabla^2 v^*$, i.e.,

$$v^*_{yy}(r,y) \geq 0, \quad \forall\, (r,y) \in [0,\infty)^2, \tag{7.6}$$

$$v^*_{rr}(r,y)v^*_{yy}(r,y) - (v^*_{ry}(r,y))^2 \geq 0, \quad \forall\, (r,y) \in [0,\infty)^2. \tag{7.7}$$

If $(r,y) \in \overline{R}^*_{\ell}$, then (5.6) and Lemma 7.1 give

$$v^*_{yy}(r,y) \geq 0, \; v^*_{ry}(r,y) \leq 0, \tag{7.8}$$

and Lemma 7.4 gives

$$\gamma(r,y) \triangleq v^*_{rr}(r,y) + v^*_{ry}(r,y) \geq 0, \tag{7.9}$$

$$\delta(r,y) \triangleq v^*_{ry}(r,y) + v^*_{yy}(r,y) \geq 0. \tag{7.10}$$

These inequalities also hold on $\overline{\mathcal{R}}_m^*$ because of Lemma 6.3, and they hold on $\overline{\mathcal{R}}_r^*$ because of (6.3). We thus have (7.6), and to obtain (7.7) we write

$$v_{rr}^* v_{yy}^* - (v_{ry}^*)^2 = \gamma\delta - (\gamma + \delta + 1)v_{ry}^*.$$

Having established the convexity of v^*, we let $x_1, x_2 \in \mathbb{R}^n$, $y_1, y_2 \in [0,\infty)$ and $\lambda \in [0,1]$ be given. Because $\|\cdot\|$ is convex, we have

$$\|\lambda x_1 + (1-\lambda)x_2\| \le \lambda\|x_1\| + (1-\lambda)\|x_2\|.$$

By Lemma 7.2, v^* is increasing in its first argument. Therefore,

$$\begin{aligned} v^*(\|\lambda x_1 + (1-\lambda)x_2\|, \lambda y_1 + (1-\lambda)y_2) &\le v^*(\lambda\|x_1\| + (1-\lambda)\|x_2\|, \lambda y_1 + (1-\lambda)y_2) \\ &\le \lambda v^*(\|x_1\|, y_1) + v^*((1-\lambda)\|x_2\|, (1-\lambda)y_2), \end{aligned}$$

where the second inequality follows from the convexity of v^*. □

8. Acknowledgement.

This work was supported by the National Science Foundation under grant DMS 90–02588 and by the Air Force Office of Scientific Research under grant AFOSR 89–0075.

9. References.

(1980) Beneš, V.E., Shepp, L.A. & Witsenhausen, H.S., Some solvable stochastic control problems, *Stochastics* 4, 39–83.

(1982) Elliott, R.J., *Stochastic Calculus and Applications*, Springer–Verlag, New York.

(1983) Harrison, J.M., Sellke, T.M. & Taylor, A.J., Impulse control of Brownian motion, *Math. Operations Research* 8, 454–466.

(1983) Harrison, J.M. & Taksar, M.I., Instantaneous control of a Brownian motion, *Math. Operations Research* 8, 439–453.

(1978) Harrison, J.M. & Taylor, A.J., Optimal control of a Brownian storage system, *Stoch. Proc. Appl.* 6, 179–194.

(1983) Jacka, S.D., A finite fuel stochastic control problem, *Stochastics* 10, 103–113.

(1983) Karatzas, I., A class of singular stochastic control problems, *Adv. Appl. Prob.* 15, 225–254.

(1985) Karatzas, I., Probabilistic aspects of finite–fuel stochastic control, *Proc. Nat'l. Acad. Sciences USA* 82, 5579–5581.

(1986) Karatzas, I. & Shreve, S.E., Equivalent models for finite–fuel stochastic control, *Stochastics* 18, 245–276.

(1986) Lehoczky, J.P. & Shreve, S.E., Absolutely continuous and singular stochastic control, *Stochastics* 17, 91–109.

(1984) Lions, P.L. & Sznitman, A.S., Stochastic differential equations with reflecting boundary conditions, *Comm. Pure Appl. Math.* 37, 511–537.

(1984) Shreve, S.E., Lehoczky, J.P. & Gaver, D.P., Optimal consumption for general diffusions with absorbing and reflecting barriers, *SIAM J. Control Optimization* 22, 55–75.

(1989) Soner, H.M. & Shreve, S.E., Regularity of the value function for a two–dimensional singular stochastic control problem, *SIAM J. Control Optimization* 27, 876–907.

1991) Soner, H.M. & Shreve, S.E., A free boundary problem related to singular stochastic control: the parabolic case, *Comm. Partial Diff. Equations* 16, 373–424.

(1987) Sun, M., Singular control problems in bounded intervals, *Stochastics* 21, 303–344.

(1985) Taksar, M.I., Average optimal singular control and a related stopping problem, *Math. Operations Research* 10, 63–81.

Numerical Methods in Ergodic Optimal Stochastic Control and Application *

F. Campillo
INRIA Sophia Antipolis
Route des Lucioles
F-06565 Valbonne Cedex

E. Pardoux
Université de Provence
UFR MIM
F-13331 Marseille Cedex 3
and INRIA

Abstract

In Campillo *[4] we presented a numerical algorithm for the computation of the optimal feedback law in an ergodic stochastic optimal control problem. This method, based on the discretization of the associated Hamilton-Jacobi-Bellman equation, can be used only in low dimensions (2,4, or 6 in a parallel computer). For higher dimensional problems, we propose here to use a stochastic gradient algorithm in order to find the optimal feedback within a given subclass of parametrized controls. As is [4], we apply these techniques to the control of semi-active suspensions for road vehicles.*

1 Introduction

In this paper we consider numerical procedures for stochastic control problems. Given a real case study (here we consider semi-active control of vehicle suspensions) we can use different methods. For low dimensional problems, we can use optimal methods : we discretize the Hamilton-Jacobi-Bellman equation (this approach is proposed in [4]). In higher dimensions this approach is cumbersome or even impossible to implement, in this case we can look for the optimal feedback in a given subclass of parametrized controls using a stochastic gradient algorithm. The aim of this paper is to compute the stochastic gradient in the simplest two-dimensional model relevant in our application, and to give some numerical results.

In section 2 we introduce the stochastic control problem of ergodic type in $\mathbb{R}^2$ motivated by the application to the control of suspension systems (see Bellizzi et al [1]).

In section 3, we derive the stochastic gradient for the above problem.

2 A stochastic control problem

2.1 The problem

Let $\{X_t(\theta); t \geq 0\}$ be the solution of the following stochastic system in $\mathbb{R}^2$

$$dX_t^1(\theta) = X_t^2(\theta)\, dt\,, \quad (1)$$

$$dX_t^2(\theta) = -\left[u(\theta, X_t(\theta))\, X_t^2(\theta) + \beta\, X_t^1(\theta) + \gamma\, \mathrm{sign}(X_t^2(\theta))\right] dt + \sigma\, dW_t \quad (2)$$

where

*Supported by RENAULT (contract II6.10.601/INRIA/17)

- β, γ are strictly positive constants,
- W is a standard Brownian motion,
- $u(\theta,x)$ is a feedback control parametrized by $\theta \in \Theta$ (Θ is a compact subset of $\mathbb{R}^d$, for some $d \geq 1$). $u(\theta,x)$ takes values in $U = [u_1,u_2]$, $0 < u_1 < u_2 < \infty$.

The components of $x \in \mathbb{R}^2$ (resp. of $X_t(\theta)$) are denoted by x^1 and x^2 (resp. by $X_t^1(\theta)$ and $X_t^2(\theta)$). For each $\theta \in \Theta$, we consider the following long time average cost functional

$$J(\theta) \triangleq \lim_{T\to\infty} \frac{1}{T} E \int_0^T f(\theta, X_t(\theta))\, dt \tag{3}$$

with

$$f(\theta,x) \triangleq \left[u(\theta,x)\,x^2 + \beta\, x^1 + \gamma\, \mathrm{sign}(x^2)\right]^2 .$$

A class of feedback controls : Let $\mathcal{U}$ denote the set of feedback functions $u(\theta,x)$ such that

- $u : \Theta \times \mathbb{R}^2 \to U = [u_1,u_2]$,
- $\forall x \in \mathbb{R}^2$, $\theta \to u(\theta,x)$ is C^1,
- $\forall \theta \in \Theta$, $x \to u(\theta,x)\,x^2$ is C^1 on $\mathbb{R} \times \mathbb{R} \setminus \{0\}$ with bounded derivatives.

This last condition allows u to be discontinuous at $x^2 = 0$; it is the case for the optimal control which we have computed in a previous work [4].

Note that in (2), the drift coefficient is C^1 on $\mathbb{R} \times \mathbb{R} \setminus \{0\}$, and it is discontinuous across $x^2 = 0$ with a jump of amplitude -2γ.

Lemma 2.1 *For any $\theta \in \Theta$, the system (1,2) admits a unique strong solution.*

Proof In view of Yamada–Watanabe's result (Karatzas–Shreve [8] proposition V–3.20), it is sufficient to prove that

(i) existence of a weak solution,

(ii) pathwise uniqueness

hold for system (1,2).

Part *(i)* has already been proved in Campillo et al [5] : using a Girsanov transformation, we remove the discontinuous terms in (1,2). We can rewrite (1,2) as

$$\begin{aligned} dX_t^1 &= X_t^2\, dt \\ dX_t^2 &= a(X_t^1, X_t^2)\, dt + b(X_t^2)\, dt + \sigma\, dW_t \end{aligned}$$

where a is Lipschitz continuous and b is decreasing. Now let X and X' be two solutions of this system and let $\bar{X} = X - X'$. We get

$$\begin{aligned} \frac{d}{dt}\bar{X}_t^1 &= \bar{X}_t^2 \\ \frac{d}{dt}\bar{X}_t^2 &= \left[a(X_t^1, X_t^2) - a(X'^1_t, X'^2_t) + b(X_t^2) - b(X'^2_t)\right] \end{aligned}$$

so

$$\begin{aligned}\frac{d}{dt}|\bar{X}_t|^2 &= 2\,\bar{X}_t^1\,\frac{d\bar{X}_t^1}{dt} + 2\,\bar{X}_t^2\,\frac{d\bar{X}_t^2}{dt}\\ &= 2\,[\bar{X}_t^1\,\bar{X}_t^2 + \bar{X}_t^2\,(a(X_t^1,X_t^2) - a(X'^1_t,X'^2_t)) + \underbrace{\bar{X}_t^2\,(b(X_t^2) - b(X'^2_t))}_{\leq 0}] \leq C\,|\bar{X}_t|^2\,.\end{aligned}$$

Hence, using Gronwall's lemma we prove the pathwise uniqueness. ∎

Note that the Lemma also follows from the monotonicity of the drift, and the results in Gyöngy-Krylov [6] and Jacod [7].

2.2 An example : a semi-active suspension system

In this section we present a damping control method for a nonlinear suspension of a road vehicle (comprising a spring, a shock absorber, a mass, and taking into account the dry friction, cf. figure 1). The aim is to improve the ride comfort.

Among alternatives to classical suspension systems (passive systems) we distinguish between active and semi-active techniques. An active suspension system consists of force elements in addition to a spring and a damper assembly. Force elements continuously vary the force according to some control law. In general, an active system is expensive, complicated, and requires an external power source. In contrast, a semi-active system requires no hydraulic power supply, and its hardware implementation is simpler and cheaper than a fully active system. A semi-active suspension system acts only on damping or spring laws, so it can only dissipate or store energy.

Here we consider a system with control on the damping law, the forces in the damper are generated by modulating its orifice for fluid flow. We use the simplest model which consists in a one degree-of-freedom model.

The equation of motion for a one degree-of-freedom model is (cf. figure 1 for the exact definition of the terms)

$$m\,\ddot{y} + c\,\dot{y} + k_s\,y + F_s\,\text{sign}(\dot{y}) = -m\,\ddot{e}\,. \tag{4}$$

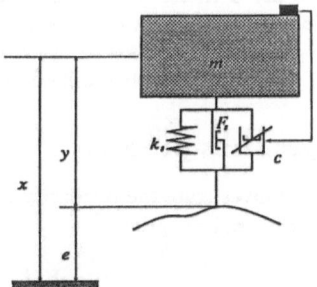

a	absolute displacement of mass m
y	relative displacement ($y = a - e$)
e	stochastic input (surface road acceleration)
m	sprung mass
c	shock-absorber damping constant (controlled)
k_s	spring constant
F_s	dry friction constant

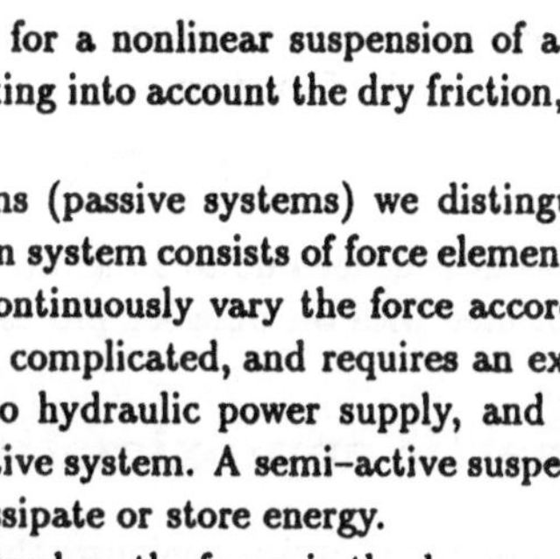

Figure 1: One degree-of-freedom model.

$\ddot{e}$ denotes the input acceleration, i.e. it models the roughness of the road surface. The restoring force $k_s\,y + F_s\,\text{sign}(\dot{y})$, has a linear part $k_s\,y$, and a nonlinear part $F_s\,\text{sign}(\dot{y})$ which describes the dry friction force. The damping force is $c\,\dot{y}$ where $c > 0$ is the instantaneous damping coefficient (the control is acting on this term).

The problem is to compute a feedback law $c = c(y, \dot{y})$ such that the solution of the system (4) minimizes a criterion — related to the vibration comfort

$$J(u) \triangleq \lim_{T\to\infty} \frac{1}{T} E \int_0^T |\ddot{a}|^2\,dt = \lim_{T\to\infty} \frac{1}{T} E \int_0^T |\ddot{y} + \ddot{e}|^2\,dt\,.$$

$\ddot{e}$ is supposed to be a white Gaussian noise process, $\ddot{e} = -\sigma\, dW/dt$ where W is a standard Wiener process.

Using

$$u = \frac{c}{m},\ \beta = \frac{k_s}{m},\ \gamma = \frac{F_s}{m},\ \text{and } X \triangleq \begin{pmatrix} y \\ \dot{y} \end{pmatrix},$$

equation (4) can be rewritten as (1), (2).

2.3 The "optimal" approach

This approach — presented in [4] — consists in discretizing the Hamilton-Jacobi-Bellman (HJB) equation associated with the ergodic stochastic control problem for the diffusion (1), (2) with the cost function (3) (but now the feedback function u is not parametrized)

The HJB equation is of the form

$$\min_{u\in[u_1,u_2]} (\mathcal{L}^u v(x) + f(u,x)) = \rho\,, \quad \forall x \in \mathbb{R}^2$$

where $v : \mathbb{R}^2 \to \mathbb{R}$ is defined up to an additive constant, ρ is a constant and $\mathcal{L}^u$ is the infinitesimal generator associated with the diffusion process (1), (2) (with $u(\theta, x)$ replaced by $u \in \mathcal{U}$).

3 Stochastic approximation algorithms

The problem is to find θ^* which minimizes the cost function (3). In [4] we have already proved that the cost function is of the form

$$J(\theta) = \int_{\mathbb{R}^2} f(\theta, x)\, \mu_\theta^X(dx) \tag{5}$$

where μ_θ^X is the unique invariant measure of the process $X_t(\theta)$. As pointed out in Ladelli [9], the gradient of $J(\theta)$ is *not* equal to

$$\int_{\mathbb{R}^2} \nabla_\theta f(\theta, x)\, \mu_\theta^X(dx)\,.$$

It is possible to use a Kiefer-Wolfowitz-type algorithm in order to minimize $J(\theta)$, thus avoiding the computation of the gradient of J. However, it seems that stochastic gradient algorithms often converge faster than Kiefer-Wolfowitz algorithms. Motivated by this remark, we shall compute here the gradient of $J(\theta)$, which involves differentiating $X_t(\theta)$.

3.1 The gradient process

In this section we investigate the regularity of the process $X_t(\theta)$ with respect to θ and we establish an equation for the gradient process

$$Y_t(i, \theta) \triangleq \frac{\partial}{\partial \theta_i} X_t(\theta)\,.$$

This derivation will give rise to the local time of the process $\{X_t^2(\theta)\}$ at 0.

Following Protter [11] we have the

Lemma 3.1 *The processes X_t^2 admits a local time at 0, which is the unique adapted, continuous, nondecreasing process L_t such that $L_0 = 0$, $\int_0^\infty 1_{\mathbb{R}\setminus\{0\}}(X_s^2)\, dL_s = 0$ a.s., and*

$$|X_t^2| = |X_0^2| + \int_0^t \operatorname{sign}(X_s^2)\, dX_s^2 + 2\, L_t\,. \tag{6}$$

Moreover,

$$L_t = \lim_{\varepsilon\to 0}\text{a.s.}\ \frac{\sigma^2}{2\varepsilon}\int_0^t 1_{[-\varepsilon,\varepsilon]}(X_s^2)\, ds\,. \tag{7}$$

The following result is proved in the Appendix below.

Proposition 3.2 *For each $t \geq 0$, $X_t(\theta)$ is mean-square differentiable w.r.t. the parameter θ_i ($i = 1, \dots, d$), and $Y_t(i,\theta) \triangleq \partial X_t(\theta)/\partial\theta_i$ is the solution of the following system*

$$dY_t(\theta) = A(\theta, X_t(\theta))\, Y_t(\theta)\, dt + B\, Y_t(\theta)\, dL_t + C(\theta, X_t(\theta))\, dt\ , \tag{8}$$

with

$$A(\theta,x) \triangleq \begin{pmatrix} 0 & 1 \\ -\beta - u_{x^1}(\theta,x)x^2 & -u(\theta,x) - u_{x^2}(\theta,x)x^2 \end{pmatrix}$$

$$B \triangleq \begin{pmatrix} 0 & 0 \\ 0 & -\frac{2\gamma}{\sigma^2} \end{pmatrix}, \qquad C(\theta,x) \triangleq \begin{pmatrix} 0 \\ -u_\theta(\theta,x)x^2 \end{pmatrix}.$$

$Y_0 \equiv 0$, *and where L is the local time of X^2 at 0. We denote*

$$Y_t(\theta) = \left[Y_t(1,\theta)|\cdots|Y_t(d,\theta)\right]\ .$$

For the sake of simplicity, from now on we suppose that θ is scalar, so we drop the subscript i. We have an explicit representation of $Y_t(\theta)$ in terms of $X_t(\theta)$. Let $\{\Phi_t(\theta)\,,\, t \geq 0\}$ be the solution of

$$d\Phi_t(\theta) \triangleq A(\theta, X_t(\theta))\, \Phi_t(\theta)\, dt + B\, \Phi_t(\theta)\, dL_t\ , \qquad \Phi_0(\theta) = I\ , \tag{9}$$

then

$$Y_t(\theta) = \int_0^t \Phi_t(\theta)\, \Phi_s(\theta)^{-1}\, C(\theta, X_s(\theta))\, ds\ . \tag{10}$$

3.2 Asymptotic properties of $(X_t(\theta), Y_t(\theta))$

We note

$$\xi_t(\theta) = \begin{pmatrix} X_t(\theta) \\ Y_t(\theta) \end{pmatrix}. \tag{11}$$

In [4] we showed that $X_t(\theta)$ admits a unique invariant measure μ_θ^X. We extend the process $X_t(\theta)$ for all $t \in \mathbb{R}$, such that it is stationary and μ_θ^X is the law of $X_t(\theta)$ for all $t \in \mathbb{R}$. We can then solve equation (9) and define $\Phi_t(\theta)$ for all $t \in \mathbb{R}$. It is easily seen that the Markov process $\xi_t(\theta)$ possesses a unique invariant measure μ_θ iff the following integral converges a.s.

$$\int_{-\infty}^0 \Phi_t(\theta)^{-1}\, C(\theta, X_t(\theta))\, dt\ .$$

A sufficient condition for that fact is given in the following lemma

Lemma 3.3 *Suppose that there exists $C > 0$ such that for all (θ, x)*

$$\beta + u_{x_1}(\theta,x)\, x^2 \geq C > 0\ ,$$
$$u(\theta,x) + u_{x^2}(\theta,x)\, x^2 \geq C > 0\ ,$$

then there exists $\lambda < 0$ such that

$$\limsup_{t\to\infty} \frac{1}{t} \log \left|\Phi_{-t}(\theta)^{-1}\right| \leq \lambda \text{ a.s.} \tag{12}$$

Proof With the hypotheses of the Lemma and using the properties of the solution of (9), we can prove that

$$\Phi_t(\theta) \underset{t\to\infty}{\longrightarrow} 0 \text{ a.s.} \tag{13}$$

Moreover there exists C such that $|\Phi_t(\theta)| \le C$, $\forall t \ge 0$, and by bounded convergence $E|\Phi_t(\theta)| \to 0$, $t \to \infty$. Hence, given $\alpha < 1$, there exists $t > 0$ such that

$$E|\Phi_t(\theta)| = \alpha \,. \tag{14}$$

Now, using the ergodicity property of $\Phi_t(\theta)$ as in Bougerol [2,3]

$$\begin{aligned}
\lim_{T\to\infty} \frac{1}{T} \log|\Phi_T(\theta)| &= \lim_{T\to\infty} \frac{1}{T} E(\log|\Phi_T(\theta)|) \\
&= \frac{1}{t} \lim_{n\to\infty} \frac{1}{n} E(\log|\Phi_{nt}(\theta)|) \\
&\le \frac{1}{t} \lim_{n\to\infty} \frac{1}{n} \sum_{k=1}^{n} E\left(\log\left|\Phi_{kt,(k-1)t}(\theta)\right|\right) \\
&= \frac{1}{t} E(\log|\Phi_t(\theta)|) \\
&\le \frac{1}{t} \log E|\Phi_t(\theta)| \\
&= \frac{\log\alpha}{t} < 0
\end{aligned}$$

where $\{\Phi_{s,t}(\theta),\, t \ge s\}$ is the solution of (9) satisfying $\Phi_{s,s}(\theta) = I$.

Note that $\Phi_t(\theta)$ and $\Phi_{-t}(\theta)^{-1}$ are equal in law, so $E|\Phi_t(\theta)| = E|\Phi_{-t}(\theta)^{-1}|$, and similarly as above

$$\lim_{T\to\infty} \frac{1}{T} \log\left|\Phi_{-T}(\theta)^{-1}\right| = \lim_{T\to\infty} \frac{1}{T} E\left(\log\left|\Phi_{-T}(\theta)^{-1}\right|\right) = \lim_{T\to\infty} \frac{1}{T} E(\log|\Phi_T(\theta)|) < 0 \text{ a.s.}$$

■

3.3 The gradient of the cost functional

Let

$$g(\theta,x) \triangleq u(\theta,x)\,x^2 + \beta\,x^1 + \gamma\,\mathrm{sign}(x^2)$$

so that $f(\theta,x) = [g(\theta,x)]^2$. We have

$$\begin{aligned}
f_\theta(\theta,x) &= 2\,g(\theta,x)\,u_\theta(\theta,x)\,x^2 \,, \\
f_{x^1}(\theta,x) &= 2\,g(\theta,x)\,[u_{x^1}(\theta,x)\,x^2 + \beta] \,,
\end{aligned}$$

and, in the sense of distributions, $\delta(x^2)$ denoting the Dirac measure "in the variable x^2,

$$\begin{aligned}
f_{x^2}(\theta,x) &= 2\,g(\theta,x)\,[u_{x^2}(\theta,x)\,x^2 + u(\theta,x) + 2\,\gamma\,\delta(x^2)] \\
&= 2\,g(\theta,x)\,[u_{x^2}(\theta,x)\,x^2 + u(\theta,x)] + 4\,\beta\,\gamma\,x^1\,\delta(x^2) \,.
\end{aligned}$$

Let

$$\begin{aligned}
\tilde f_{x^1}(\theta,x) &= f_{x^1}(\theta,x) \\
\tilde f_{x^2}(\theta,x) &= 2\,g(\theta,x)\,[u_{x^2}(\theta,x)\,x^2 + u(\theta,x)]
\end{aligned}$$

and

$$F(\theta,x,y) \triangleq f_\theta(\theta,x) + \tilde f_x(\theta,x)\,y \,.$$

Formally, if we take the derivative of the cost functional (3) with respect to θ, and if we interchange the derivation and the limit as $T \to \infty$, we get formally

$$\lim_{T\to\infty} \frac{1}{T} E \left[\int_0^T F(\theta, X_t(\theta), Y_t(\theta))\, dt + \int_0^T 4\beta\gamma X_t^1(\theta) Y_t^2(\theta)\, \delta(X_t^2(\theta))\, dt \right]$$

and one can show rigorously that the gradient of the cost functional is given by

$$\nabla_\theta J(\theta) = \lim_{T\to\infty} \frac{1}{T} E \left[\int_0^T F(\theta, X_t(\theta), Y_t(\theta))\, dt + \int_0^T 4\beta\gamma X_t^1(\theta) Y_t^2(\theta)\, dL_t \right] . \tag{15}$$

It is possible to prove that the process $(X_t(\theta), Y_t(\theta))$ admits a unique invariant measure μ_θ which is regular with respect to the parameter θ, from which one can conclude that the limit (15) is well defined.

3.4 Stochastic gradient algorithm

In order to minimize (5), we want to find $\theta^* \in \Theta$ such that

$$\nabla_\theta J(\theta)|_{\theta=\theta^*} = 0 . \tag{16}$$

The associated stochastic gradient algorithm is the following : given $\Delta t > 0$ and $t_k \triangleq k\,\Delta t$, we solve equations (1),(2),(8) with

$$\theta = \theta_k \text{ for } t_k \le t < t_{k+1} ,$$

and θ_k is given by

$$\theta_{k+1} = \theta_k - \rho_k \left[F(\theta_k, X_{t_k}(\theta_k), Y_{t_k}(\theta_k))\, \Delta t + 4\gamma\beta X_{t_k}^1(\theta_k) Y_{t_k}^2(\theta_k)\, \Delta L_k \right] , \tag{17}$$

where

$$\Delta L_k = L_{t_{k+1}} - L_{t_k} ,$$

and where the sequence of positive gains $\{\rho_k\}$ satisfies appropriate conditions.

4 Computational aspects and Numerical Results

4.1 Time discretization

We approximate X_t by X_t^n given by the following Euler scheme

$$X_{k+1}^{n,1} = X_k^{n,1} + X_k^{n,2}\, \Delta t , \tag{18}$$

$$X_{k+1}^{n,2} = X_k^{n,2} - \left(u(\theta, X_t^n) X_k^{n,2} + \beta X_k^{n,1} + \gamma\, \mathrm{sign}(X_k^{n,2}) \right) \Delta t + \sigma\, \Delta W_k^n \tag{19}$$

where

$$\Delta t \triangleq t_{k+1} - t_k , \quad \Delta W_k^n \triangleq W_{t_{k+1}} - W_{t_k} \sim \mathcal{N}(0, t_{k+1} - t_k) .$$

For Y_t we also use an Euler scheme

$$Y_{k+1}^{n,1} = Y_k^{n,1} + Y_k^{n,2}\, \Delta t , \tag{20}$$

$$\begin{aligned} Y_{k+1}^{n,2} = {} & Y_k^{n,2} - (u_\theta(\theta, X_k^n) + u_x(\theta, X_k^n) Y_k^n)\, X_k^{n,2}\, \Delta t \\ & - u(\theta, X_k^n)\, Y_k^{n,2}\, \Delta t - \beta Y_k^{n,1}\, \Delta t - \frac{2\gamma}{\sigma^2} Y_k^{n,2}\, \Delta L_k^n \end{aligned} \tag{21}$$

where ΔL_k^n is an approximation of $L_{t_{k+1}} - L_{t_k}$ given by

$$\Delta L_k^n = \begin{cases} |X_{k+1}^{n,2}| & \text{if } X_k^{n,2} X_{k+1}^{n,2} < 0 , \\ 0 & \text{otherwise,} \end{cases}$$

(cf. proposition 4.2).

Proposition 4.1 *For all $t \geq 0$ and all $\theta \in \Theta$*

$$X_t^n(\theta) \underset{n\to\infty}{\overset{L^2}{\rightarrow}} X_t(\theta) .$$

Proof We fixe $\theta \in \Theta$. We use the notation

$$dX_t = h(X_t)\,dt + G\,dW_t \tag{22}$$

where

$$h(x) = h_1(x) + h_2(x)\,, \quad h_2(x) = \begin{pmatrix} 0 \\ -\gamma\,\mathrm{sign}(x_2) \end{pmatrix}$$

h_1 is Lipschitz continuous and h_2 is discontinuous and *monotonic*. Let $\Delta t = 1/n$, $t_k = k\,\Delta t$ and $\phi_n(t) = t_k$ if $t \in [t_k, t_{k+1}[$. Let W_t^n be the polygonal interpolation of the Wiener process W_t, that is

$$W_t^n = W_{t_k} + (t - t_k)\frac{W_{t_{k+1}} - W_{t_k}}{\Delta t}\,, \quad t \in [t_k, t_{k+1}[$$

Then the Euler scheme (18) reads

$$dX_t^n = h(X_{\phi_n(t)}^n)\,dt + G\,dW_t^n\,. \tag{23}$$

Because X_t^n is not adapted to the filtration of the Wiener process and for technical simplification, we can replace W_t^n is this last equation by

$$W_t^n = W_{t_k} + (t - t_k)\frac{W_{t_{k+1}} - W_{t_k}}{\Delta t}\,, \quad t \in [t_{k-1}, t_k[$$

(with the convention $W_t = 0$ for $t \leq 0$) in this case X_t^n is adapted.

The difference between (22) and (23) gives

$$d[X_t - X_t^n] = [h(X_t) - h(X_t^n)]\,dt + [h(X_t^n) - h(X_{\phi_n(t)}^n)]\,dt + G\,d[W_t - W_t^n]$$

and by Itô formula

$$\begin{aligned} |X_t - X_t^n|^2 &= 2\int_0^t (h(X_s) - h(X_s^n), X_s - X_s^n)\,ds + 2\int_0^t \left(h(X_s^n) - h(X_{\phi_n(s)}^n), X_s - X_s^n\right)ds \\ &\quad +2\,(X_t - X_t^n, G\,(W_t - W_t^n)) + |G\,(W_t - W_t^n)|^2 \\ &\quad +2\int_0^t (G\,(W_s - W_s^n), h(X_s) - h(X_s^n))\,ds\,. \end{aligned} \tag{24}$$

Now we have the following results

- By Lispchitz continuity of h_1 and monotonicity of h_2

$$\begin{aligned} &\int_0^t (h(X_s) - h(X_s^n), X_s - X_s^n)\,ds \\ &= \int_0^t (h_1(X_s) - h_1(X_s^n), X_s - X_s^n)\,ds + \underbrace{\int_0^t (h_2(X_s) - h_2(X_s^n), X_s - X_s^n)\,ds}_{\leq 0} \\ &\leq \int_0^t (h_1(X_s) - h_1(X_s^n), X_s - X_s^n)\,ds \\ &\leq C\int_0^t |X_s - X_s^n|^2 ds\,. \end{aligned}$$

- By Lispchitz continuity of h_1

$$\int_0^t \left(h(X_t^n) - h(X_{\phi_n(t)}^n), X_s - X_s^n\right) ds$$
$$\leq C \int_0^t |X_s - X_s^n|^2 ds + C \int_0^t \left(h(X_s^n) - h(X_{\phi_n(s)}^n), X_s - X_s^n\right) ds$$
$$\leq C \int_0^t |X_s - X_s^n|^2 ds + C \int_0^t |X_s^n - X_{\phi_n(s)}^n|^2 ds + C \int_0^t |\text{sign}(X_s^{n,2}) - \text{sign}(X_{\phi_n(s)}^{n,2})|^2 ds$$

and

$$E|\text{sign}(X_t^{n,2}) - \text{sign}(X_{\phi_n(t)}^{n,2})|^2 \leq C\, P(|X_s^{n,2}| \leq \alpha) + C\, E\left(1_{(|X_s^{n,2}|>\alpha)} |\text{sign}(X_t^{n,2}) - \text{sign}(X_{\phi_n(t)}^{n,2})|^2\right)$$

and this last term tends to 0 as $n \to \infty$.

- For $0 < \rho < 1$

$$2\, |(X_t - X_t^n, G(W_t - W_t^n))| \leq \rho |X_t - X_t^n|^2 + \frac{\sigma^2}{\rho} |W_t - W_t^n|^2$$

- By Hölder inequality

$$E\, 2 \int_0^t (G(W_s - W_s^n), h(X_s) - h(X_s^n))\, ds \leq C \left(E \int_0^t |W_s - W_s^n|^2 ds\right)^{1/2}$$

So from (24) and Gronwall's Lemma

$$E|X_t - X_t^n|^2 \leq \left(\varepsilon_n + C \int_0^t P(|X_s^{n,2}| \leq \alpha)\, ds\right) e^{Ct} \tag{25}$$

where $\varepsilon_n \to 0$, as $n \to \infty$.

It is simple to see that the sequence (X^n, W^n) is tight, so there exists a subsequence (also denoted (X^n, W^n)) and a process (Z, W) such that $(X^n, W^n) \to (Z, W)$ weakly. The process Y_t is of the from

$$Y_t = X_0 + \int_0^t \chi_s\, ds + G\, W_t$$

where W is a Wiener process, so we have $P(|Y_s^2| \leq \alpha) \to 0$ as $\alpha \to 0$. Then form (25) we deduce

$$\limsup_{n\to\infty} E|X_t - X_t^n|^2 \leq C \int_0^t P(|Y_s^2| \leq 2\alpha)\, ds\, e^{Ct}$$

which tends to 0 as $\alpha \to 0$. Then by uniqueness of the limit we prove that $X^n \to X$ in L^2. ■

4.2 Approximation of the local time

From (6)

$$L_t = \frac{1}{2}|X_t^2| - \frac{1}{2}|X_0^2| - \frac{1}{2}\int_0^t \text{sign}(X_s^2)\, dX_s^2\,. \tag{26}$$

We approximate L_t the following way : in (26) we replace X_t^2 by the polygonal interpolation of the discrete process $X_k^{n,2}$ given in (19)

$$X_t^{n,2} \triangleq \sum_{k\geq 0} \frac{(t_{k+1}^n - t)\, X_{n_k}^{n,2} + (t - t_k^n)\, X_{n_{k+1}}^{n,2}}{t_{k+1}^n - t_k^n} 1_{[t_k^n, t_{k+1}^n)}(t)$$

with $t_k^n = k/n$. So we get the approximation

$$L_t^n = \frac{1}{2}|X_t^{n,2}| - \frac{1}{2}|X_0^{n,2}| - \frac{1}{2}\int_0^t \text{sign}(X_{\phi_n(s)}^{n,2})\, dX_s^{n,2}\,, \tag{27}$$

where $\phi_n(s) = t_k^n$ if $s \in [t_k^n, t_{k+1}^n)$.

Proposition 4.2

$$L_{t_{k+1}^n}^n - L_{t_k^n}^n = \begin{cases} |X_{t_{k+1}^n}^{n,2}| & \textit{if } X_{t_{k+1}^n}^{n,2} X_{t_k^n}^{n,2} < 0\,, \\ 0 & \textit{otherwise.} \end{cases}$$

Proof Let $t \in [t_k^n, t_{k+1}^n)$, $\Delta t_k^n = t_{k+1}^n - t_k^n$, $\Delta X_k^{n,2} = X_{t_{k+1}^n}^{n,2} - X_{t_k^n}^{n,2}$,

$$\begin{aligned}\frac{dL_t^n}{dt} &= \frac{1}{2}\frac{d}{dt}|X_t^{n,2}| - \frac{1}{2}\operatorname{sign}(X_{t_k^n}^{n,2})\frac{\Delta X_k^{n,2}}{\Delta t_k^n}\,, \\ &= \frac{1}{2}\operatorname{sign}(X_t^{n,2})\frac{\Delta X_k^{n,2}}{\Delta t_k^n} - \frac{1}{2}\operatorname{sign}(X_{t_k^n}^{n,2})\frac{\Delta X_k^{n,2}}{\Delta t_k^n}\,, \\ &= \frac{1}{2}\frac{\Delta X_k^{n,2}}{\Delta t_k^n}\left[\operatorname{sign}(X_t^{n,2}) - \operatorname{sign}(X_{t_k^n}^{n,2})\right]\,,\end{aligned}$$

then

$$\operatorname{sign}(X_t^{n,2}) - \operatorname{sign}(X_{t_k^n}^{n,2}) = 2\operatorname{sign}(X_{t_{k+1}^n}^{n,2})\,1_{(X_t^{n,2}X_{t_k^n}^{n,2}<0)}\,,$$

so

$$\frac{dL_t^n}{dt} = \operatorname{sign}(X_{t_{k+1}^n}^{n,2})\frac{\Delta X_k^{n,2}}{\Delta t_k^n}\,1_{(X_t^{n,2}X_{t_k^n}^{n,2}<0)}\,,$$

and

$$L_t^n = \sum_{k\geq 0}\operatorname{sign}(X_{t_{k+1}^n}^{n,2})\frac{\Delta X_k^{n,2}}{\Delta t_k^n}\lambda(s\in[t_k^n, t_{k+1}^n)\cap[0,t];\, X_s^{n,2}X_{t_k^n}^{n,2}<0)$$

where λ is Lebesgue measure. Finally,

$$L_{t_{k+1}^n}^n - L_{t_k^n}^n = \operatorname{sign}(X_{t_{k+1}^n}^{n,2})\frac{\Delta X_k^{n,2}}{\Delta t_k^n}\lambda(s\in[t_k^n, t_{k+1}^n);\, X_s^{n,2}X_{t_k^n}^{n,2}<0)$$

and

$$\lambda(s\in[t_k^n, t_{k+1}^n);\, X_s^{n,2}X_{t_k^n}^{n,2}<0) = \begin{cases} X_{t_{k+1}^n}^{n,2}\dfrac{\Delta t_k^n}{\Delta X_k^{n,2}} & \text{if } X_{t_{k+1}^n}^{n,2}X_{t_k^n}^{n,2}<0\,, \\ 0 & \text{otherwise}\,. \end{cases}$$

■

Proposition 4.3 *For all t*

$$L_t^n \underset{n\to\infty}{\overset{L^2}{\to}} L_t\,.$$

Proof Using the definitions (26), (27) of L_t and L_t^n, it is sufficient to prove that

$$\int_0^t \operatorname{sign}(X_s^2)\,dX_s^2 \underset{n\to\infty}{\overset{L^2}{\to}} \int_0^t \operatorname{sign}(X_{\phi_n(s)}^{n,2})\,dX_s^{n,2}\,.$$

Using the notations of the proof of Proposition 4.1, we have

$$\begin{aligned}&E\left|\int_0^t \operatorname{sign}(X_s^2)\,dX_s^2 - \int_0^t \operatorname{sign}(X_{\phi_n(s)}^{n,2})\,dX_s^{n,2}\right|^2 \\ &\quad\leq C\int_0^t E|\phi(X_{\phi_n(s)}^n) - \phi(X_s)|^2\,ds \\ &\quad\quad + C\int_0^t E|\operatorname{sign}(X_{\phi_n(s)}^{n,2}) - \operatorname{sign}(X_s^2)|^2\,ds + C\,E\left|\int_0^t \operatorname{sign}(X_{\phi_n(s)}^{n,2})\,d[W_s^n - Ws]\right|^2\end{aligned}$$

where $\phi(x) = \operatorname{sign}(x^2)\,h(x)$ satisfies $|\phi(x)| \leq C(1+|x|)$. So the result follows from the same arguments used in the proof of Proposition 4.1. ■

4.3 Numerical results

We consider the example of section 2.2. We use the following parameters

m	110
K	26000
F	85
σ	0.5

We use a feedback control of the form

$$u(\theta, x) = \theta^1 + \left[-\theta^2 x^1 \operatorname{sign}(x^2)\right]^+$$

where θ^1 and θ^2 are positive.

We do not exactly use the algorithm (17), but the following modification

$$\theta_{k+1} = \theta_k - \begin{pmatrix} \rho_k^1 & 0 \\ 0 & \rho_k^2 \end{pmatrix} \left[F\left(\theta_k, X_{t_k}(\theta_k), Y_{t_k}(\theta_k)\right) \Delta t + 4\beta\gamma X_{t_k}^1(\theta_k) Y_{t_k}^2(\theta_k) \Delta L_k\right] ,$$

that is we do not exactly use the direction of the gradient but another direction of minimization which is more convenient in practice. Moreover we do not take $\rho_k^i = 1/i$ (which is not very good in practice) but

$$\rho_k^i = a_i + \frac{b_i}{\max(1, k - k_i)} , \qquad i = 1, 2 ,$$

where a_i, b_i, k_i are given. In figure 2 we have an example of trajectories : $k \to \hat{\theta}_k^1$ and $k \to \hat{\theta}_k^2$.

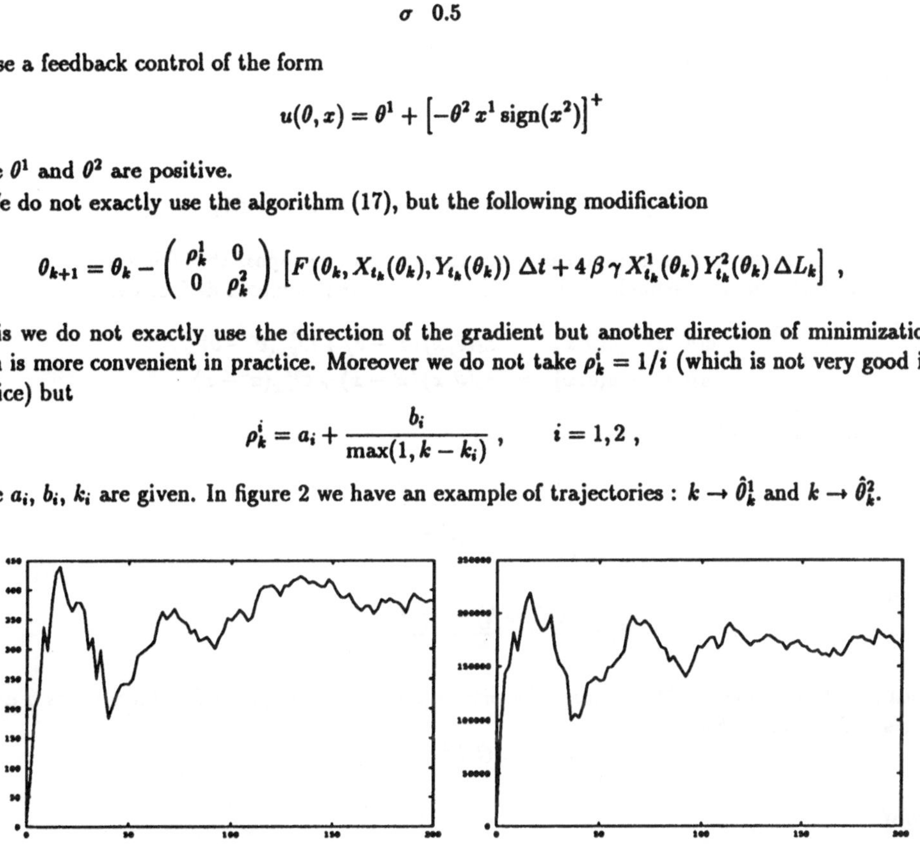

$k \to \theta_k^1$ $\qquad$ $k \to \theta_k^2$

Figure 2: Case 3, the plot of θ^1

5 Conclusion

We have studied in this paper an ergodic stochastic control problem, where the control is to be chosen in a parametrized family of feedback laws. We have computed the gradient of its cost functional. The corresponding stochastic gradient algorithm has been implemented on an applied problem, and has proved to be efficient. It would be worthwhile to prove the convergence of that algorithm towards a global minimum of our cost functional. Unfortunately, the best results that we know in that direction are those of M. Métivier and P. Priouret [10], which required the cost functional to have a single local (and hence global) minimum. There is no reason for that assumption to hold in our case. However, using ideas from simulated annealing, one might perhaps show that our algorithm (or possibly a modified version of it) converges to a global minimum. Such a study is far beyond the scope of the present paper.

The authors wish to thank Prof. P. Bougerol for some helpful remarks used in the proof of Lemma 3.3.

Appendix : Regularity of $X_t(\theta)$

We use the notation of Lemma 2.1,

$$\begin{aligned} dX_t^1(\theta) &= X_t^2(\theta)\,dt\,, \\ dX_t^2(\theta) &= a(\theta, X_t(\theta))\,dt + b(X_t^2(\theta))\,dt + \sigma\,dW_t\,, \end{aligned}$$

where $b(x^2) = -\gamma\,\mathrm{sign}(x^2)$ is non increasing. In this section we suppose that θ is scalar and that the function $a(\theta, x)$ is regular enough in (θ, x) : differentiable with bounded derivatives,

$$\begin{aligned} a(\theta', x) - a(\theta, x) &= a_\theta(\theta, x)\,(\theta' - \theta) + O^1_{\theta,x}(\theta' - \theta)\,, \\ a(\theta, x') - a(\theta, x) &= a_x(\theta, x)\,(x' - x) + O^2_{\theta,x}(x' - x)\,, \end{aligned}$$

with $a_\theta(\theta, x)$, $a_x(\theta, x)$ bounded, continuous and

$$\begin{aligned} &\sup_x O^1_{\theta,x}(\theta)/|\theta| \to 0 \text{ as } |\theta| \to 0\,, \\ &\sup_\theta O^2_{\theta,x}(x)/|x| \to 0 \text{ as } |x| \to 0\,. \end{aligned}$$

Define

$$\bar{X}_t^i = \frac{X_t^i(\theta + h) - X_t^i(\theta)}{h}\,, \quad i = 1, 2\,.$$

Lemma 5.1 *The process X_t^θ is continuous with respect to θ, more precisely : there exists a certain constant $C > 0$ such that*

$$\left|\bar{X}_t\right|^2 \le C\,e^{Ct}\,, \quad \forall t \ge 0\,. \tag{28}$$

Proof

$$\begin{aligned} \frac{d}{dt}|\bar{X}_t|^2 &= 2\,\bar{X}_t^1\,\frac{d}{dt}\bar{X}_t^1 + 2\,\bar{X}_t^2\,\frac{d}{dt}\bar{X}_t^2 \\ &= 2\,\bar{X}_t^1\,\bar{X}_t^2 + 2\,\bar{X}_t^2\,\frac{a(\theta+h, X_t(\theta+h)) - a(\theta, X_t(\theta))}{h} + \underbrace{2\,\bar{X}_t^2\,\frac{b(X_t^2(\theta+h)) - b(X_t^2(\theta))}{h}}_{\le 0}\,, \\ &\le C_1\,|\bar{X}_t|^2 + C_2\,. \end{aligned}$$

The result follows from Gronwall's lemma. ∎

Proof of Proposition 3.2

$$\begin{aligned} d\bar{X}_t^1 &= \bar{X}_t^2\,dt\,, \\ d\bar{X}_t^2 &= \frac{a(\theta+h, X_t(\theta+h)) - a(\theta, X_t(\theta))}{h}\,dt + \frac{b(X_t^2(\theta+h)) - b(X_t^2(\theta))}{h}\,dt \end{aligned}$$

and

$$\begin{aligned} dY_t^1 &= Y_t^2\,dt\,, \\ dY_t^2 &= (a_\theta(\theta,X_t(\theta)) + a_x(\theta,X_t(\theta))\,Y_t)\,dt - \frac{2\gamma}{\sigma^2}Y_t^2\,dL_t\,, \end{aligned}$$

so

$$\begin{aligned} d[\bar X_t^1 - Y_t^1] &= [\bar X_t^2 - Y_t^2]\,dt\,, \\ d[\bar X_t^2 - Y_t^2] &= \left[\frac{a(\theta+h,X_t(\theta)) - a(\theta,X_t(\theta))}{h} - a_\theta(\theta,X_t(\theta))\right]dt \\ &\quad + \left[\frac{a(\theta+h,X_t(\theta+h)) - a(\theta+h,X_t(\theta))}{h} - a_x(\theta+h,X_t(\theta))\,\bar X_t\right]dt \\ &\quad + [a_x(\theta+h,X_t(\theta)) - a_x(\theta,X_t(\theta))]\,\bar X_t\,dt \\ &\quad + a_x(\theta,X_t(\theta))\,[\bar X_t - Y_t]\,dt \\ &\quad + \frac{b(X_t^2(\theta+h)) - b(X_t^2(\theta))}{h}\,dt + \frac{2\gamma}{\sigma^2}Y_t^2\,dL_t\,. \\ &= \frac{O^1_{\theta,X_t(\theta)}(h)}{h}\,dt + \frac{O^2_{\theta+h,X_t(\theta)}(h\,\bar X_t)}{h}\,dt \\ &\quad + [a_x(\theta+h,X_t(\theta)) - a_x(\theta,X_t(\theta))]\,\bar X_t\,dt \\ &\quad + a_x(\theta,X_t(\theta))\,[\bar X_t - Y_t]\,dt \\ &\quad + \frac{b(X_t^2(\theta+h)) - b(X_t^2(\theta))}{h}\,dt + \frac{2\gamma}{\sigma^2}Y_t^2\,dL_t\,. \end{aligned}$$

Then

$$\begin{aligned} d|\bar X_t - Y_t|^2 &= 2(\bar X_t^1 - Y_t^1)\,d[\bar X_t^1 - Y_t^1] + 2(\bar X_t^2 - Y_t^2)\,d[\bar X_t^2 - Y_t^2] \\ &= 2(\bar X_t^1 - Y_t^1)(\bar X_t^2 - Y_t^2)\,dt \\ &\quad + 2(\bar X_t^2 - Y_t^2)\frac{O^1_{\theta,X_t(\theta)}(h)}{h}\,dt + 2(\bar X_t^2 - Y_t^2)\frac{O^2_{\theta+h,X_t(\theta)}(h\,\bar X_t)}{h}\,dt \\ &\quad + 2(\bar X_t^2 - Y_t^2)\,[a_x(\theta+h,X_t(\theta)) - a_x(\theta,X_t(\theta))]\,\bar X_t\,dt \\ &\quad + 2(\bar X_t^2 - Y_t^2)\,a_x(\theta,X_t(\theta))\,[\bar X_t - Y_t]\,dt \\ &\quad + 2(\bar X_t^2 - Y_t^2)\frac{b(X_t^2(\theta+h)) - b(X_t^2(\theta))}{h}\,dt \\ &\quad + 4(\bar X_t^2 - Y_t^2)\frac{\gamma}{\sigma^2}Y_t^2\,dL_t\,, \end{aligned}$$

so we have the following inequality

$$\begin{aligned} |\bar X_t - Y_t|^2 &\le C\int_0^t |\bar X_s - Y_s|^2\,ds \\ &\quad + C\int_0^t \left|\frac{O^1_{\theta,X_s(\theta)}(h)}{h}\right|^2 ds + C\int_0^t \left|\frac{O^2_{\theta+h,X_s(\theta)}(h\,\bar X_t)}{h}\right|^2 ds \\ &\quad + \int_0^t |\bar X_s|^2\,|a_x(\theta+h,X_s(\theta)) - a_x(\theta,X_s(\theta))|^2\,ds \\ &\quad + C\left|\int_0^t (\bar X_s^2 - Y_s^2)\frac{b(X_s^2(\theta+h)) - b(X_s^2(\theta))}{h}\,ds + \int_0^t \frac{2\gamma}{\sigma^2}(\bar X_s^2 - Y_s^2)\,Y_s^2\,dL_s\right|. \end{aligned}$$

Let Δ_t^1 denotes the last term of the right hand side of this last inequality.

$$\Delta_t^1 \le \gamma\left|\int_0^t (\bar X_s^2 - Y_s^2)\frac{\operatorname{sign}(X_s^2(\theta+h)) - \operatorname{sign}(X_s^2(\theta))}{h}\,ds\right.$$

$$
\begin{aligned}
&- \int_0^t (\bar{X}_s^2 - Y_s^2) \frac{\operatorname{sign}_\varepsilon(X_s^2(\theta+h)) - \operatorname{sign}_\varepsilon(X_s^2(\theta))}{h} ds \Big| \\
&+\gamma \left| \int_0^t (\bar{X}_s^2 - Y_s^2) \frac{\operatorname{sign}_\varepsilon(X_s^2(\theta+h)) - \operatorname{sign}_\varepsilon(X_s^2(\theta))}{h} ds - \int_0^t \frac{2\gamma}{\sigma^2} (\bar{X}_s^2 - Y_s^2) Y_s^2 \, dL_s \right| \\
\leq\ & 2 \int_0^t |\bar{X}_s - Y_s|^2 ds \\
&+2 \int_0^t \left| \frac{\operatorname{sign}(X_s^2(\theta+h)) - \operatorname{sign}(X_s^2(\theta))}{h} - \frac{\operatorname{sign}_\varepsilon(X_s^2(\theta+h)) - \operatorname{sign}_\varepsilon(X_s^2(\theta))}{h} \right|^2 ds \\
&+\gamma \left| \int_0^t (\bar{X}_s^2 - Y_s^2) \frac{\operatorname{sign}_\varepsilon(X_s^2(\theta+h)) - \operatorname{sign}_\varepsilon(X_s^2(\theta))}{h} ds - \int_0^t \frac{2\gamma}{\sigma^2} (\bar{X}_s^2 - Y_s^2) Y_s^2 \, dL_s \right| .
\end{aligned}
$$

Let Δ_t^2 denotes the last term of the right hand side of this last inequality.

For h small enough

$$
\begin{aligned}
&\int_0^t (\bar{X}_s^2 - Y_s^2) \frac{\operatorname{sign}_\varepsilon(X_s^2(\theta+h)) - \operatorname{sign}_\varepsilon(X_s^2(\theta))}{h} ds \\
&= \int_0^t (\bar{X}_s^2 - Y_s^2) \frac{\operatorname{sign}_\varepsilon(X_s^2(\theta+h)) - \operatorname{sign}_\varepsilon(X_s^2(\theta))}{h} 1_{[-\varepsilon,\varepsilon]}(X_s^2(\theta)) \, ds \\
&= 2 \int_0^t (\bar{X}_s^2 - Y_s^2) \bar{X}_s^2 \frac{1}{2\varepsilon} 1_{[-\varepsilon,\varepsilon]}(X_s^2(\theta)) \, ds \\
&\underset{\varepsilon \to 0}{\overset{\text{a.s.}}{\to}} \frac{2}{\sigma^2} \int_0^t (\bar{X}_s^2 - Y_s^2) \bar{X}_s^2 \, dL_s \\
&\underset{h \to 0}{\simeq} \frac{2}{\sigma^2} \int_0^t (\bar{X}_s^2 - Y_s^2) Y_s^2 \, dL_s .
\end{aligned}
$$

Then from (29),

$$
|\bar{X}_t - Y_t|^2 \leq C \int_0^t |\bar{X}_s - Y_s|^2 ds + \nu_t(\theta, h, \varepsilon) ,
$$

where

$$
\lim_{h \to 0} \lim_{\varepsilon \to 0} \nu_s(\theta, h, \varepsilon) = 0 \quad \text{a.s.}
$$

Then using Gronwall's inequality

$$
|\bar{X}_t - Y_t|^2 \leq \nu_t(\theta, h, \varepsilon) + C \int_0^t \nu_s(\theta, h, \varepsilon) e^{C(t-s)} ds
$$

so

$$
|\bar{X}_t - Y_t|^2 \to 0 .
$$

References

[1] S. BELLIZZI, R. BOUC, F. CAMPILLO, and E. PARDOUX. Contrôle optimal semi-actif de suspension de véhicule. In A. Bensoussan and J.L. Lions, editors, *Analysis and Optimization of Systems*, pages 689–699, INRIA, Antibes, Lecture Notes in Control and Information Sciences 111, 1988.

[2] P. BOUGEROL. Comparaison des exposants de lyapounov des processus markoviens multiplicatifs. *Annales de l'Institut Henri Poincaré - Probabilités et Statistiques*, 24(4):439–489, 1988.

[3] P. BOUGEROL. Théorèmes limite pour les systèmes linéaires à coefficients markoviens. *Probability Theory and Related Fields*, 78:193–221, 1988.

[4] F. CAMPILLO. Optimal ergodic control for a class of nonlinear stochastic systems. In *Proceedings of 28th Conference on Decision and Control*, pages 1190–1195, IEEE, Tampa, Florida, December 1989.

[5] F. CAMPILLO, F. LE GLAND, and E. PARDOUX. Approximation of a stochastic ergodic control problem. In J. Descusse, M. Fliess, A. Isidori, and D. Leborgne, editors, *New Trends in Nonlinear Control Theory*, pages 379–395, CNRS, Nantes 1988, Lecture Notes in Control and Information Sciences 122, Springer–Verlag, 1989.

[6] J. GYÖNGY and N. KRYLOV. On stochastic equations with respect to semimartingales i. *Stochastics*, 4:1–21, 1980.

[7] J. JACOD. Une condition d'existence et d'unicité pour les solutions fortes d'équations différentielles stochastiques. *Stochastics*, 4:23–38, 1980.

[8] I. KARATZAS and S.E. SHREVE. *Brownian Motion and Stochastic Calculus*. Volume 113 of *Graduate Texts in Mathematics*, Srpinger–Verlag, New–York, 1988.

[9] L. LADELLI. *Théorèmes limites pour les chaînes de Markov : Application aux algorithmes stochastiques*. PhD thesis, Université Paris VI, 1989.

[10] M. METIVIER and P. PRIOURET. Théorèmes de convergence presque sûre pour une classe d'algorithmes stochastiques à pas décroissant. *Probability Theory and Related Fields*, 74:403–428, 1987.

[11] P. PROTTER. *Stochastic integration and differential equations : a new approach*. Volume 21 of *Applications of Mathematics*, Springer Verlag, 1990.

EXPONENTIAL TRIANGULAR COOLING SCHEDULES FOR SIMULATED ANNEALING ALGORITHMS : A CASE STUDY

O. Catoni
DIAM - Intelligence Artificielle et Mathématiques
Laboratoire de Mathématiques, UA 762 du CNRS
Département de Mathématiques et d'Informatique
Ecole Normale Supérieure
45, rue d'Ulm 75230 Paris Cedex 05 France.

Introduction

This paper is a case study of the probability of convergence in a finite large number of steps of simulated annealing algorithms. The choice of the model has been led by two considerations : that it retain the important features of the general case, and be easy to solve. All the results presented here have already been established in the general case of simulated annealing algorithms with reversible transitions. They are to appear in a paper called "Rough Large Deviations Estimates for Simulated Annealing Algorithms : Application to Exponential Schedules" [6]. Solving the general case requires indulging in quite complex large deviations techniques. We hope that the line of reasoning will be more clear in this simple case study.

Here we study simulated annealing algorithms from the point of view of optimization. In our case study, we are given an energy function with only one global minimum, and our aim is to find this global minimum. We will restrict the transitions to be star shaped around the global minimum. Moreover each state will be a local minimum. Thus our model, though very simple and unrealistic, can be considered as a simplification of a "real" problem where all the states in any basin of attraction have been regrouped into a single generic state.

In simulated annealing algorithms, a finite state space is explored by a non-stationary Markov chain. As the transition matrix evolves with time, so does the invariant probability distribution. The evolution of the transitions with time is chosen so that the invariant measure should concentrate on the states of minimum energy. Since the energy can be chosen to be any arbitrary function on a finite set, we have a general purpose optimization technique (though it is clear that its performance will strongly depend on the shape of the function).

Most theoretical studies about simulated annealing are linked with estimates of the second eigenvalue of the transition matrix (and its evolution with time). This kind of approach is the one naturaly inspired by the consideration of the stationary case : the speed of convergence of the powers of a Markov irreducible matrix is given by the powers of its second eigenvalue. In the non-stationary case of simulated annealing we have to consider a product of matrices which are different from one another. The aim of this paper is to show that in this more complex setting all the eigenvalues of the transition matrix can enter significantly into the game.

Our plan will be the following. The quantity of interest will be the probability to be in the state of minimal energy at time N. We call this quantity the probability of convergence in N steps. Simulated annealing algorithms are governed by a control parameter, the so called temperature. We will first give an upper bound for the probability of convergence in N steps whatever the control parameters may have been during these N steps. In this sense we study annealing in finite time, because we allow the whole sequence of control parameters to depend on N. In an infinite time study we would have fixed the control parameters first and then let N go to infinity.

Then we will show that this bound is sharp up to a multiplicative constant. For this we build a control sequence, a so called cooling schedule, for which the bound is reached (up to a multiplicative constant). The striking fact is that this control sequence depends on N. We show that if we choose any infinite control sequence once and for all, and then let N tend to infinity, the complement of the probability of convergence in N steps does not tend to zero according to the best possible power of $1/N$. Another striking fact is that the control sequence depends on all the eigenvalues of the transition matrix. Thus it is not a good cooling schedule for any practical application, because it is very unlikely that such precise features of the energy landscape should be known whereas the global minimum should not !

For this reason we turn towards the problem of optimization of an "unknown" function : we fix the number of steps N, then the control sequence (depending on N) and we let the energy function vary in some large set defined by a few equations and look at the infimum of the probability of convergence.

In this context we show that for some particular kind of constraints defining the set of admissible energy functions, exponential cooling schedules are "nearly optimal", i.e. are doing a good job, whatever the admissible energy function may be. Thus adopting the point of view of a finite time study and looking at the worst probability of convergence when the energy function vary in some set, we can give a rigorous mathematical justification to the use of exponential cooling schedules.

1 Description of the model

The state space $L = \{0, 1, \dots, r\}$ will be finite. The time will be discrete. On $L \times L$ we consider a star shaped Markov matrix $q : L \times L :\longrightarrow [0, 1]$. We assume that

$$\begin{aligned} q(0,i) > 0, \quad q(i,0) > 0, \quad & i > 0, \\ q(i,j) = 0 \quad & \text{otherwise.} \end{aligned} \tag{1}$$

We consider on $L \times L$ the set $\mathcal{V}$ of all cost functions $V : L \times L \longrightarrow \mathbb{R}_+$ such that $V(0,i) > V(i,0), i > 0$. We consider also the set $\mathcal{E}$ of all control sequences $(\epsilon_n)_{n \in \mathbb{N}^*}$ such that $0 \leq \epsilon_{n+1} \leq \epsilon_n \leq 1$, $n > 0$.

To any $V \in \mathcal{V}$ and any $\epsilon \in \mathcal{E}$ we associate the simulated annealing algorithm $X^{V,\epsilon}$

which is a non-stationary Markov chain with transitions

$$P(X_n = j \mid X_{n-1} = i) = p^{V,\epsilon_n}(i,j), \tag{2}$$

where

$$\begin{cases} p^{V,\alpha}(i,j) &= q(i,j)\alpha^{V(i,j)}, \quad i \neq j, \\ p^{V,\alpha}(i,i) &= 1 - \sum_{j,j\neq i} q(i,j)\alpha^{V(i,j)}. \end{cases} \tag{3}$$

For any $V \in \mathcal{V}$ we define the potential U associated with V to be

$$\begin{aligned} U(0) &= 0, \\ U(i) &= V(0,i) - V(i,0), \quad i > 0. \end{aligned} \tag{4}$$

Proposition 1.1 *1. The matrix q is reversible with invariant probability measure μ^1 given by*

$$\begin{cases} \mu^1(i) &= \left(1 + \sum_{j=1}^r \frac{q(0,j)}{q(j,0)}\right)^{-1} \frac{q(0,i)}{q(i,0)}, \quad i > 0, \\ \mu^1(0) &= \left(1 + \sum_{j=1}^r \frac{q(0,j)}{q(j,0)}\right)^{-1} \end{cases} \tag{5}$$

2. The matrix p^α is reversible with invariant probability measure

$$\mu^\alpha(i) = \left(\sum_{j=0}^r \alpha^{U_j}\mu^1(j)\right)^{-1} \alpha^{U_i}\mu^1(i). \tag{6}$$

3. In the case of "general position", more precisily when the $V(i,0)$ are distinct for $i > 0$, and for α sufficiently small, p^α has $r+1$ distinct eigenvalues $1, 1-e_1(\alpha), \dots, 1-e_r(\alpha)$ with $e_1(\alpha) < \cdots < e_r(\alpha)$ and

$$\lim_{\alpha\to 0} \frac{e_j(\alpha)}{q(j,0)\alpha^{V(j,0)}} = 1. \tag{7}$$

Proof :

Part 1) and 2) are straightforward.It is not hard to see that the characteristic polynomial of p^α is

$$\begin{aligned} (X-1) \quad &\left(\prod_{j=1}^r (X - 1 + q(j,0)\alpha^{V(j,0)} + q(0,j)\alpha^{V(0,j)})\right. \\ &\left. + \sum_{j=1}^r q(0,j)\alpha^{V(0,j)} \prod_{h\notin\{0,j\}} (X - 1 + q(h,0)\alpha^{V(h,0)} + q(0,h)\alpha^{V(0,h)})\right). \end{aligned} \tag{8}$$

For α small enough this polynomial takes values of opposite signs for $X = 1 - q(j,0)\alpha^{V(j,0)} - q(0,j)\alpha^{V(0,j)}$ and for $X = 1 - q(j,0)\alpha^{V(j,0)} - 3q(0,j)\alpha^{V(0,j)}$. This proves part 3).

End of the proof of proposition 1.1.

2 An upper bound for the convergence rate.

As we have already mentioned in the introduction, the quantity of interest will be the probability of convergence in N steps. We will use the notation

$$\nu_n^{(V,\epsilon)}(i,j) = P(X_n^{V,\epsilon} = j \mid X_0^{V,\epsilon} = i). \quad (9)$$

For any $j > 0$, we will give two kinds of lower bounds for $\nu_n^{V,\epsilon}(i,j)$. One is in term of ϵ_n, and the other is independent of the control sequence. This will allow us to generalize our study to convergence sets of the form $\{j \mid U(j) \geq \delta\}$ for some positive δ. The second result is of the form

$$\inf_{1\geq\epsilon_1\geq\epsilon_2\geq\ldots\geq\epsilon_n\geq 0} \max_{i\in L} P(U(X_n) \geq \delta \mid X_0 = i) \geq \frac{\text{constant}}{n^{D_\delta^{-1}}}, \quad (10)$$

where D_δ is a constant depending on the cost function. We will call D_δ the difficulty of $\mathcal{V}$.

We will contrast this second result with a third "infinite time" result : for any choice of an infinite sequence $(\epsilon_n)_{n\in N^*}$ there is a constant K such that

$$\limsup_{n\to+\infty} n^{D_{\infty,\delta}^{-1}} \max_{i\in L} P(U(X_n) \geq \delta \mid X_0 = i) \geq K. \quad (11)$$

In general the "infinite time difficulty" $D_{\infty,\delta}$ is larger than the finite time difficulty. The first theorem is a comparison with the invariant probability measure at time n.

Theorem 2.1 *We have for any starting probability distribution* $\rho : L \to [0,1]$

$$\sum_{i\in L} \rho(i)\nu_n^{V,\epsilon}(i,j) \geq \inf_{i\in L}\left(\frac{\rho(i)}{\mu^1(i)}\right)\mu^1(j)\epsilon_n^{U_j} \quad (12)$$

Proof

The proof is by induction. Let us put

$$\sum_{i\in L} \rho(i)\nu_n^{V,\epsilon}(i,j) = \mu_n(j) \quad (13)$$

and $\epsilon_0 = 1$. We have

$$\begin{aligned} \mu_n(j) &= \sum_{i\in L} \mu_{n-1}(i) p^{V,\epsilon_n}(i,j) \\ &\geq \inf_{h\in L} \frac{\rho(h)}{\mu^1(h)} \sum_{i\in L} \mu^1(i)\epsilon_{n-1}^{U(i)} p^{V,\epsilon_n}(i,j) \\ &\geq \inf_{h\in L} \frac{\rho(h)}{\mu^1(h)} \sum_{i\in L} \mu^1(i)\epsilon_n^{U(i)} p^{V,\epsilon_n}(i,j) \\ &\geq \inf_{h\in L} \frac{\rho(h)}{\mu^1(h)} \mu_n^1(j)\epsilon_n^{U(j)}, \end{aligned} \quad (14)$$

because $\mu^1(i)\epsilon_n^{U(i)}$ is p^{V,ϵ_n} invariant.

End of the proof of theorem 2.1

Let us come now to the finite time result :

Theorem 2.2 *There are $a > 0$ and $b > 0$ such that for any n, any $V \in \mathcal{V}$, any $1 \geq \epsilon_1 \geq \cdots \geq \epsilon_n \geq 0$*

$$\max_i \nu_n^{V,\epsilon}(i,j) \geq \frac{a}{(bn)^{\frac{U(j)}{V(j,0)}}}. \tag{15}$$

Corollary: *For any $\delta > 0$, let us put*

$$D_\delta = \max_{j, U(j) \geq \delta} \frac{V(j,0)}{U(j)}, \tag{16}$$

and call it the difficulty of level δ of the energy landscape. Then for some constants $a > 0$ and $b > 0$, independent of V and of the control sequence ϵ,

$$\max_{i \in L} P(U(X_n^{V,\epsilon}) \geq \delta \mid X_0^{V,\epsilon} = i) \geq \frac{a}{(bn)^{D_\delta^{-1}}}. \tag{17}$$

Proof

Let us put in theorem 2.1 $\rho = \mu^1$. Let us put again $\mu_n = \sum_{i \in L} \mu^1(i) \nu_n^{V,\epsilon}(i,j)$. We have $\max_{i \in L} \nu_n^{V,\epsilon}(i,j) \geq \mu_n(j)$ and

$$\mu_n(j) \geq \mu^1(0) q(0,j) \epsilon_n^{V(0,j)} + \mu_{n-1}(j)(1 - q(j,0)\epsilon_n^{V(j,0)}). \tag{18}$$

It is clear that if the sequence $(R_n)_{n \in N}$ is such that $R_0 \leq \mu_0(j)$ and

$$R_n = \inf_{\epsilon_n} \mu^1(0) q(0,j) \epsilon_n^{V(0,j)} + R_{n-1}(1 - q(j,0)\epsilon_n^{V(j,0)}) \tag{19}$$

then, for all n, $\mu_n(j) \geq R_n$. This infimum is equal to

$$R_{n-1} - (1 - \frac{V(j,0)}{V(0,j)}) R_{n-1} q(j,0) \left(\frac{R_{n-1} V(j,0) q(j,0)}{\mu^1(0) V(0,j) q(0,j)} \right)^{V(j,0)/U(j)}, \tag{20}$$

thus

$$R_n - R_{n-1} = A \frac{U(j)}{V(0,j)} R_{n-1}^{V(0,j)/U(j)}, \tag{21}$$

with

$$A = q(j,0) \left(\frac{V(j,0) q(j,0)}{\mu^1(0) V(0,j) q(0,j)} \right)^{V(j,0)/U(j)}$$

Thus

$$\begin{aligned} R_n^{-V(j,0)/U(j)} - R_{n-1}^{-V(j,0)/U(j)} &\leq \frac{V(j,0)}{V(0,j)} A \left(1 - \frac{U(j)}{V(0,j)} A R_{n-1}^{V(j,0)/U(j)} \right)^{-V(0,j)/U(j)} \\ &\leq \frac{V(j,0)}{V(0,j)} A \left(1 - \frac{U(j)}{V(0,j)} A R_0^{V(j,0)/U(j)} \right)^{-V(0,j)/U(j)} . \end{aligned} \tag{22}$$

Take

$$R_0 = \min(\mu^1(j), \left(\frac{V(0,j)}{2U(j)A}\right)^{U(j)/V(j,0)}),$$

then

$$R_n^{-V(j,0)/U(j)} - R_{n-1}^{-V(j,0)/U(j)} \leq \frac{V(j,0)}{V(0,j)} 2^{V(0,j)/U(j)} A, \tag{23}$$

thus

$$R_n \geq \left(R_0^{-V(j,0)/U(j)} + 2^{V(0,j)/U(j)} An\right)^{-U(j)/V(j,0)} \tag{24}$$

Although it requires some care, it is elementary to deduce the theorem from this expression, replacing R_0 and A by their values in terms of V and q.

End of the proof of theorem 2.2

Let us contrast theorem 2.2 with an "infinite time" result:

Theorem 2.3 *For any $V \in \mathcal{V}$, any $\epsilon \in \mathcal{E}$, any $\delta > 0$,*

$$\limsup_{n \to +\infty} n^{D_{\infty,\delta}^{-1}} \max_{i \in L} P(U(X_n^{V,\epsilon}) \geq \delta \mid X_0 = i) > 0, \tag{25}$$

where

$$D_{\infty,\delta} = \frac{\max_{j, U(j) \geq \delta} V(j,0)}{\min_{j, U(j) \geq \delta} U(j)}. \tag{26}$$

Proof

Let us put $\rho = \mu^1$ in theorem 2.1. Choose j and h in L such that $U(j) = \min_{s, U(s) \geq \delta} U(s)$ and $V(h,0) = \max_{s, U(s) \geq \delta} V(s,0)$. We have

$$\max_{i \in L} P(U(X_n^{V,\epsilon}) \geq \delta \mid X_0 = i) \geq \max(\mu^1(j)\epsilon_n^{U(j)}, \prod_{k=1}^{n}(1 - q(h,0)\epsilon_k^{V(h,0)})). \tag{27}$$

Let us consider the control sequence $(\alpha_n)_{n \in N^*}$ defined by

$$\mu^1(j)\alpha_n^{U(j)} = \prod_{k=1}^{n}(1 - q(h,0)\alpha_k^{V(h,0)}). \tag{28}$$

This sequence is obviously decreasing, hence $\alpha \in \mathcal{E}$. It satisfies the induction relation

$$\alpha_n^{U(j)} = \alpha_{n-1}^{U(j)}(1 - q(h,0)\alpha_n^{V(h,0)}), \tag{29}$$

from which it is easy to deduce that for some constant K

$$\mu^1(j)\alpha_n^{U(j)} \geq \frac{K}{n^{U(j)/V(h,0)}}, \tag{30}$$

using the same line of reasoning as in the previous theorem. Then we can distinguish between two cases:

1. Either $\{n \mid \epsilon_n \geq \alpha_n\}$ is infinite, in which case the theorem holds,
2. either there is N such that for $n \geq N$ we have $\epsilon_n \leq \alpha_n$. In this case

$$\prod_{k=1}^{n}(1 - q(h,0)\epsilon_k^{V(h,0)}) \geq \prod_{k=1}^{N-1}\left(\frac{1 - q(h,0)\epsilon_k^{V(h,0)}}{1 - q(h,0)\alpha_k^{V(h,0)}}\right)\frac{K}{n^{U(j)/V(h,0)}}, \tag{31}$$

thus the theorem holds also in this case.

3 Localization of the quasi-equilibrium equations

In this section we will establish a converse to theorem 2.1. By quasi-equilibrium here, we mean quasi-equilibrium up to a multiplicative constant. We will say that $X^{V,\epsilon}$ is in quasi-equilibrium (asymptotically) if for some constant $K < +\infty$ we have $P(X_n = i)/\mu^{\epsilon_n}(i) \leq K$ for $i \in L$ and $n \in \mathbb{N}$.

A known condition for quasi-equilibrium ([19], [7], [8]) is that

$$\lim_{n\to+\infty}(\ln \epsilon_n^{-1} - \ln \epsilon_{n-1}^{-1})\epsilon_n^{-\max_{j>0} V(j,0)} = 0, \tag{32}$$

in which case we have even $\lim_{n\to+\infty} P(X_n = i)/\mu^{\epsilon_n}(i) = 1$. Here we localize the quasi-equilibrium relation, that is, we give a different equation for each $i \in L$. The result is the following:

Theorem 3.1 *For any $\eta > 0$, any $V \in \mathcal{V}$, any $m \in \mathbb{N}$, any $\epsilon \in \mathcal{E}$, any $j > 0$, if*

$$(U(j) \wedge \eta)(\ln \epsilon_{l+1}^{-1} - \ln \epsilon_l^{-1}) \leq \frac{q(j,0)}{4}\epsilon_l^{V(j,0)}, \quad l < m, \tag{33}$$

then for any $i \in L$

$$\nu_n^{V,\epsilon}(i,j) \leq (\frac{q(j,0)}{q(0,j)}\epsilon_1^{-(U(j)\wedge\eta)} + 2)(1 + q(j,0)/2)\frac{q(0,j)}{q(j,0)}\epsilon_m^{U(j)\wedge\eta}, \quad n \geq m. \tag{34}$$

Remark : take good notice of the fact that there are three time variables, l, m and n and that equation (34) holds not only at time m but at any further time n.

Proof

Considering the last jump of $X^{V,\epsilon}$ out of state 0 and writing $\epsilon_k^{V(0,j)} = \epsilon_k^{V(j,0)}\epsilon_k^{U(j)}$, we see that

$$\begin{aligned} \nu_n(i,j) &= \nu_0(i,j)\prod_{s=1}^{n}(1 - q(j,0)\epsilon_s^{V(j,0)}) \\ &\quad + \sum_{k=1}^{n} \nu_{k-1}(i,0)\frac{q(0,j)}{q(j,0)}\epsilon_k^{U(j)} q(j,0)\epsilon_k^{V(j,0)} \prod_{s=k+1}^{n}(1 - q(j,0)\epsilon_s^{V(j,0)}). \end{aligned} \tag{35}$$

We can replace in the right-hand member $U(j)$ by $U(j) \wedge \eta$, since $\epsilon_k^{U(j)} \leq \epsilon_k^{U(j)\wedge\eta}$. Putting $\epsilon_0 = \epsilon_1$ we get that

$$\begin{aligned} \nu_n(i,j) &\leq \frac{q(0,j)}{q(j,0)}\epsilon_m^{U(j)\wedge\eta}\Bigg(\nu_0(i,j)\frac{q(j,0)}{q(0,j)}\epsilon_1^{-(U(j)\wedge\eta)}\prod_{s=1}^{m}(1 - q(j,0)\epsilon_s^{V(j,0)})\left(\frac{\epsilon_{s-1}}{\epsilon_s}\right)^{U(j)\wedge\eta} \\ &\quad + \sum_{k=1}^{m} q(j,0)\epsilon_k^{V(j,0)}\prod_{s=k+1}^{m}(1 - q(j,0)\epsilon_s^{V(j,0)})\left(\frac{\epsilon_{s-1}}{\epsilon_s}\right)^{U(j)\wedge\eta}\prod_{s=m+1}^{n}(1 - q(j,0)\epsilon_s^{V(j,0)}) \\ &\quad + \sum_{k=m+1}^{n} q(j,0)\epsilon_k^{V(j,0)}\prod_{s=k+1}^{n}(1 - q(j,0)\epsilon_s^{V(j,0)})\Bigg). \end{aligned} \tag{36}$$

The quasi-equilibrium equations imply that

$$\left(\frac{\epsilon_{s-1}}{\epsilon_s}\right)^{U(j)\wedge\eta} \leq 1 + \frac{q(j,0)}{2}\epsilon_{s-1}^{V(j,0)}, \tag{37}$$

hence that

$$\prod_{s=k+1}^{m}(1-q(j,0)\epsilon_s^{V(j,0)})\left(\frac{\epsilon_{s-1}}{\epsilon_s}\right)^{U(j)\wedge\eta} \leq (1+\frac{q(j,0)}{2})\prod_{s=k+1}^{m}(1-\frac{q(j,0)}{2}\epsilon_s^{V(j,0)}). \quad (38)$$

Thus we have

$$\begin{aligned} \nu_n(i,j) \leq\ & \frac{q(0,j)}{q(j,0)}\epsilon_m^{U(j)\wedge\eta}(1+\frac{q(j,0)}{2}) \\ & \times\left(\frac{q(j,0)}{q(0,j)}\epsilon_1^{-(U(j)\wedge\eta)}+2\sum_{k=1}^{n}\frac{q(j,0)}{2}\epsilon_k^{V(j,0)}\prod_{s=k+1}^{n}(1-\frac{q(j,0)}{2}\epsilon_s^{V(j,0)})\right). \quad (39)\end{aligned}$$

Noticing that

$$\sum_{k=1}^{n}\frac{q(j,0)}{2}\epsilon_k^{V(j,0)}\prod_{s=k+1}^{n}(1-\frac{q(j,0)}{2}\epsilon_s^{V(j,0)}) \leq 1 \quad (40)$$

ends the proof.

End of the proof of theorem 3.1

4 Almost optimal cooling schedules

The aim of this section is to show that D_δ^{-1} is sharp in theorem 2.2. For this purpose we will build a cooling schedule for which the complement of the probability of convergence in N steps decreases according to this power of $1/N$. From theorem 2.3 we already know that this schedule has to be a *triangular schedule*, that is a schedule of the form $(\epsilon_n^N)_{1\leq n\leq N}$: for each given total number of steps N we are allowed to choose the whole sequence $\epsilon_1^N,\ldots,\epsilon_N^N$. We will prove the following theorem

Theorem 4.1 *For any $\delta > 0$ and any $V \in \mathcal{V}$ there is $(\epsilon_n^N)_{1\leq n\leq N}$ and $a > 0$ such that*

$$\max_{i\in L} P(U(X_N^{V,\epsilon^N}) \geq \delta \mid X_0 = i) \leq \frac{a}{N^{D_\delta^{-1}}}, \quad (41)$$

Proof

For any given λ, $0<\lambda<1$, we will find N_λ and $\epsilon_1^{N_\lambda},\ldots,\epsilon_{N_\lambda}^{N_\lambda}$ such that

$$\nu_{N_\lambda}^{V,\epsilon^{N_\lambda}}(i,j) \leq \lambda, \quad j \in L, U(j) \geq \delta. \quad (42)$$

We will verify that $\lambda \leq aN_\lambda^{-D_\delta^{-1}}$ for some constant a. For this purpose we will use the quasi-equilibrium equations of theorem 3.1. We have one equation for each $j = 1,\ldots,r$. We will ask that these equations are satisfied as long as

$$(\frac{q(j,0)}{q(0,j)}\epsilon_1^{-(U(j)\wedge\eta)}+2)(1+\frac{q(j,0)}{2})\frac{q(0,j)}{q(j,0)}\epsilon_l^{U(j)\wedge\eta} \geq \lambda. \quad (43)$$

Thus the life-times of the equations will be different, depending on $U(j)$. For each period of time we will build some leading equation which implies all the quasi-equilibrium equations which are still alive. The first leading equation will be of the form

$$W_1(\ln \epsilon_{l+1}^{-1} - \ln \epsilon_l^{-1}) \leq B\epsilon_l^{H_1} \tag{44}$$

where $H_1 = \max_{j,U(j)\geq\delta} V(j,0)$, and where $W_1 = \eta$. Then when all equations corresponding to those j's such that $V(j,0) = H_1$, will have died, we will change the leading equation into

$$W_2(\ln \epsilon_{l+1}^{-1} - \ln \epsilon_l^{-1}) \leq B\epsilon_l^{H_2} \tag{45}$$

with some suitable $H_2 < H_1$ and $W_2 < W_1$. Solving these equations successively will give a control sequence which is piecewise of the form $\epsilon_l = (A_k\, l + E_k)^{-H_k^{-1}}$ for a decreasing sequence of characteristic depths $H_1 > H_2 > \cdots > H_u$.

Let us come to the details of the proof:

Let us put

$$\begin{aligned}
H_1 &= \max\{V(j,0) \mid j, U(j) \geq \delta\},\\
W_1 &= \min\{U(j) \mid j, U(j) \geq \delta, V(j,0) \geq H_1\}\\
&\dots\\
H_k &= \max\{V(j,0) \mid j, \delta \leq U(j) < W_{k-1}\},\\
W_k &= \min\{U(j) \mid j, U(j) \geq \delta, V(j,0) \geq H_k\}\\
&\dots
\end{aligned} \tag{46}$$

where we stop at the first k for which $\{j \mid \delta \leq U(j) < W_k\}$ becomes empty.

Let us take $\eta = W_1$ in theorem 3.1 and let us put $B = \min_{j,U(j)\geq\delta} \frac{q(j,0)}{4}$. All the equations in theorem 3.1 are satisfied when

$$W_1(\ln \epsilon_{l+1}^{-1} - \ln \epsilon_l^{-1}) \leq B\epsilon_l^{H_1}, \tag{47}$$

which is itself satisfied when

$$\epsilon_l = \left(\frac{BH_1}{W_1}\, l + \epsilon_0^{-H_1}\right)^{-1/H_1}, \tag{48}$$

where $\epsilon_0 > 0$ is some arbitrary constant.

Let us put

$$C = \max_{j,U(j)\geq\delta}\left(\frac{q(j,0)}{q(0,j)}\epsilon_0^{-(U(j)\wedge\eta)} + 2\right)\frac{q(0,j)}{q(j,0)}\left(1 + \frac{q(j,0)}{2}\right). \tag{49}$$

We will keep this leading equation as long as

$$C\epsilon_{l+1}^{W_1} \geq \lambda, \tag{50}$$

that is on the intervall $(1, m_1($ with

$$m_1 = \left[\frac{W_1}{BH_1}\left(\frac{\lambda}{C}^{-H_1/W_1} - \epsilon_0^{-H_1}\right)\right], \tag{51}$$

where $\lceil x \rceil$ is the lowest integer superior or equal to x. Applying theorem 3.1 we see that for any j sucht that $U(j) \geq W_1$, any control sequence beginning as (48), we have

$$\nu_n^{V,\epsilon}(i,j) \leq \lambda, \quad n \geq m_1. \tag{52}$$

For $n > m_1$ equations for j such that $W_1 \leq U(j)$ disappear from the set of quasi-equilibrium equations, and we can take as leading equation

$$W_1(\ln \epsilon_{l+1}^{-1} - \ln \epsilon_l^{-1}) \leq B\epsilon_l^{H_2}. \tag{53}$$

This equation is satisfied when

$$\epsilon_l = \left(\frac{BH_2}{W_1} l + \frac{\lambda}{C}^{-H_2/W_1}\right)^{-1/H_2}, \tag{54}$$

which we will use on the intervall $(m_1, m_2($ with

$$m_2 = m_1 + \left\lceil \frac{W_1}{BH_2}\left(\frac{\lambda}{C}^{-H_2/W_2} - \frac{\lambda}{C}^{-H_2/W_1}\right)\right\rceil. \tag{55}$$

Carrying on this process we define

$$m_k = m_1 + \sum_{s=2}^{k} \left\lceil \frac{W_{s-1}}{BH_s}\left(\left(\frac{\lambda}{C}\right)^{-H_s/W_s} - \left(\frac{\lambda}{C}\right)^{-H_s/W_{s-1}}\right)\right\rceil \tag{56}$$

and

$$\epsilon_l = \left(\frac{BH_s}{W_{s-1}} l + \left(\frac{\lambda}{C}\right)^{-H_s/W_{s-1}}\right)^{-1/H_s} \vee \left(\frac{\lambda}{C}\right)^{1/W_s}, \quad m_{s-1} \leq l \leq m_s, \quad s = 2, \ldots, u. \tag{57}$$

We have assumed here that $H_u > 0$. The case $H_u = 0$ is elementary but has to be treated separately. It can be found in [6].

For this cooling schedule the quasi-equilibrium equations in theorem 3.1 are sastisfied as long as $C\epsilon_l^{U(j)\wedge W_1} \geq \lambda$, $j, U(j) \geq \delta$, therefore

$$\nu_{m(u)}^{V,\epsilon}(i,j) \leq \lambda, \quad j, U(j) \geq \delta. \tag{58}$$

We put $m_u = N_\lambda$. We have

$$N_\lambda \leq a\left(\frac{\lambda}{C}\right)^{-\max_s H_s/W_s} = a\left(\frac{\lambda}{C}\right)^{-D_\delta} \tag{59}$$

with

$$a = \frac{u}{B}\max_s \frac{W_s}{H_s}. \tag{60}$$

This ends the proof of theorem 4.1

Theorem 4.1 shows all the benefits resulting from the use of triangular schedules.

To each quasi-equilibrium equation corresponds one different eigenvalue of the transition matrix at low temperature, in the case of general position (see proposition 1.1). Depending on the cost function V it is possible that all the quasi-equilibrium equations become in turn the leading equation (it is the case when $V(1,0) > V(2,0) > \cdots > V(r,0)$ and $U(1) > U(2) > \cdots > U(r)$). Thus our almost optimal control sequence may take into account all the eigenvalues of the transition matrix (at low temperatures). Moreover we have seen that the quasi-equilibrium equations are satisfied on different periods of time. The result is that quasi-equilibrium *is not maintained* up to the end of the algorithm. It is only maintained in the beginning period $(0, m_1)$. To the contrary, at the end (at time N), $\nu_n(i,j)$ is of the same order of magnitude λ for all states j such that $U(j) \geq \delta$, that is for all states outside the convergence set.

We insist on the fact that the introduction of triangular control sequences is not a mathematical joke about the different notions of limit. It has the concrete meaning that if you are allowed twice as much computation time, you'd better share your extra time between low and high temperatures than increase only the time spent at low temperatures, as you would do by continuing the same control sequence ϵ.

5 Performance of exponential cooling schedules on a poorly known energy function

As explained before, the control sequence built in the previous section is useless in any practical situation because so much knowledge is required about the cost function.

Here we will try to satisfy the quasi-equilibrium equations of theorem 3.1 *uniformly with respect to the cost function V* .

The reasoning is the following. We put

$$C = \max_{j,j>0}\left(\frac{q(j,0)}{q(0,j)}\epsilon_1^{-\eta} + 2\right)\left(1 + \frac{q(j,0)}{2}\right)\frac{q(0,j)}{q(j,0)} \vee e \tag{61}$$

(so that $C \geq e$, which will be helpful in the following) and we want

$$(U(j) \wedge \eta)(\ln \epsilon_{l+1}^{-1} - \ln \epsilon_l^{-1}) \leq \frac{q(j,0)}{4}\epsilon_l^{V(j,0)}, \quad j, U(j) \geq \delta \tag{62}$$

to hold as long as $C\ \epsilon_{l+1}^{(U(j)\wedge\eta)} \geq \lambda$.

This gives the relation

$$(U(j) \wedge \eta) \leq \frac{\ln \frac{C}{\lambda}}{\ln \epsilon_{l+1}^{-1}}. \tag{63}$$

Hence it is sufficient to have

$$\frac{\ln \frac{C}{\lambda}}{\ln \epsilon_{l+1}^{-1}}(\ln \epsilon_{l+1}^{-1} - \ln \epsilon_l^{-1}) \leq \frac{q(j,0)}{4}\epsilon_l^{V(j,0)} \tag{64}$$

as long as $C\epsilon_{l+1}^{(U(j)\wedge\eta)} \geq \lambda$. Now we will assume that

$$V(j,0) \leq D_\delta(U(j) \wedge \eta), \tag{65}$$

which is the case when

$$\max_{j,U(j)\geq\delta} V(j,0) \leq D_\delta \eta \tag{66}$$

Equation (66) defines a subset $\mathcal{W}_{\delta,\eta}$ of $\mathcal{V}$. For any $V \in \mathcal{W}_{\delta,\eta}$ a sufficient set of quasi-equilibrium equations is

$$\frac{\ln\frac{C}{\lambda}}{\ln\epsilon_{l+1}^{-1}}(\ln\epsilon_{l+1}^{-1} - \ln\epsilon_l^{-1}) \leq \frac{q(j,0)}{4}\left(\frac{\lambda}{C}\right)^{D_\delta} \tag{67}$$

as long as $C\epsilon_{l+1}^{(U(j)\wedge\eta)} \geq \lambda$.

Now putting $B = \min_{j,U(j)\geq\delta}\frac{q(j,0)}{4}$ we see that this set of equations is a consequence of the single equation

$$\frac{\ln\frac{C}{\lambda}}{\ln\epsilon_{l+1}^{-1}}(\ln\epsilon_{l+1}^{-1} - \ln\epsilon_l^{-1}) \leq B\left(\frac{\lambda}{C}\right)^{D_\delta} \tag{68}$$

as long as $C\epsilon_{l+1}^{\delta} \geq \lambda$.

Thus we can choose

$$\ln\epsilon_l^{-1} = A\exp(\xi l) \tag{69}$$

with

$$\xi = B\frac{\left(\frac{\lambda}{C}\right)^{D_\delta}}{\ln\frac{C}{\lambda}} \tag{70}$$

and N_λ defined by

$$\epsilon_{N_\lambda}^{\delta} \leq \frac{\lambda}{C}. \tag{71}$$

In these formulas A is an arbitrary positive constant and C depends on A, namely

$$C = \max_{j>0}\frac{q(0,j)}{q(j,0)}(1+\frac{q(j,0)}{2})(\frac{q(j,0)}{q(0,j)}e^{\eta A}+2) \vee e. \tag{72}$$

With this choice of cooling schedule

$$\nu_{N_\lambda}^{V,N_\lambda}(i,j) \leq \lambda, \quad j,U(j)\geq\delta,\ i\in L \tag{73}$$

for any $V \in \mathcal{W}_{\delta,\eta}$.

Let us explicit the stopping rule. It can be written as

$$N_\lambda \geq \frac{1}{\xi_\lambda}\left(\ln_2\frac{C}{\lambda} - \ln(\delta A)\right) \tag{74}$$

where $\ln_2 x = \ln(\ln x)$. Writing

$$\ln\frac{B}{\xi} = D_\delta\ln\left(\frac{C}{\lambda}\right) + \ln_2\left(\frac{C}{\lambda}\right) \tag{75}$$

and remembering that we have chosen $C \geq e$, we see that $\ln(C/\lambda) \geq 0$ and that $\ln_2(C/\lambda) \geq 0$, hence that sufficient stopping rules are

$$N_\lambda \xi_\lambda \geq \ln_2 \frac{B}{\xi_\lambda} - \ln(\delta A D_\delta), \tag{76}$$

or

$$N_\lambda \xi_\lambda \geq \ln \frac{B}{\xi_\lambda} - \ln(\delta A). \tag{77}$$

This last stopping rule is interesting because it is independent of $V \in \mathcal{W}_{\delta,\eta}$.

With either of these stopping rules

$$\lim_{\lambda \to 0} \frac{\ln N_\lambda}{\ln \lambda^{-1}} = D_\delta, \tag{78}$$

thus $P(U(X_N^{V,\epsilon^N}) \geq \delta)$ is of order $N^{-D_\delta^{-1}}$ for any $V \in \mathcal{W}_{\delta,\eta}$.

The important fact is that, using the second stopping rule, we can build ϵ_n^N, parametrized only by η (and constants linked with (L, q)) such that *for any* $V \in \mathcal{W}_{\delta,\eta}$ the convergence probability tends to one with the optimal rate (in the sense of log equivalence).

Let us sum up the results of this section into a theorem.

Theorem 5.1 *There is a positive constant B such that for any $\delta > 0$, $\eta \geq \delta$, $A > 0$, there is a constant C such that for any $V \in \mathcal{W}_{\delta,\eta}$, any triangular control sequence*

$$\epsilon_n^N = \exp(-A e^{n \xi_N}) \tag{79}$$

with

$$\xi_N \leq B \frac{\left(\frac{\lambda}{C}\right)^{D_\delta}}{\ln \frac{C}{\lambda}} \tag{80}$$

and

$$N \geq \frac{1}{\xi_N} \left(\ln_2 \frac{C}{\lambda} - \ln(\delta A) \right) \tag{81}$$

for some $\lambda > 0$ depending on N, we have

$$\max_{i \in N} P(X_N^{V,\epsilon^N} = j \mid X_0 = i) \leq \lambda, \quad j, U(j) \geq \delta. \tag{82}$$

Corollary :
If

$$N = \frac{1}{\xi_N} \left(\ln_2 \frac{B}{\xi_N} - \ln(\delta A D_\delta) \right) \tag{83}$$

or

$$N = \frac{1}{\xi_N} \left(\ln \frac{B}{\xi_N} - \ln(\delta A) \right) \tag{84}$$

the hypothesis of the theorem are fulfilled for N large enough and we have

$$\lim_{N \to +\infty} \frac{-\ln P(U(X_N^{V,\epsilon^N}) \geq \delta \mid X_0 = i)}{\ln N} = D_\delta^{-1} \tag{85}$$

for any starting point i.

Exponential control sequences are widely used in practice. We see here that A should be chosen of the order of $D_\delta / \max_j V(j, 0)$ (from the expression of C) and that the rate ξ *should be a function of the total number of steps* N. We see also that we can choose ξ_N independently of $V \in \mathcal{W}_{\delta,\eta}$ (the second formula), for which the logarithm of the complement of the probability of convergence is equivalent to its optimal value for any $V \in \mathcal{W}_{\delta,\eta}$, *despite the fact that* D_δ *may in this case be unknown.*

Conclusion

This case study had not the pretention to describe a realistic optimization problem, but only to raise some questions and make some features of the behaviour of simulated annealing algorithms easier to grasp.

Interestingly enough, the influence of the choice of the control sequence on the probability of convergence is almost as complex as in the general case, although the difficulty of the proofs is kept to a reasonable level.

The main points to be retained are the following:

- The quasi-equilibrium equations can be localized and each localized equation corresponds in the case of general position to a different eigenvalue of the transition matrix.
- Triangular control sequences should be prefered : supplementary computation time should be shared between high temperatures and low temperatures, which requires changing the whole control sequence from the beginning.
- Exponential control sequences ($\ln \epsilon_n^{-1} = A \exp(\xi n)$) can be justified on a mathematically rigorous basis and are more robust to "unexpected" energy functions than piecewise logarithmic ones ($\ln \epsilon_n^{-1} = A \ln(nB)$ piecewise). They are more efficient than the global logarithmic control sequences which are common in the literature ($\ln \epsilon_n^{-1} = A \ln(nB)$ with the same constants from beginning to end).

This paper is a good introduction to the general case of simulated annealing with reversible transitions [6], the proofs being indeed in the same spirit.

I have been introduced to simulated annealing algorithms by Robert Azencott, and I am grateful to him for all the useful discussions we had on this subject.

References

[1] Azencott, R. (1988). Simulated annealing. *Séminaire Bourbaki 40 ième année, 1987 - 1988* **697**.

[2] Catoni, (1988) O. Grandes déviations et décroissance de la température dans les algorithmes de recuit. *C. R. Acad. Sci. Paris.* **307** série I p. 535 – 538.

[3] Catoni, O. (1990a) Sharp large deviations estimates for simulated annealing algorithms. *Preprint, submitted to Ann. Inst. H. Poincaré Probab. Statist.*.

[4] Catoni, O. (1990b) Applications of sharp large deviations estimates to optimal cooling schedules, *Preprint, submitted to Ann. Inst. H. Poincaré Probab. Statist.*

[5] Catoni, O. (1990c) Large deviations for annealing, *Thesis held at university Paris-Sud Orsay.*

[6] Catoni, (1990) Rough Large Deviations for Simulated Annealing. Application to Exponential Schedules. *Preprint, submitted to Annals of Probability.*

[7] Chiang, T. S. and Chow, Y. (1987) On the convergence rate of annealing processes, *Siam J. Control and Optimization* vol 26, No 6, November 1988.

[8] Chiang, T.-S. and Chow, Y. (1989) A limit theorem for a class of inhomogeneous Markov processes, *The Annals of Probability*, vol. 17, No 4, 1483-1502.

[9] Chiang, T.-S. and Chow, Y. (1990) The asymptotic behaviour of simulated annealing processes with absorption, *Preprint, Institute of Mathematics, Academia Sinica, Taipei, Taiwan.*

[10] Chiang, T.-S. and Chow, Y. (1990) On occupation times of annealing processes, *Preprint, Institute of Mathematics, Academia Sinica, Taipei, Taiwan 11529 Rep. of China.*

[11] Dobrushin R. L. (1956) Central limit theorems for non-stationary Markov chains, I, II (english translation) *Theory Probab. Appl.*, **1**, 65-80, 329-383.

[12] Freidlin M. I. and Wentzell A. D. (1984) Random perturbations of dynamical systems, *Springer Verlag.*

[13] Geman, S. and Geman, D. (1984). Stochastic relaxation, Gibbs distribution, and the Bayesian restoration of images, *I.E.E.E. Transactions on Pattern Analysis and Machine Intelligence.* **6**, 721- 741.

[14] Gidas, B. (1985) Non-stationary Markov chains and convergence of the annealing algorithms, *J. Statist. Phys.* **39**, 73-131.

[15] Hajek, B. (1987) Cooling schedules for optimal annealing, *Preprint submitted to Math. Oper. Res.*

[16] Holley, R. A., Kusuoka, S. and Stroock, D. W. (1989) Asymptotics of the spectral gap with applications to the theory of simulated annealing. *J. Funct. Anal.* **83** 333-347.

[17] Holley, R. and Stroock, D. (1987) Simulated annealing via Sobolev inequalities. *Preprint.*

[18] Hwang, C. R. and Sheu, S. J. (1986) Large time behaviours of perturbed diffusion Markov processes with applications I, II and III, *Preprints, Institute of mathematics, Academia Sinica, Tapei, Taiwan.*

[19] Hwang, C. R. and Sheu, S. J. (1988a) Singular perturbed Markov chains and and exact behaviours of simulated annealing process, *Preprint, Institute of mathematics, Academia Sinica, Taipei, Taiwan.*

[20] Hwang, C. R. and Sheu, S. J. (1988b) On the weak reversibility condition in simulated annealing. *Preprint, Institute of mathematics, Academia Sinica, Taipei, Taiwan.*

[21] Iosifescu, M. and Theodorescu, R. (1969) Random processes and learning, *Springer Verlag.*

[22] Isaacson, D. L. and Madsen, R. W. (1973) Strongly ergodic behaviour for non-stationary Markov processes. *Ann. Probab.* **1** 329-335.

[23] Kirkpatrick, S. Gelatt, C. D. and Vecchi, M. P. (1983) Optimization by simulated annealing. *Science.* **220** 621-680.

[24] Miclo L., (1991) Evolution de l'énergie libre. Applications à l'étude de la convergence des algorithmes de recuit simulé. *Thesis, University Paris 6, Paris, France.*

[25] Tsitsiklis, J. N. (1988) A survey of large time asymptotics of simulated annealing algorithms, in "Stochastic differential systems, stochastic control theory and applications", Fleming, W. and Lions, P. L. ed., *IMA Vol. Math. Appl.* **10** Springer Verlag.

[26] Tsitsiklis, J. N. (1989) Markov chains with rare transitions and simulated annealing, *Math. Op. Res.*, **14**, *1*.

A Numerical Method for a Calculus of Variations Problem with Discontinuous Integrand

Paul Dupuis*
Department of Mathematics and Statistics
University of Massachusetts at Amherst
Amherst, Massachusetts 01003

1 Introduction.

Consider the calculus of variations problem

$$V(x) = \inf \left[\int_0^T L(\phi, \dot{\phi}) ds + g(\phi(T)) \right], \tag{1.1}$$

where $L(\cdot, \cdot)$ maps $\mathbb{R}^n \times \mathbb{R}^n$ to $\mathbb{R}$, $g(\cdot)$ maps $\mathbb{R}^n$ to $\mathbb{R}$, and the infimum is over all absolutely continuous functions $\phi : [0, T] \to \mathbb{R}^n$ satisfying $\phi(0) = x$. Such problems, together with variations and generalizations, arise in a wide variety of settings. Well known examples are classical mechanics and geometric optics (e.g., [7], [2]). A more recent example is the theory of large deviations of stochastic processes [6]. In all these examples there is some dynamical model for a "physical" system which determines the function $L(\cdot, \cdot)$, and minimizing (or nearly minimizing) paths $\phi(\cdot)$ have important physical interpretations. For example, in the case of classical Hamiltonian mechanics the relevant laws of motion define $L(\cdot, \cdot)$, and the path $\phi(\cdot)$ describes the trajectory followed by a particle subject to those laws. In the case of geometric optics $L(\cdot, \cdot)$ is defined in terms of the local speed of light. Thus, in the typical formulation of a calculus of variations problem appropriate to some given applied problem the function $L(\cdot, \cdot)$ is obtained during the modeling stage.

Clearly the function $L(\cdot, \cdot)$ reflects properties of the "medium" in which the underlying dynamical system evolves. Typically, the definition is local in the sense that $L(x, \cdot)$ reflects the properties of the medium at x. If $L(x, \beta)$ possesses some kind of continuity in x, then it seems reasonable to believe this reflects a type of spatial continuity in the properties of the medium, e.g. the speed of propagation varies continuously from point to point in the geometrical optics problem. However, in many problems of interest such continuity properties may be violated. For example, it may be the case that there are two (or more)

*This research was supported in part by a grant from the National Science Foundation (NSF-DMS-8902333) disjoint regions $R^{(1)}$ and $R^{(2)}$ interior to which the physical properties of the media vary continuously, with a smooth boundary or interface separating the regions. It is simple to produce examples of this type from classical mechanics or geometrical optics. An example from large deviation theory will be described in Section 2.

In such a case one must rethink the modeling of the original physical problem as a calculus of variations problem. Clearly, the modeling appropriate for a single region of continuous behavior should be appropriate for defining or identifying $L(x,\cdot)$ when x is in the relative interior of either of the regions $R^{(1)}$ or $R^{(2)}$. However, there is still the question of the proper definition of $L(x,\cdot)$ for points x on the interface. The mathematical problem (1.1) will be well posed under just appropriate measurability assumptions on $L(\cdot,\cdot)$. But from the point of view of modeling, certain additional properties can be expected (or perhaps should even be demanded). For example, regardless of how $L(x,\cdot)$ is defined on the interface, it should lead to a value function $V(x)$ which has desirable stability properties under approximations, discretizations, small perturbations, etc. This turns out to impose restrictions on the form of $L(x,\cdot)$. Perhaps the most naive thing one could imagine would be to define $L(x,\cdot)$ on the interface by simply extending the definition on either $R^{(1)}$ or $R^{(2)}$ (assuming this is possible). Then, as will be shown in Section 2, the resulting value $V(x)$ is not stable if we attempt to approximate by discretizing in time.

In this paper we present what appears to be a "natural" definition of the integrand on the interface, and describe an associated numerical procedure. By natural, what is meant is that the definition: (a) occurs in applications; (b) leads to a value that is stable under discretizations; (c) can be shown to be the *only* definition on the boundary that is stable under a wide class of approximations; (d) leads to a simple numerical scheme for approximating the resulting value $V(x)$. Given (b) and (c) it is perhaps not very surprising that (d) should be true.

The main contribution of the paper is the convergence proof for the numerical scheme, and the main intended application is the numerical solution of problems from large deviation theory. However, properties (b) and (c) strongly suggest the definition on the boundary would be correct for many problems of this type, and thus the numerical scheme may be of wider interest.

An outline of the paper is as follows. In Section 2 we define the integrand for the boundary and discuss the properties (a) and (b). In Section 3 we describe a numerical scheme for the resulting value. The basic mathematical results in the paper are the convergence proof given in Section 4. Some extensions are described in Section 5. All the discussion in Sections 2-4 is restricted to the case of one boundary. This is because a key element in the definition of the integrand on the boundary is the identification of necessary and sufficient conditions for stability of a related process. In the case of a single boundary the related process is one dimensional and the necessary and sufficient conditions can be given explicitly. In Section 5 we make some remarks on the possibilities when there are two or more intersecting smooth boundaries.

2 Problem statement and background.

We now define the calculus of variations problem. For each $i = 1,2$, let $L^{(i)} : I\!R^n \to I\!R$ be convex and continuous. We assume that both of the $L^{(i)}(\cdot)$ satisfy the following superlinear growth condition:

$$\lim_{c\to\infty} \inf_{\beta:|\beta|=c} L^{(i)}(\beta)/c = +\infty. \tag{2.1}$$

It seems likely that the convexity assumption can be dropped at the expense of complicating the formula for $L^{(0)}(\cdot)$ given below. In the absence of a particular application we will not pursue such a generalization.

Example 2.1 For each $i = 1, 2$, let $b^{(i)} \in I\!\!R^n$ and let $a^{(i)}$ be a symmetric positive definite $n \times n$–matrix. An example of functions $L^{(i)}(\cdot)$ satisfying (2.1) is

$$L^{(i)}(\beta) = \frac{1}{2}(\beta - b^{(i)}) \left[a^{(i)}\right]^{-1} (\beta - b^{(i)})',$$

where $'$ denotes transpose. ■

We next define

$$L^{(0)}(\beta) = \inf \left\{\rho^{(1)}L^{(1)}(\beta^{(1)}) + \rho^{(2)}L^{(2)}(\beta^{(2)})\right\}, \tag{2.2}$$

where the infimum is over $(\rho^{(1)}, \rho^{(2)}) \in I\!\!R^2$ and $(\beta^{(1)}, \beta^{(2)}) \in I\!\!R^{2n}$ satisfying

$$\rho^{(1)} + \rho^{(2)} = 1, \rho^{(1)} \geq 0, \rho^{(2)} \geq 0 \tag{2.3}$$

$$\beta_1^{(1)} \geq 0, \beta_1^{(2)} \leq 0 \tag{2.4}$$

$$\rho^{(1)}\beta^{(1)} + \rho^{(2)}\beta^{(2)} = \beta. \tag{2.5}$$

Define

$$L(x, \beta) = \begin{cases} L^{(1)}(\beta) & \text{if} \quad x_1 < 0 \\ L^{(0)}(\beta) & \text{if} \quad x_1 = 0 \\ L^{(2)}(\beta) & \text{if} \quad x_1 > 0. \end{cases} \tag{2.6}$$

Note that the function $L(x, \beta)$ may be discontinuous in x. It follows directly from the definition that $L^{(0)}(\beta) \leq L^{(1)}(\beta)$ for β satisfying $\beta_1 \geq 0$ and $L^{(0)}(\beta) \leq L^{(2)}(\beta)$ for β satisfying $\beta_1 \leq 0$. It also follows that $L^{(0)}(\cdot)$ satisfies the superlinearity condition (2.1).

Let $g : I\!\!R^n \to I\!\!R$ be a bounded and continuous function. The value function of interest is

$$V(x) = \inf \left[\int_0^T L(\phi, \dot{\phi})ds + g(\phi(T))\right], \tag{2.7}$$

where the infimum is over all absolutely continuous functions $\phi : [0, T] \to I\!\!R^n$ satisfying $\phi(0) = x$. Note that only those values of $L^{(0)}(\beta)$ for β satisfying $\beta_1 = 0$ affect the value of $V(x)$. Suppose we define $\int_0^T L(\phi, \dot{\phi})ds = +\infty$ if ϕ is not absolutely continuous. Since $L^{(0)}(\beta) \leq L^{(1)}(\beta) \wedge L^{(2)}(\beta)$ for β satisfying $\beta_1 = 0$, $\int_0^T L(\phi, \dot{\phi})ds$ is lower semicontinuous in the space $C([0, T] : I\!\!R^n)$ of continuous functions from $[0, T]$ to $I\!\!R^n$ [8]. The superlinearity of the integrand then guarantees the infimum is achieved in (2.7). (Note that $L(\cdot, \cdot)$ is not necessarily lower semicontinuous, although it may be redefined to be so without changing the value $V(x)$.)

There are a number of interesting generalizations. For example, one may consider cases where the boundary is no longer flat (although still smooth), or where the $L^{(1)}(\cdot)$ and $L^{(2)}(\cdot)$ are allowed to depend on x in some continuous fashion. These generalizations involve more or less straightforward adaptations of the arguments given here. A much more difficult generalization is to the case of several intersecting boundaries of discontinuity. Some remarks along these lines are in Section 5.

Origins of the problem. The calculus of variations problem (2.7) can be motivated in several ways.

Continuous time limit of discrete time problems. Consider the following discrete time control problem. Let $\Delta > 0$ be such that T/Δ is an integer. We consider dynamics defined by

$$\phi_{i+1}^{\Delta} = \phi_i^{\Delta} + \Delta u_i, \qquad \phi_0^{\Delta} = x, \tag{2.8}$$

where $\{u_i, i = 0, ..., (T/\Delta) - 1\}$ is any sequence in $\mathbb{R}^n$, and the cost function

$$V^{\Delta}(x) = \inf_{\{u_i, i=0,...,(T/\Delta)-1\}} \sum_{i=0}^{(T/\Delta)-1} \Delta \left[1_{\{(\phi_i^{\Delta})_1 \le 0\}} L^{(1)}(u_i) + 1_{\{(\phi_i^{\Delta})_1 > 0\}} L^{(2)}(u_i) \right] + g\left(\phi_{(T/\Delta)}^{\Delta}\right). \tag{2.9}$$

This is a discrete time optimal control problem in which the running cost suffers a discontinuity as the state ϕ_i^{Δ} of the process crosses the boundary $\{x : x_1 = 0\}$. Substituting $u_i = (\phi_{i+1}^{\Delta} - \phi_i^{\Delta})/\Delta$ into (2.9), we see that (2.9) superficially resembles (2.7), at least when the process is away from the boundary $\{x : x_1 = 0\}$. Perhaps the most naive guess for the limit of $V^{\Delta}(x)$ would be the function $\tilde{V}(x)$ defined by (2.7) with the given integrand replaced by $\left[1_{\{(\phi)_1 \le 0\}} L^{(1)}(\dot{\phi}) + 1_{\{(\phi)_1 > 0\}} L^{(2)}(\dot{\phi})\right]$. The interesting fact is that $V(x)$, including the rather involved definition of the integrand $L(x, \beta)$ for points x such that $x_1 = 0$, is the correct continuous time limit of the problem (2.9) as $\Delta \to 0$. Note that the discrete time problems are defined simply in terms of $L^{(1)}(\cdot)$ and $L^{(2)}(\cdot)$. Thus, if $V(x)$ is truly the continuous time limit of the $V^{\Delta}(x)$, then the function $L^{(0)}(\cdot)$ must naturally reflect the complicated interactions of the prelimit process ϕ_i^{Δ} as it "selects" between all "feasible" combinations of the two costs $L^{(1)}(\cdot)$ and $L^{(2)}(\cdot)$ whenever it is close to the boundary.

It is simple to heuristically derive (2.7) as the limit of (2.9), and we do so now. A rigorous proof can be based on the methods used below to prove the convergence of numerical schemes for approximating the solution to (2.7). Clearly the only issue is the minimal cost associated to the process ϕ_i^{Δ} while in a neighborhood of the boundary. Fix β satisfying $\beta_1 = 0$, $\delta > 0$, and $\phi_0^{\Delta} = 0$, and assume the discrete time process ends up at β/δ at discrete time δ/Δ (which we assume is an integer). Under the natural scaling of time defined by the parameter Δ, this corresponds to the continuous time process starting at 0 at time 0 and ending at β/δ at time δ. Since $L^{(0)}(\cdot)$ is convex and $L^{(0)}(\beta) \le L^{(1)}(\beta) \wedge L^{(2)}(\beta)$, the "cheapest" way in which $\phi(\cdot)$ can arrive at β/δ at time δ after starting at 0 at time 0, is to satisfy $\dot{\phi}(t) = \beta$ on $[0, \delta]$. Thus, the minimal cost for the continuous time problem is $\delta L^{(0)}(\beta)$. We would like to show

$$\sum_{i=0}^{(\delta/\Delta)-1} \Delta \left[1_{\{(\phi_i^{\Delta})_1 \le 0\}} L^{(1)}(u_i) + 1_{\{(\phi_i^{\Delta})_1 > 0\}} L^{(2)}(u_i) \right] \to \delta L^{(0)}(\beta) \tag{2.10}$$

as $\Delta \to 0$, when the $\{u_i, i = 0, ..., (\delta/\Delta) - 1\}$ (or equivalently the $\{\phi_i^{\Delta}, i = 1, ..., (\delta/\Delta)\}$) are chosen to minimize the left hand side of (2.10) and satisfy $\phi_0^{\Delta} = 0$, $\phi_{(\delta/\Delta)}^{\Delta} = \beta/\delta$.

A little thought suggests the choice of the control at discrete time i should depend only on whether $(\phi_i^{\Delta})_1 \le 0$ or $(\phi_i^{\Delta})_1 > 0$. Suppose we use $u_i = \tilde{\beta}^{(1)}$ when $(\phi_i^{\Delta})_1 \le 0$ and

$u_i = \tilde\beta^{(2)}$ when $(\phi_i^\Delta)_1 > 0$. We can rewrite the left hand side of (2.10) as

$$\delta\left[r_\Delta^{(1)} L^{(1)}(\tilde\beta^{(1)}) + r_\Delta^{(2)} L^{(2)}(\tilde\beta^{(2)})\right],$$

where

$$r_\Delta^{(1)} = \frac{1}{\delta}\sum_{i=0}^{(\delta/\Delta)-1} \Delta 1_{\{(\phi_i^\Delta)_1 \le 0\}} \qquad \left(\text{respectively } r_\Delta^{(2)} = \frac{1}{\delta}\sum_{i=0}^{(\delta/\Delta)-1} \Delta 1_{\{(\phi_i^\Delta)_1 > 0\}}\right)$$

is the fraction of time ϕ_i^Δ spends in $\{x : x_1 \le 0\}$ (respectively $\{x : x_1 > 0\}$) for $i \in \{0, ..., (\delta/\Delta) - 1\}$. Since we impose the condition $\phi_{\delta/\Delta}^\Delta = \delta\beta$,

$$r_\Delta^{(1)}\tilde\beta^{(1)} + r_\Delta^{(2)}\tilde\beta^{(2)} = \delta\beta. \tag{2.11}$$

Clearly, $r_\Delta^{(1)}$ and $r_\Delta^{(2)}$ satisfy

$$r_\Delta^{(1)} + r_\Delta^{(2)} = 1, r_\Delta^{(1)} \ge 0, r_\Delta^{(2)} \ge 0. \tag{2.12}$$

It follows from the conditions $\phi_{(\delta/\Delta)}^\Delta = \delta\beta$ and $\beta_1 = 0$ that without loss of generality we may assume $\tilde\beta_1^{(1)} \ge 0$ and $\tilde\beta_1^{(2)} \le 0$. The only difference between the conditions imposed on $\rho^{(1)}, \rho^{(2)}, \beta^{(1)}$ and $\beta^{(2)}$ ((2.3), (2.4), and (2.5)) and the conditions imposed on the $r_\Delta^{(1)}, r_\Delta^{(2)}, \tilde\beta^{(1)}$ and $\tilde\beta^{(2)}$ is that the $r_\Delta^{(i)}$ must take values in the set $\{0, \Delta/\delta, 2\Delta/\delta, ..., 1\}$, while the $\rho^{(i)}$ may take values in the interval $[0, 1]$. This naturally imposes restrictions on the $\tilde\beta^{(i)}$ through (2.11). However, as $\Delta \to 0$ the set of allowed values of $r_\Delta^{(1)}, r_\Delta^{(2)}, \tilde\beta^{(1)}, \tilde\beta^{(2)}$ converges up to a dense subset of the set of all $\rho^{(1)}, \rho^{(2)}, \beta^{(1)}, \beta^{(2)}$ satisfying (2.3)–(2.5), and the continuity of the functions $L^{(i)}(\cdot), i = 1, 2$ gives (2.10).

Large deviation theory. Let $b^{(i)}$ and $a^{(i)}$ be as in Example 2.1. Consider the stochastic differential equation

$$dx^\varepsilon(t) = b(x^\varepsilon(t))dt + \varepsilon^{1/2}\sigma(x^\varepsilon(t))dw(t),$$

where

$$b(x) = 1_{\{x_1 \le 0\}} b^{(1)} + 1_{\{x_1 > 0\}} b^{(2)}$$
$$\sigma(x) = 1_{\{x_1 \le 0\}} \sigma^{(1)} + 1_{\{x_1 > 0\}} \sigma^{(2)}.$$

Such models involving discontinuous coefficients and generalizations appear in the areas of communication theory and the control of small noise diffusions. Suppose $a^{(i)} = \sigma^{(i)}(\sigma^{(i)})'$, and define $L^{(i)}(\cdot)$ as in Example 2.1. Then for such processes, it turns out that $\int_0^T L(\phi, \dot\phi)ds$ is what is known as the *large deviation rate function* [12] appropriate to this model [3], [4]. Very loosely speaking, this means $\int_0^T L(\phi, \dot\phi)ds$ characterizes the probability that a sample path of $x^\varepsilon(\cdot)$ will be in a small neighborhood of the path $\phi(\cdot)$, in the sense that

$$P_{\phi(0)}\{x^\varepsilon(\cdot) \text{ is "close" to } \phi(\cdot) \text{ over } [0, T]\} \approx \exp -\frac{1}{\varepsilon}\int_0^T L(\phi, \dot\phi)ds.$$

Here P_x denotes probability conditioned on $x^\varepsilon(0) = x$.. An example of a precise statement which follows from the identification of $\int_0^T L(\phi, \dot\phi)ds$ as the appropriate rate function is as follows. Let G be a closed subset of $\mathbb{R}^n$ which is the closure of its interior and let

$$g(x) = \begin{cases} 0 & x \in G \\ +\infty & x \notin G. \end{cases} \tag{2.13}$$

(Up to this point in the paper we have assumed g is continuous, and the convergence proof in Section 4 also makes this assumption. However, continuity of g is not necessary, and we remark in Section 5 on the slight modifications needed to cover the present case.) Then,

$$\lim_{\varepsilon\to 0} \varepsilon \log P_x \{x^\varepsilon(T) \in G\} = -V(x).$$

Estimates of other probabilities are also of interest, e.g. $P_x\{x^\varepsilon(t) \in G \text{ for all } t \in [0,T]\}$, $P_x\{x^\varepsilon(t) \notin G \text{ for some } t \in [0,T]\}$. These correspond to calculus of variations problems that are slightly different from (2.7), but the design and proof of convergence of numerical schemes may also be treated by the method of this paper. The expression (2.2) and the constraints (2.3)-(2.5) have interesting interpretations in terms of the associated stochastic process, and a discussion on these points may be found in [3].

3 Description of the numerical scheme.

We next describe a numerical scheme for approximating the solution to (2.7). Many variations on the given scheme are possible and of interest [9]. We let $h > 0$ denote the discretization parameter. The approximations will be defined on the grid $G_h = h\mathbb{Z}^n$. The spacing in the time variable will be given by Δt^h. The sequence Δt^h is assumed to satisfy $\Delta t^h/h \to 0$ and $\Delta t^h/h^2 \to \infty$, and for notational convenience we also assume $T/\Delta t^h \in \mathbb{Z}$.

Following the ideas of [10,9], we wish to construct a Markov chain which is "locally consistent" with the dynamics $\dot\phi = u$ (i.e. we interpret (2.7) as an optimal control problem, with control u, the given dynamical equation, and running cost $L(\phi, u)$). The proof of convergence given in Section 4 is complicated by the fact that when we rewrite the original calculus of variations problem as an optimal control problem we must deal with a potentially unbounded control space. It is this fact which forces the condition $\Delta t^h/h \to 0$. If there is *a priori* knowledge available concerning bounds on the optimal control then this information may be exploited, both to simplify the proof of convergence and to improve the selection of discretization in the time variable.

Let e_k denote the unit vector in the kth coordinate direction, and let $Z^h \equiv h/\Delta t^h \to \infty$. Define $\{\xi_i^h, i < \infty\}$ to be the controlled Markov chain with state space G_h and transition probability

$$p^h(x,y|u) = \begin{cases} [(u)_k \vee 0]/Z^h & \text{if } y = x + he_k \\ -[(u)_k \wedge 0]/Z^h & \text{if } y = x - he_k \\ 1 - \left(\sum_{k=1}^n |(u)_k|/Z^h\right) & \text{if } y = x \\ 0 & \text{otherwise.} \end{cases} \tag{3.1}$$

In this equation $(u)_k$ denotes the kth component of u. Here and below we will use the parentheses to indicate that the subscript denotes a component, and not the discrete time variable i.

The choice (3.2) corresponds to approximating the dynamics $\dot\phi = u$ by using a "one-sided" difference scheme. For the chain $\{\xi_i^h, i < \infty\}$ the control space is restricted to $\mathcal{U}^h = \{u : |(u)_k| \le Z^h/n, k = 1,...,n\}$. This limitation on the control guarantees that

$p^h(x,y|u)$ is a probability transition function. For this choice of $p^h(x,y|u)$, $\{\xi_i^h, i < \infty\}$ satisfies the "local consistency" conditions

$$E\left[\xi_{i+1}^h - \xi_i^h | \xi_j^h, u_j, j \le i, u_i = u\right] = u\Delta t^h, \tag{3.2}$$

$$E\left[\left(\xi_{i+1}^h - \xi_i^h - u\Delta t^h\right)^2 | \xi_j^h, u_j, j \le i, u_i = u\right] = O(h^2). \tag{3.3}$$

With $\{\xi_i^h, i < \infty\}$ as given above, we define an approximation scheme that exploits the similarity between a suitably interpolated version of $\{\xi_i^h, i < \infty\}$ and the dynamics $\dot{\phi} = u$, at least when $u \in \mathcal{U}^h$. Let

$$V^h(x) = \inf E_x \left\{ \sum_{i=0}^{(T/\Delta t^h)-1} \Delta t^h \left[1_{\{(\xi_i^h)_1 \le 0\}} L^{(1)}(u_i) + 1_{\{(\xi_i^h)_1 > 0\}} L^{(2)}(u_i) \right] + g\left(\xi_{(T/\Delta t^h)}^h\right) \right\}. \tag{3.4}$$

Here E_x indicates expectation conditioned on $\xi_0^h = x$, and the infimum is over all control sequences $\{u_i, i = 0, ..., (T/\Delta t^h) - 1\}$ satisfying

$$u_i \in \mathcal{U}^h$$
$$u_i \text{ is } \sigma\left(\xi_{k+1}^h, u_k, k < i\right) \text{ measurable.}$$

Let $V^h(x,i)$ denote the cost corresponding to the analogous problem that starts at discrete time i and conditions on $\xi_i^h = x$. The computational scheme for approximating $V(x)$ is given by the dynamic programming equation satisfied by $V^h(x,i)$:

$$V^h(x,i) = \begin{cases} \inf_{u \in \mathcal{U}^h} \left[\Delta t^h L^{(1)}(u) + \sum_{y \in G_h} p^h(x,y|u) V^h(y,i+1)\right] & \text{if} \quad x_1 \le 0 \\ \inf_{u \in \mathcal{U}^h} \left[\Delta t^h L^{(2)}(u) + \sum_{y \in G_h} p^h(x,y|u) V^h(y,i+1)\right] & \text{if} \quad x_1 > 0 \end{cases}$$

for $i = 0, ..., (T/\Delta t^h) - 1$, and

$$V^h(x, T/\Delta t^h) = g(x).$$

Note that the scheme does not explicitly involve the complicated definition of $L(\cdot,\cdot)$ on the boundary. In some cases (e.g. the case of quadratic $L^{(i)}$ of Example 2.1) the minimization may be simplified considerably.

4 Proof of convergence.

Theorem 4.1 *Define $V(x)$ by (2.7), where we assume $L(\cdot,\cdot)$ and $L^{(0)}(\cdot)$ are defined by (2.2) through (2.6), that the functions $L^{(1)}(\cdot)$ and $L^{(2)}(\cdot)$ are convex, continuous and satisfy (2.1), and that $g(\cdot)$ is bounded and continuous. Then for $V^h(x)$ defined as in Section 3 we have $V^h(x) \to V(x)$.*

Remarks on the proof. We wish to show $\lim_h V^h(x) = V(x)$. The starting point is the representation (3.4) of $V^h(x)$ as a functional of an optimally controlled Markov chain $\{\xi_i^h, i < \infty\}$, which has been chosen so that its dynamics mimic those of $\dot{\phi} = u$. Let

Ω denote the underlying probability space with generic element ω. In the proof that $\liminf_h V^h(x) \geq V(x)$ we essentially show that for almost all $\omega \in \Omega$ the cost

$$\liminf_h \sum_{i=0}^{(T/\Delta t^h)-1} \Delta t^h \left[1_{\{(\xi_i^h)_1 \leq 0\}} L^{(1)}(u_i) + 1_{\{(\xi_i^h)_1 > 0\}} L^{(2)}(u_i)\right] + g\left(\xi^h_{(T/\Delta t^h)}\right)$$

is bounded below by one of the form

$$\int_0^T L(\phi, u)ds + g(\phi(T)), \dot{\phi} = u, \phi(0) = x,$$

where u may or may not be optimal. Owing to the definition of $V(x)$, this gives the lower bound. The main point one must resolve is to show $L^{(0)}(\cdot)$ provides a lower bound for the cost for a sample path of the suitably interpolated version of $\{\xi_i^h, i < \infty\}$ when this process is near the boundary $\{x : x_1 = 0\}$. The key is to show that if the process stays near this set, and if the average (in time) control selected by the process fails to satisfy $(u)_1 \geq 0$ (respectively $(u)_1 \leq 0$) when the process is in $\{x : x_1 \leq 0\}$ (respectively $\{x : x_1 > 0\}$), then we obtain a contradiction. This is accomplished by using an argument based on Lyapunov functions. Thus the controls selected by the process must satisfy a condition analogous to (2.4) when the process stays in a neighborhood of the boundary, and this implies the needed bound from below.

In order to show $\limsup_h V^h(x) \leq V(x)$, we take a path $\phi(\cdot)$ which minimizes in (2.7) and define $u(t) = \dot{\phi}(t)$. We then describe how a nonanticipative control scheme based on $u(\cdot)$ may be defined for the chain $\{\xi_i^h, i < \infty\}$, which when applied yields a cost no greater than $V(x)$ plus an error that vanishes as $h \to 0$. As in the case of the lower bound, the main difficulty is with any sections of $\phi(\cdot)$ that lie on the set $\{x : x_1 = 0\}$. The definition of $L^{(0)}(\cdot)$, and in particular the constraint (2.4), are the keys to the proper definition of the control scheme near the boundary.

Lower bound. We first prove

$$\liminf_h V^h(x) \geq V(x) \tag{4.1}$$

for $V^h(x)$ defined by (3.4). Of course we have actually specified $V^h(x)$ only on G_h. What we mean precisely by (4.1) is

$$\liminf_{\delta \to 0} \liminf_{h \to 0} \inf_{y \in G_h : |y - x| \leq \delta} V^h(y) \geq V(x),$$

and analogously for the upper bound statement.

For $h > 0$, let $u^h(\cdot, i)$ be a minimizing feedback control for the chain, given that the chain starts at x. Define the continuous time interpolations of the process and control by

$$\left.\begin{array}{rcl} \xi^h(t) & = & \xi_i^h \\ \tilde{u}^h(t) & = & u^h(\xi_i^h, i) \end{array}\right\} \text{ for } t \in [i\Delta t^h, (i+1)\Delta t^h).$$

The space of *relaxed controls* appropriate to this problem is defined as the set $\mathcal{M}$ of nonnegative Borel measures m on $\mathbb{R}^n \times [0, T]$ satisfying $m(\mathbb{R}^n \times [0, t]) = t$. A fact we will use is that for each $t \in [0, T]$ there exists a Borel probability measure $m_t(\cdot)$ on $\mathbb{R}^n$ such that for each Borel set B, $m.(B)$ is Borel measurable and $m_t(d\alpha)dt = m(d\alpha dt)$. We may consider $\tilde{u}^h(\cdot)$ as taking values in the space of relaxed controls by identifying

$$m^h(B \times [0,t]) = \int_0^t 1_{\{\tilde{u}^h(s)\in B\}} ds$$

for all $t \in [0,T]$ and Borel $B \subset \mathbb{R}^n$.

We first show the sequence $\{m^h(\cdot), h > 0\}$ is tight as a sequence of measure valued random variables. To this end, we note that equation (2.1) implies the existence of a convex function $l : \mathbb{R} \to \mathbb{R}$ such that $l(c)/c \to \infty$ as $c \to \infty$, and

$$\min_{i=0,1,2} L^{(i)}(\beta) \geq l(|\beta|)$$

for all $\beta \in \mathbb{R}^n$. Let

$$B \equiv \sup_{x,h} V^h(x) \leq T \left[\max_{i=0,1,2} L^{(i)}(0)\right] + \sup_x g(x) < \infty.$$

Then

$$\sup_{x,h} E_x \left[\int_{\mathbb{R}^n \times [0,T]} 1_{\{|\alpha|\geq c\}} l(|\alpha|) m^h(d\alpha ds)\right] \leq B - \inf_x g(x) < \infty,$$

which implies that

$$\sup_{x,h} E_x \left[\int_{\mathbb{R}^n \times [0,T]} 1_{\{|\alpha|\geq c\}} |\alpha| m^h(d\alpha ds)\right] \to 0 \tag{4.2}$$

as $c \to \infty$. By [11, Theorem 1.6.1] the collection $\{m^h(\cdot), h > 0\}$ is tight.

According to equation (3.2),

$$\left\{\xi^h(i\Delta t^h) - \xi^h(0) - \int_{\mathbb{R}^n \times [0, i\Delta t^h]} \alpha m^h(d\alpha ds), i = 0, 1, ..., T/\Delta t^h\right\}$$

is a martingale, and by (3.3)

$$E_x \left[\xi^h(T) - \xi^h(0) - \int_{\mathbb{R}^n \times [0,T]} \alpha m^h(d\alpha ds)\right]^2 = \frac{1}{\Delta t^h} O(h^2) \to 0.$$

By the martingale inequality (Corollary 1 to Theorem 9.4.1 of [1]),

$$\sup_{0\leq t\leq T} \left|\xi^h(t) - \xi^h(0) - \int_{\mathbb{R}^n \times [0,t]} \alpha m^h(d\alpha ds)\right| \to 0 \tag{4.3}$$

in probability. Retain h to index a subsequence such that $\{m^h(\cdot)\}$ converges weakly to the limit $m(\cdot)$. By the Skorokhod Representation Theorem [11], and by the uniform integrability given in (4.2), we may assume $(\xi^h(\cdot), m^h(\cdot))$ converges almost surely to a pair $(\xi(\cdot), m(\cdot))$ in $D([0,T] : \mathbb{R}^n) \times \mathcal{M}$, and that this pair satisfies the equation

$$\xi(t) - x = \int_{\mathbb{R}^n \times [0,t]} \alpha m(d\alpha ds)$$

$$\left(\text{equivalently } \dot{\xi}(t) = \int_{\mathbb{R}^n} \alpha m_t(d\alpha) \text{ a.s. in } t, \xi(0) = x \right)$$

almost surely in $\omega \in \Omega$.

We next define random measures $\nu^h(\cdot)$ on $\mathbb{R}^n \times D([0,T] : \mathbb{R}^n) \times [0,T]$ by

$$\nu^h(A \times B \times C) = \int_C I_{\{u^h(s) \in A, \xi^h(\cdot) \in B\}} ds.$$

Note that $\nu^h(A \times D([0,T] : \mathbb{R}^n) \times C) = m^h(A \times C)$. Thus the tightness of the sequence $\{(\xi^h(\cdot), m^h(\cdot)), h > 0\}$ implies the tightness of $\{\nu^h(\cdot), h > 0\}$. We also define

$$S^{(1)} = \{(\phi(\cdot), s) \in D([0,T] : \mathbb{R}^n) \times [0,T] : (\phi(s))_1 \le 0\}$$

$$S^{(2)} = \{(\phi(\cdot), s) \in D([0,T] : \mathbb{R}^n) \times [0,T] : (\phi(s))_1 > 0\}$$

$$\nu^{(i),h}(S) = \nu^h \left(S \cap (\mathbb{R}^n \times S^{(i)}) \right), \quad i = 1, 2.$$

The measures $\nu^{(1),h}(\cdot)$ and $\nu^{(2),h}(\cdot)$ record the control effort that is applied, when it is applied, and also distinguish between when the state $\xi^h(s)$ is in $\{x : (x)_1 \le 0\}$ and $\{x : (x)_1 > 0\}$. Clearly $\nu^h(\cdot) = \nu^{(1),h}(\cdot) + \nu^{(2),h}(\cdot)$.

We now apply the Skorohhod Representation and extract a weakly converging subsequence from

$$\left\{ \left(\xi^h(\cdot), m^h(\cdot), \nu^h(\cdot), \nu^{(1),h}(\cdot), \nu^{(2),h}(\cdot) \right), h > 0 \right\}$$

with limit

$$\left(x(\cdot), m(\cdot), \nu(\cdot), \nu^{(1)}(\cdot), \nu^{(2)}(\cdot) \right).$$

It follows easily from $\xi^h(\cdot) \to x(\cdot)$ and the weak convergence that

$$\nu^{(1)}(\mathbb{R}^n \times \{x(\cdot)\}^c \times [0,T]) = \nu^{(2)}(\mathbb{R}^n \times \{x(\cdot)\}^c \times [0,T]) = 0$$

w.p.1. We may therefore conclude the existence of measures $\nu_s^{(1)}(\cdot)$ and $\nu_s^{(2)}(\cdot)$, $0 \le s \le T$, such that $\nu^{(i)}(A \times \{x(\cdot)\} \times [0,t]) = \int_0^t \nu_s^{(i)}(A) ds, i = 1, 2$, for all Borel sets A and $t \in [0,T]$.

The measures $\nu_s^{(1)}(\cdot)$ and $\nu_s^{(2)}(\cdot)$ possess the following properties. Almost surely in s, and w.p.1,

$$\begin{aligned} (x(s))_1 < 0 &\Rightarrow \int_{\mathbb{R}^n} \nu_s^{(1)}(d\alpha) = 1 \\ (x(s))_1 > 0 &\Rightarrow \int_{\mathbb{R}^n} \nu_s^{(2)}(d\alpha) = 1 \end{aligned} \tag{4.4}$$

$$\int_{\mathbb{R}^n} \nu_s^{(1)}(d\alpha) + \int_{\mathbb{R}^n} \nu_s^{(2)}(d\alpha) = 1 \tag{4.5}$$

$$(x(s))_1 = 0 \Rightarrow \int_{\mathbb{R}^n} (\alpha)_1 \nu_s^{(1)}(d\alpha) \ge 0 \text{ and } \int_{\mathbb{R}^n} (\alpha)_1 \nu_s^{(2)}(d\alpha) \le 0 \tag{4.6}$$

$$\int_{\mathbb{R}^n} \alpha \nu_s^{(1)}(d\alpha) + \int_{\mathbb{R}^n} \alpha \nu_s^{(2)}(d\alpha) = \int_{\mathbb{R}^n} \alpha m_s(d\alpha) = \dot{x}(s). \tag{4.7}$$

Equation (4.4) follows easily from the definitions of the $\nu^{(i),h}(\cdot)$ and the weak convergence, while (4.5) and (4.7) follow from the relationship $\nu^h(A \times D([0,T] : \mathbb{R}^n) \times C) = m^h(A \times C)$. The only property that is not obvious is (4.6). We will first prove the lower bound assuming (4.6), and then show (4.6).

Now fix an ω for which there is convergence via the Skorokhod representation. Then,

$$\begin{aligned}
\liminf_{h\to 0} \int_0^T \int_{\mathbb{R}^n} & \left[1_{\{(\xi^h(s))_1\le 0\}} k^{(1)}(\alpha) + 1_{\{(\xi^h(s))_1>0\}} k^{(2)}(\alpha)\right] m^h(d\alpha ds) \\
&= \liminf_{h\to 0} \int_0^T \int_{\mathbb{R}^n} k^{(1)}(\alpha)\nu^{(1),h}(d\alpha \times D([0,T]:\mathbb{R}^n)\times ds) \\
&\quad + \int_0^T \int_{\mathbb{R}^n} k^{(2)}(\alpha)\nu^{(2),h}(d\alpha \times D([0,T]:\mathbb{R}^n)\times ds) \\
&\ge \int_0^T \int_{\mathbb{R}^n} k^{(1)}(\alpha)\nu^{(1)}(d\alpha \times D([0,T]:\mathbb{R}^n)\times ds) \\
&\quad + \int_0^T \int_{\mathbb{R}^n} k^{(2)}(\alpha)\nu^{(2)}(d\alpha \times D([0,T]:\mathbb{R}^n)\times ds) \\
&= \int_0^T \int_{\mathbb{R}^n} \left[k^{(1)}(\alpha)\nu_s^{(1)}(d\alpha) + k^{(2)}(\alpha)\nu_s^{(2)}(d\alpha)\right] ds.
\end{aligned}$$

The set $\{s : (x(s))_1 = 0, (\dot{x}(s))_1 \neq 0\}$ is a set of measure zero. Therefore the definition of $k(\cdot,\cdot)$, the convexity of the $k^{(i)}(\cdot)$, and the properties of the $\nu_s^{(i)}(\cdot)$ given in (4.4)-(4.7) imply

$$\int_{\mathbb{R}^n} \left[k^{(1)}(\alpha)\nu_s^{(1)}(d\alpha) + k^{(2)}(\alpha)\nu_s^{(2)}(d\alpha)\right] \ge k(x(s),\dot{x}(s))$$

a.s. in s. We also have

$$g(\xi^h(T)) \to g(x(T)).$$

Assembling the inequalities and using the definition of $V(x)$, we conclude that

$$\liminf_{h\to 0}\left[\int_0^T \int_{\mathbb{R}^n} \left[1_{\{(\xi^h(s))_1\le 0\}} k^{(1)}(\alpha) + 1_{\{(\xi^h(s))_1>0\}} k^{(2)}(\alpha)\right] m^h(d\alpha ds) + g(\xi^h(T))\right] \ge V(x). \tag{4.8}$$

By Fatou's lemma we have

$$\liminf_{h\to 0} V^h(x) \ge V(x)$$

for the convergent subsequence. The proof of the lower bound can now be completed by using an argument by contradiction.

We next prove (4.6). Consider any subsequence from

$$\left\{\left(\xi^h(\cdot), m^h(\cdot), \nu^h(\cdot), \nu^{(1),h}(\cdot), \nu^{(2),h}(\cdot)\right), h>0\right\}$$

that converges w.p.1 to a limit

$$\left(x(\cdot), m(\cdot), \nu(\cdot), \nu^{(1)}(\cdot), \nu^{(2)}(\cdot)\right).$$

Fix $\gamma > 0$. In order to proceed with the proof of (4.6) we must first build an approximation to the function $F : \mathbb{R} \to \mathbb{R}$ given by

$$F(z) = \begin{cases} |z| & \text{if } |z| \le \gamma \\ \gamma & \text{if } |z| > \gamma. \end{cases}$$

For each $\eta > 0$, let $F^\eta(\cdot)$ be a function such that

$$\begin{aligned}
|F^\eta(z)| &\le 2\gamma && \text{for all } z \\
F^\eta(z) &= F(z) && \text{for } z \in [-\gamma, \gamma] \\
|F^\eta_{zz}(z)| &< B && \text{for } z \notin [-\gamma/2, 2\gamma/2],
\end{aligned}$$

where $B < \infty$ depends on $\eta > 0$. Let

$$f^\eta(z) = \begin{cases} F_z^\eta(z) & \text{if } z \neq 0 \\ -1 & \text{if } z = 0, \end{cases}$$

and let $f^\eta(z) \to 0$ as $\eta \to 0$ for $z \notin [-\gamma, \gamma]$. By local consistency, for all small $h > 0$ we have

$$\begin{aligned} 4\gamma &\geq E_x \sum_{j=0}^{T/\delta-1} F^\eta((\xi_{j+1}^h)_1) - F^\eta((\xi_j^h)_1) \\ &\geq E_x \sum_{j=0}^{T/\delta-1} \delta f^\eta((\xi_j^h)_1) \left[(u_j^h)_1 + o(|u_j^h|)\right] + B\delta o(|u_j^h|). \end{aligned}$$

As noted previously, the boundedness of the running costs and the superlinearity of the integrand implies

$$\limsup_{h\to 0} E_x \sum_{j=0}^{T/\delta-1} \delta|u_j^h| = \limsup_{h\to 0} E_x \int_0^T \int_{\mathbb{R}^n} |\alpha| m^h(d\alpha ds) < \infty.$$

Sending $h \to 0$, we have

$$\begin{aligned} 4\gamma \geq E_x\bigg[\int_0^T \int_{\mathbb{R}^n} &(\alpha)_1 f^{(2),\eta}(x(s))\nu^{(2)}(d\alpha \times D([0,T] : \mathbb{R}^n) \times ds) \\ + \int_0^T \int_{\mathbb{R}^n} &(\alpha)_1 f^{(1),\eta}(x(s))\nu^{(1)}(d\alpha \times D([0,T] : \mathbb{R}^n) \times ds)\bigg], \end{aligned}$$

where $f^{(1),\eta}(\cdot)$ (respectively $f^{(2),\eta}(\cdot)$) is a continuous extension of $f^\eta(\cdot)$ from $(-\infty, 0)$ to $(-\infty, 0]$ (respectively $(0, \infty)$ to $[0, \infty)$).

Sending $\eta \to 0$, we have

$$E_x\left[\int_{\mathbb{R}^n} (\alpha)_1 \nu^{(2)}(d\alpha \times S^\gamma) - \int_{\mathbb{R}^n} (\alpha)_1 \nu^{(1)}(d\alpha \times S^\gamma)\right] \leq 4\gamma,$$

where $S^\gamma = \{(\psi(\cdot), s) \in D([0,T] : \mathbb{R}^n) \times [0,T] : |(\psi(s))_1| \leq \gamma\}$. A very similar proof that is based on approximating the function

$$F(z) = \begin{cases} z & \text{if } |z| \leq \gamma \\ \gamma & \text{if } z > \gamma \\ -\gamma & \text{if } z < -\gamma \end{cases}$$

shows that

$$\left|E_x\left[\int_{\mathbb{R}^n} (\alpha)_1 \nu^{(2)}(d\alpha \times S^\gamma) + \int_{\mathbb{R}^n} (\alpha)_1 \nu^{(1)}(d\alpha \times S^\gamma)\right]\right| \leq 4\gamma.$$

Adding and substracting these equations gives

$$E_x \int_{\mathbb{R}^n} (\alpha)_1 \nu^{(1)}(d\alpha \times S^\gamma) \geq -8\gamma, \quad E_x \int_{\mathbb{R}^n} (\alpha)_1 \nu^{(2)}(d\alpha \times S^\gamma) \leq 8\gamma.$$

Thus for each $\theta > 0$, both the quantities

$$P_x\left\{\int_{\mathbb{R}^n}(\alpha)_1\nu^{(1)}(d\alpha\times S^\gamma)\leq -\theta\right\},\quad P_x\left\{\int_{\mathbb{R}^n}(\alpha)_1\nu^{(2)}(d\alpha\times S^\gamma)\geq \theta\right\}$$

tend to zero as $\gamma \to 0$. From the definitions of the sets S^γ and the measures $\nu^{(1)}(\cdot)$ and $\nu^{(2)}(\cdot)$,

$$\begin{aligned}&\int_{\mathbb{R}^n}(\alpha)_1\nu^{(1)}(d\alpha\times\{(\psi(\cdot),s)\in D([0,T]:\mathbb{R}^n)\times[0,T]:(\psi(s))_1=0\})\geq 0\\&\int_{\mathbb{R}^n}(\alpha)_1\nu^{(2)}(d\alpha\times\{(\psi(\cdot),s)\in D([0,T]:\mathbb{R}^n)\times[0,T]:(\psi(s))_1=0\})\leq 0\end{aligned}\tag{4.9}$$

w.p.1.

The argument that lead to (4.9) can be repeated with s restricted to any interval $[a,b]\subset[0,T]$ with the same conclusion. Thus we can assume it holds simultaneously for all such intervals with rational endpoints. Using the definitions of $\nu_s^{(1)}(\cdot)$ and $\nu_s^{(2)}(\cdot)$, this implies

$$\int_{\mathbb{R}^n}(\alpha)_1\nu_s^{(1)}(d\alpha)\geq 0 \text{ and } \int_{\mathbb{R}^n}(\alpha)_1\nu_s^{(2)}(d\alpha)\leq 0$$

whenever $(x(s))_1=0$ a.s. in s and with probability one. This proves (4.6).

Upper bound. We next prove $\limsup_h V^h(x)\leq V(x)$. Fix x and let $\phi(\cdot)$ be a minimizing path in (2.7). We will consider the case where $(\phi(0))_1=x_1<0$ and $(\phi(T))_1>0$. Define $t_1=\inf\{t\in[0,T]:(\xi(t))_1=0\}$ and $t_2=\sup\{t\in[0,T]:(\xi(t))_1=0\}$. Then $0<t_1\leq t_2<T$. We further restrict to the case where $t_1<t_2$. Proofs for the other possible cases are obvious adaptations of the argument for this case. Owing to the convexity of the $L^{(i)}(\cdot), i=0,1,2$ and the inequality $L^{(0)}(\beta)\leq L^{(1)}(\beta)\wedge L^{(2)}(\beta)$ for β satisfying $\beta_1=0$, we may assume the existence of vectors β_a,β_b and β_c such that $\dot\phi(t)=\beta_a$, $\dot\phi(t)=\beta_b$, and $\dot\phi(t)=\beta_c$ almost surely on the intervals $[0,t_1]$, $[t_1,t_2]$, and $[t_2,T]$, respectively. Let $\varepsilon>0$ and note that $(\beta_b)_1=0$. Exploiting the continuity of the $L^{(i)}(\cdot)$ there exist $\rho^{(1)},\rho^{(2)},\beta_b^{(1)}$ and $\beta_b^{(2)}$ which satisfy

$$\rho^{(1)}+\rho^{(2)}=1,\rho^{(1)}>0,\rho^{(2)}>0\tag{4.10}$$

$$(\beta_b^{(1)})_1>0,(\beta_b^{(2)})_1<0\tag{4.11}$$

$$\rho^{(1)}\beta_b^{(1)}+\rho^{(2)}\beta_b^{(2)}=\beta_b.\tag{4.12}$$

$$\rho^{(1)}L^{(1)}(\beta_b^{(1)})+\rho^{(2)}L^{(2)}(\beta_b^{(2)})\leq L^{(0)}(\beta_b)+\varepsilon.\tag{4.13}$$

Note that for $h^0>0$ sufficiently small $\{\beta_a,\beta_b^{(1)},\beta_b^{(2)},\beta_c\}\subset\mathcal{U}^{h^0}$. For the remainder of the proof we assume $0<h<h^0$.

We next describe a collection of control schemes for the discrete time Markov chains defined in Section 3. We begin with the control $u_0^h=\beta_a$, and continue with $u_i^h=\beta_a$ until the first time J^h such that $(\xi_{J^h}^h)_1=0$. For $i\in\{J^h,J^h+1,\ldots,J^h+[(t_2-t_1)/\Delta t^h]-1\}$ we use

$$u_i^h=\begin{cases}\beta_b^{(1)} & \text{if } (\xi_i^h)_1\leq 0\\ \beta_b^{(2)} & \text{if } (\xi_i^h)_1>0.\end{cases}$$

Finally, for $i \in \{J^h + [(t_2 - t_1)/\Delta t^h], ..., (T/\Delta t^h) - 1\}$ we use $u_i^h = \beta_c$. Let $V^h(x, u^h)$ denote the expectation

$$E_x \left\{ \sum_{i=0}^{(T/\Delta t^h)-1} \Delta t^h \left[1_{\{(\xi_i^h)_1 \le 0\}} L^{(1)}(u_i^h) + 1_{\{(\xi_i^h)_1 > 0\}} L^{(2)}(u_i^h) \right] + g\left(\xi_{(T/\Delta t^h)}^h\right) \right\}$$

when $\xi_0^h = x$ and the control scheme u^h described above is used. Since the scheme u^h is nonanticipative, $V^h(x) \le V^h(x, u^h)$. Thus, in order to prove the upper bound it suffices to demonstrate

$$\limsup_h V^h(x, u^h) \le V(x) + T\varepsilon.$$

We do this by considering the different parts of the path $\phi(\cdot)$ separately.

Part 1. Suppose the control $u_i = \beta_a$ were used for all $i = 0, ..., T/\Delta t^h - 1$. An elementary application of the martingale inequality (Corollary 1 to Theorem 9.4.1 of [1]) similar to that used in the proof of the lower bound implies

$$P_x \left\{ \sup_{0 \le t \le T} \left| \xi^h(t) - t\beta_a \right| \ge \delta \right\} \to 0$$

as $h \to 0$, for all $\delta > 0$. Therefore $(\beta_a)_1 > 0$ and the definition of J^h imply

$$J^h \Delta t^h \to t_1$$

in probability as $h \to 0$, and thus

$$\sum_{i=0}^{J^h - 1} \Delta t^h L^{(1)}(u_i^h) \to t_1 L^{(1)}(\beta_a)$$

in probability as $h \to 0$.

Part 2. Suppose we start the chain at a point satisfying $(\xi_0^h)_1 = 0$ and apply the control

$$u_i^h = \begin{cases} \beta_b^{(1)} & \text{if } (\xi_i^h)_1 \le 0 \\ \beta_b^{(2)} & \text{if } (\xi_i^h)_1 > 0 \end{cases}$$

for all i. Then the definition of the control scheme implies $\{(\xi_i^h)_1, i = 0, 1, ...\}$ is a Markov chain with state space $\{0, h\}$ and transition matrix

$$\begin{pmatrix} 1 - ((\beta_b^{(1)})_1 / Z^h) & (\beta_b^{(1)})_1 / Z^h \\ -(\beta_b^{(2)})_1 / Z^h & 1 + ((\beta_b^{(2)})_1 / Z^h) \end{pmatrix}.$$

A direct calculation (see (3.2)) gives

$$E\left[(\xi_{i+1}^h)_1 - (\xi_i^h)_1 | (\xi_i^h)_1 \right] = \Delta t^h \left[1_{\{(\xi_i^h)_1 = 0\}} (\beta_b^{(1)})_1 + 1_{\{(\xi_i^h)_1 = h\}} (\beta_b^{(2)})_1 \right] \tag{4.14}$$

Define the quantities

$$r_i^{(1),h} = \frac{1}{i}\sum_{j=0}^{i-1} 1_{\{(\xi_j^h)_1=0\}}, \qquad r_i^{(2),h} = \frac{1}{i}\sum_{j=0}^{i-1} 1_{\{(\xi_j^h)_1=h\}}.$$

Then using (4.14),

$$\begin{aligned} & E\left[\frac{1}{i\Delta t^h}\left\{(\xi_i^h)_1 - (\xi_0^h)_1 - i\Delta t^h\left[r_i^{(1),h}(\beta_b^{(1)})_1 + r_i^{(2),h}(\beta_b^{(2)})_1\right]\right\}\right]^2 \\ &= E\left[\frac{1}{i\Delta t^h}\sum_{j=0}^{i-1}\left\{(\xi_{j+1}^h)_1 - (\xi_j^h)_1 - \Delta t^h\left[1_{\{(\xi_j^h)_1=0\}}(\beta_b^{(1)})_1 + 1_{\{(\xi_j^h)_1=h\}}(\beta_b^{(2)})_1\right]\right\}\right]^2 \\ &\le \frac{h^2}{i(\Delta t^h)^2}. \end{aligned}$$

Let $\tau^h = [(t_2 - t_1)/\Delta t^h]$ and let $i = \tau^h$. Since $|(\xi_{\tau^h}^h)_1 - (\xi_0^h)_1| \to 0$ almost surely, and since $h^2/\Delta t^h \to 0$, we conclude

$$r_{\tau^h}^{(1),h}(\beta_b^{(1)})_1 + r_{\tau^h}^{(2),h}(\beta_b^{(2)})_1 \to 0 \tag{4.15}$$

in probability as $h \to 0$. Note that (4.11) implies $\rho^{(1)}$ and $\rho^{(2)}$ are uniquely determined by the equations $\rho^{(1)} + \rho^{(2)} = 1$ and $\rho^{(1)}(\beta_b^{(1)})_1 + \rho^{(2)}(\beta_b^{(2)})_1 = 0$. Combining $r_{\tau^h}^{(1),h} + r_{\tau^h}^{(2),h} = 1$ with (4.15) produces

$$(r_{\tau^h}^{(1),h}, r_{\tau^h}^{(2),h}) \to (\rho^{(1)}, \rho^{(2)})$$

as $h \to 0$. Therefore

$$\begin{aligned} \sum_{i=0}^{\tau^h-1} \Delta t^h\left[1_{\{(\xi_i^h)_1\le 0\}}L^{(1)}(u_i^h) + 1_{\{(\xi_i^h)_1>0\}}L^{(2)}(u_i^h)\right] &= \tau^h\Delta t^h\left[r_{\tau^h}^{(1),h}L^{(1)}(\beta_b^{(1)}) + r_{\tau^h}^{(2),h}L^{(2)}(\beta_b^{(2)})\right] \\ &\to (t_2 - t_1)\left[\rho^{(1)}L^{(1)}(\beta_b^{(1)}) + \rho^{(2)}L^{(2)}(\beta_b^{(2)})\right] \\ &\le (t_2 - t_1)\left[L^{(0)}(\beta_b) + \varepsilon\right], \end{aligned}$$

and

$$\sum_{i=0}^{\tau^h-1} \Delta t^h\left[1_{\{(\xi_i^h)_1\le 0\}}\beta_b^{(1)} + 1_{\{(\xi_i^h)_1>0\}}\beta_b^{(2)}\right] \to (t_2 - t_1)\beta_b.$$

Let $s^h = (J^h + [(t_2 - t_1)/\Delta t^h])\Delta t^h$. By combining the results of Parts 1 and 2 and using the Markov property, we obtain

$$(\xi^h(s^h))_1 \to 0 \text{ and } s^h \to t_2$$

in probability.

Part 3. Using the martingale inequality in the same way as it was used in Part 1,

$$P_x\left\{\sup_{s^h\le t\le T}\left|\xi^h(t) - \xi^h(s^h) - (t - s^h)\beta_c\right| \ge \delta\right\} \to 0$$

as $h \to 0$, for all $\delta > 0$. Since $(\beta_c)_1 > 0$ and $(\xi^h(s^h))_1 \to 0$,

$$\sum_{i=J^h+[(t_2-t_1)/\Delta t^h]}^{T/\Delta t^h-1} \Delta t^h \left[1_{\{(\xi_i^h)_1\le 0\}}L^{(1)}(u_i^h) + 1_{\{(\xi_i^h)_1>0\}}L^{(2)}(u_i^h)\right] \to (T-t_2)L^{(2)}(\beta_c)$$

and

$$\xi^h(T) \to t_1\beta_a + (t_2-t_1)\beta_b + (T-t_2)\beta_c = \phi(T).$$

Combined with the conclusions of Parts 1 and 2 this implies

$$\sum_{i=0}^{(T/\Delta t^h)-1} \Delta t^h \left[1_{\{(\xi_i^h)_1\le 0\}}L^{(1)}(u_i^h) + 1_{\{(\xi_i^h)_1>0\}}L^{(2)}(u_i^h)\right] + g(\xi^h(T))$$
$$\to t_1L^{(1)}(\beta_a) + (t_2-t_1)\left[L^{(0)}(\beta_b)+\varepsilon\right] + (T-t_2)L^{(2)}(\beta_c) + g(\phi(T))$$
$$\le V(x) + T\varepsilon$$

in probability. The upper bound now follows by extracting convergent subsequences, using the dominated convergence theorem, and an argument by contradiction. ∎

5 Concluding remarks.

Many of the assumptions of Theorem 4.1 can be weakened easily with only a little additional complication in the proof.

For example, the assumption that the functions $L^{(1)}(\cdot)$ and $L^{(2)}(\cdot)$ are continuous can be weakened. Assume that $L^{(1)}(\cdot)$ and $L^{(2)}(\cdot)$ are convex and lower semicontinuous, but not necessarily finite at all points (this is the case when the calculus of variations problem occurs in the study of large deviations properties of degenerate diffusions). Assume also that there exist $\beta^{(i)}, i = 1,2$, such that $L^{(i)}(\beta^{(i)}) < \infty$ and $\beta_1^{(1)} > 0$, $\beta_1^{(2)} < 0$. This condition is quite weak and natural, and corresponds to Hypothesis (H) in [3]. Under the condition, the conclusion of Theorem 4.1 remains valid. Only two slight alterations are needed in the proof. In the proof of the lower bound, tightness and a uniform integrability condition for the random measures $m^h(\cdot)$ were shown to follow from $\sup_{x,h} V^h(x) < \infty$. While $\sup_{x,h} V^h(x) < \infty$ still holds, its verification now must be given in terms of the vectors $\beta^{(i)}$ described above. In the proof of upper bound, we must be a little more careful regarding the existence of vectors $\beta_b^{(i)}$ which satisfy (4.10)-(4.11) since the functions $L^{(i)}(\cdot)$ are no longer assumed continuous. However, this is easily overcome under the conditions given above.

Another generalization that is easily dealt with is a weakening of the conditions on $g(\cdot)$. For example, if G is a closed set that is the closure of its interior and if we define $g(\cdot)$ by (2.13), the conclusion of Theorem 4.1 again remains valid. The only adjustment in the proof of the lower bound is again a modification of the argument used to get $\sup_{x,h} V^h(x) < \infty$. For the upper bound, it may be the case that the optimizing path in (2.7) terminates at a point on the boundary of G: $\phi(T) \in \partial G$. In this case, in the construction of the control scheme for the chain we must work with nearly optimal (rather than optimal) paths which terminate strictly interior to G.

A generalization which does not involve a modest rewriting of the proof of Theorem 4.1 occurs in the case of several intersecting boundaries of discontinuity. The reason that this case is so much harder than the case of a single boundary has to do with the role that condition (2.4) plays in the definition of $L^{(0)}(\cdot)$. Consider as a concrete case the formal calculations done in Section 2 to determine the cost on the boundary so that the continuous time problem was the limit of simpler discrete time problems. To solve this problem it was necessary to learn the the optimal way in which the discrete time process could exploit the two available costs $L^{(1)}(\cdot)$ and $L^{(2)}(\cdot)$ while simultaneously maintaining an average velocity β, where β is a vector that points along the boundary. The definition of $L^{(0)}(\beta)$ provides the answer. Roughly speaking, the optimal way for the discrete time process to behave is to track $\beta^{(1)}$ when the state is in the set $\{x : x_1 \leq 0\}$ and $\beta^{(2)}$ when the state is in the set $\{x : x_1 > 0\}$. The $\rho^{(i)}$ turn out to represent respective fractions of time spent in the two sets, and the interpretations of (2.3) and (2.5) are clear. The constraint (2.4) defines necessary and sufficient conditions for the stability of the 1−component of the discrete time process, under the condition that it follows $\beta^{(1)}$ when the state is in the set $\{x : x_1 \leq 0\}$ and $\beta^{(2)}$ when the state is in the set $\{x : x_1 > 0\}$. The sufficiency implies that the limit of the costs for the discrete time process is no larger that the cost for the continuous time process. The necessity implies the limit cannot be smaller. Consider what might happen if (2.4) were dropped. It may then be the case that the minimum will be obtained at vectors which do not satisfy (2.4). However, we cannot use such vectors to determine the behavior of the discrete time process and simultaneously follow the vector β. Thus, if we omit (2.4) from the definition of $L^{(0)}(\cdot)$ the continuous time problem will in general have strictly smaller cost. The crucial role of the condition (2.4) as defining necessary and sufficient conditions for stability of a related one dimensional process is repeated in in the proof of Theorem 4.1.

This leads to the conjecture that if there is an analogue of the definition of $L^{(0)}(\cdot)$ for problems involving several intersecting boundaries then this definition must involve analogous necessary and sufficient conditions. However, it is to be expected that the processes will be of dimension greater than one. It would then seem likely that (except in special cases) the conditions themselves will not be available in any explicit from, and therefore there will not be an explicit expression for $L^{(0)}(\cdot)$. Nonetheless, there still remains an intriguing possibility. If it can be shown that such conditions *exist*, then although there may never be an explicit identification of the function $L^{(0)}(\cdot)$, the analogue of the numerical scheme discussed in this paper could still yield a method of calculating or approximating the solutions to problems like (2.7). This would be particularly useful for problems from large deviations, where such intersections of more than one surface of discontinuity arise in the asymptotic analysis of queueing systems [5].

References

[1] K.-L. Chung. *A Course in Probability Theory*. Academic Press, New York, second edition, 1974.

[2] R. Courant and D. Hilbert. *Methods of Mathematical Physics*. Volume 1, Interscience, New York, first english edition, 1937.

[3] P. Dupuis and R.S. Ellis. *Large deviations for Markov processes with discontinuous statistics, II: Random walks.* Technical Report, University of Massachusetts, 1990. To appear in *Probab. Th. Rel. Fields.*

[4] P. Dupuis and R.S. Ellis. *Large deviations for Markov processes with discontinuous statistics, III: Diffusions.* Technical Report, University of Massachusetts, 1991.

[5] P. Dupuis, H. Ishii, and H.M. Soner. A viscosity solution approach to the asymptotic analysis of queueing systems. *The Annals of Probab.*, 18:226–255, 1990.

[6] M.I. Freidlin and A.D. Wentzell. *Random Perturbations of Dynamical Systems.* Springer-Verlag, New York, 1984.

[7] H. Goldstein. *Classical Mechanics.* Addison-Wesley, Reading, Massachusetts, 1950.

[8] A.D. Ioffe and V.M. Tikhomirov. *Theory of Extremal Problems.* Volume 6 of *Studies in Mathematics and Its Applications*, North-Holland, Amsterdam, 1979.

[9] H.J. Kushner. Numerical methods for stochastic control problems in continuous time. *SIAM J. Control and Optimization*, 28:999–1048, 1990.

[10] H.J. Kushner. *Probability Methods for Approximations in Stochastic Control and for Elliptic Equations.* Academic Press, New York, 1977.

[11] H.J. Kushner. *Weak Convergence Methods and Singularly Perturbed Stochastic Control and Filtering Problems.* Volume 3 of *Systems and control*, Birkhaeuser, Boston, 1990.

[12] S.R.S. Varadhan. *Large Deviations and Applications. CBMS-NSF Regional Conference Series in Mathematics*, SIAM, Philadelphia, 1984.

Piecewise monotone filtering with small observation noise: Numerical simulations

Wendell H. Fleming*
Division of Applied Mathematics
Brown University
Providence, RI 02912

Qing Zhang
Department of Mathematics
University of Kentucky
Lexington, KY 40506

Abstract

A discrete time model for nonlinear filtering with small observation noise is considered in this paper. The observation function is assumed to be piecewise monotone with finitely many intervals of monotonicity. A sequential quadratic variation test is used to detect intervals of monotonicity of the observation function. It is shown that the original nonlinear model can be approximated by a simple linear model in the sense that the difference of the associated test statistic processes is small. Then, a diffusion process is used to approximate the linear model test statistic process. Therefore, the test statistic process for the original problem can be approximated by a diffusion process. Consequently, the associated confidence intervals and the mean decision time can be estimated and written explicitly in terms of the error probabilities. Finally, numerical simulation results are reported.

*Partially supported by NSF under grant MCS-812940, by ARO under grant DAAL 03-86-K-0171, and by AFOSR under grant AFOSR-89-0015.

1 Introduction

There is a substantial literature on the problem of optimal nonlinear filtering. In continuous time an unobserved state X_t and observation Y_t are modelled according to

$$\begin{cases} dX_t = f(X_t)dt + g(X_t)dU_t, & X_0 = x \\ dY_t = h(X_t)dt + \epsilon dV_t, & Y_0 = 0, \end{cases} \quad 0 \le t \le T \tag{1.1}$$

where U_t and V_t are independent brownian motions and T is a finite number. To find the mean square optimal estimate $\hat{X}_t$ for X_t given Y_s for $0 \le s \le t$ requires knowing the conditional distribution of X_t. Since the dynamics of the conditional distribution are governed by the nonlinear functional-partial differential equation of nonlinear filtering, the problem is inherently infinite dimensional [8]. If X_t and Y_t are of the same dimension and h is one-to-one, then X_t would be known exactly if $\epsilon = 0$. For small $\epsilon > 0$ good finite dimensional approximate filters have been described in [10]. References [3] [4] [5] are concerned with the case when h is piecewise one-to-one. Under a certain 'detectability' condition, a hypothesis test based on observations on Y_t can be used to decide the region on which h is one-to-one. Once this is done, an approximate filter of the type in [10] is used to estimate $\hat{X}_t$.

In this paper we consider the following discrete time analogue of (1.1):

$$\begin{cases} x_{k+1} = x_k + \epsilon f(x_k) + \sqrt{\epsilon} g(x_k) u_k, & x_0 = x \\ y_k = h(x_k) + \sqrt{\epsilon} v_k, & y_0 = 0, \end{cases} \quad k = 0, 1, 2, \ldots, [T/\epsilon] \tag{1.2}$$

where u_k and v_k are gaussian random variables, $x_0 = x$ is a given constant, and $[T/\epsilon]$ is the largest integer less than T/ϵ.

Actually, (1.2) approximates (1.1) in the following way. One discretizes (1.1) with time step size ϵ and replaces $X_{k\epsilon}$ by x_k, $U_{(k+1)\epsilon} - U_{k\epsilon}$ by $\sqrt{\epsilon} u_k$, $V_{(k+1)\epsilon} - V_{k\epsilon}$ by $\sqrt{\epsilon} v_k$, and $\epsilon^{-1}(Y_{(k+1)\epsilon} - Y_{k\epsilon})$ by y_k.

To illustrate the ideas without undue technical complications, we assume that all of the processes x_k, y_k, u_k, v_k in (1.2) are 1-dimensional, and that $h(x)$ has finite intervals of monotonicity. Extensions regard to the dimension of the model will be made in §3.

In [6], the nonlinear filtering in (1.2) was considered. Mean decision time and confidence intervals estimates were given in terms of the error probabilities. However, the estimates in [6] are purely theoretical and are very conservative for numerical simulations. In [5], numerical experiments in (1.2) were conducted with piecewise linear observation functions and with iid gaussian system noise.

It is the purpose of this paper to carry out numerical simulation analyses in discrete time model with general nonlinear observation functions and with correlated system noises. In this paper, correlations between $u_k, u_{k'}$ and between $v_k, v_{k'}$ will be allowed for $|k - k'| \le r$ with fixed $r < \infty$. This is done to avoid, at least to some extent, the white-noise idealization in (1.2) upon which quadratic variation tests are based.

Our method is as follows, roughly speaking. First, we test for critical point crossings of x_k on a time interval $K \le k \le K + M$, where K and $K + M$ are the time moments at which a critical point crossing may occur. Next we apply a test based on quadratic variations of y_k to decide the monotonicity intervals. On each of these monotonicity intervals, we can construct a nearly optimal filter (cf. [10]). The third step is to use these nearly optimal filters and the testing outcomes to construct our asymptotically optimal filters.

For the sequential hypothesis test in §3, we can find explicit estimates for the probabilities of incorrect decisions and for the mean decision time. This is done by a diffusion approximation technique. We first show that the general model (1.2) can be approximated by a simpler linear model. Then we show that this linear model can be approximated by a diffusion process. Therefore, the general model (1.2) can be approximated by a diffusion process. This diffusion process is then used to obtain mean decision time and error probabilities analytically.

In this paper we focus on description of the method, on the diffusion approximation technique, and on numerical results reported in §5. Many proofs of underlying mathematical results are referred to [6].

The plan of this paper is as follows. In §2 we formulate the hypothesis test problems and make assumptions of (1.2). In §3 we carry out sequential analysis on the hypothesis tests and obtain the mean decision time and error probability estimates. In §4 we construct asymptotically optimal filters based on the testing outcomes and the results of nearly optimal filtering. Finally in §5, numerical experiments in three cases are reported in detail.

2 Problem formulation

We consider the system noise u_k, v_k given by the following:

$$\begin{aligned} u_k &= a_1 u_k^0 + a_2 u_{k+1}^0 + \cdots + a_r u_{k+r-1}^0 \\ v_k &= b_1 v_k^0 + b_2 v_{k+1}^0 + \cdots + b_r v_{k+r-1}^0, \end{aligned} \tag{2.1}$$

where $\{u_k^0\}$ and $\{v_k^0\}$ are two sequences of iid gaussian $N(0,1)$, $\{u_k^0\}$ is independent of $\{v_k^0\}$, and $a_1, a_2, ..., a_r, b_1, l$ are $2r$ real numbers. Then it follows immediately that (u_k, v_k) is independent of $(u_{k'}, v_{k'})$ if $|k-k'| \geq r$. Moreover, $\{u_k\}$ is independent of $\{v_k\}$. Now if we take

$$\sigma_u^2 = a_1^2 + \cdots + a_r^2 \text{ and } \sigma_v^2 = b_1^2 + \cdots + b_r^2,$$

then $u_k \sim N(0, \sigma_u^2)$, $v_k \sim N(0, \sigma_v^2)$.

Let $\rho_u(k)$ and $\rho_v(k)$ denote the covariances between u_0 and u_k and between v_0 and v_k, respectively, i.e.,

$$\rho_u(k) := \text{cov}(u_0, u_k) = Eu_0u_k, \ \rho_v(k) := \text{cov}(v_0, v_k) = Ev_0v_k.$$

Then

$$\rho_u(k) = \begin{cases} a_1 a_{k+1} + \cdots + a_{r-k} a_r & \text{if } 0 \leq k \leq r-1 \\ 0 & \text{if } k \geq r \end{cases} \tag{2.2}$$

$$\rho_v(k) = \begin{cases} b_1 b_{k+1} + \cdots + b_{r-k} b_r & \text{if } 0 \leq k \leq r-1 \\ 0 & \text{if } k \geq r \end{cases} \tag{2.3}$$

We make the following assumptions on the functions in (2.1):

A1) $f(x)$ and $g(x)$ are bounded functions.

A2) There exist a piecewise linear function $H(x)$ and a monotone increasing function $\phi(x)$ with bounded $\phi''(x)$ and $0 < c_h^1 \leq \phi'(x) \leq c_h^2 < \infty$ such that $h(x) = H(\phi(x))$ and $H(x)$ is linear on $\phi(I_i) := (\phi(x_{i-1}^*), \phi(x_i^*))$, $i = 1, 2, ..., l+1$ for a set of points $-\infty = x_0^* < x_1^* < ... < x_l^* < x_{l+1}^* = \infty$.

A3) There exist constants c_g and C_g such that $C_g \geq |g(x)| \geq c_g > 0$, $|g(x) - g(x')| \leq C_g|x - x'|$ and $|(\phi' g)(x) - (\phi' g)(x')| \leq C_g|x - x'|$.

A4) For all $x_i \in I_i$, and $x_j \in I_j$ such that $H(\phi(x_i)) = H(\phi(x_j))$, then

$$|\alpha_i \phi'(x_i) g(x_i)|^2 \neq |\alpha_j \phi'(x_j) g(x_j)|^2, \text{ for } i \neq j, \tag{2.4}$$

where $\alpha_i = H'(x)$ for $x \in \phi(I_i)$.

Remark 2.1. Note that A2) will capture many of the functions with piecewise monotone $h(x)$. A4) is the so called 'detectability condition' which will be used in a sequential quadratic variation test to determine which intervals I_i the state x_k belongs to; see also [3].

3 Sequential quadratic variation test (QVT)

In this section, we describe a sequential hypothesis test, based on quadratic variations of the observation sequence $\{y_k\}$ to decide the intervals I_i such that $x_k \in I_i$. We start with a simple case with $f(x) = 0$, $g(x) = 1$, and $\phi(x) = x$.

QVT with $f(x) = 0$, $g(x) = 1$, and $h(x)$ piecewise linear

We first consider a simpler case $f(x) = 0$, $g(x) = 1$, and $h(x) = \sum_{i=0}^{l+1} \alpha_i x \chi_{I_i}(x)$, where $\chi_{I_i}(x)$ is the indicator function of I_i. We fix $0 \le K$ and $0 \le M \le T/\varepsilon$. Then we define $D_{K,M} = \{k : K \le k \le K+M\}$, where K and $K+M$ are the times at which x_k switches from a monotonicity interval to another. Let $\mathcal{H}_i$ denote the hypothesis

$$\mathcal{H}_i : \ x_k \in I_i = (x_{i-1}^*, x_i^*), \ k \in D_{K,M}.$$

Let $\Delta_k = y_{k+1} - y_k$ denote the consecutive increment of y_k and let

$$\sigma_i^2 = \alpha_i^2 \rho_u(0) + 2\rho_v(0) - 2\rho_v(1).$$

For each $i, j = 1, 2, \ldots l+1$ and $i \ne j$, define $S_n^{i,j}$ to be the test statistic process:

$$S_n^{i,j} = c_1^{i,j} n + c_2^{i,j} \sum_{k=K}^{K+n-1} \Delta_k^2/\varepsilon \tag{3.1}$$

where $c_1^{i,j} = \log(\sigma_j/\sigma_i)$ and $c_2^{i,j} = (1/\sigma_j^2 - 1/\sigma_i^2)/2$.

Description of the hypothesis test: First of all, we will use $S_n^{i,j}$ to determine a preference between $\mathcal{H}_i$ and $\mathcal{H}_j$ for any fixed i, j. If $\mathcal{H}_i$ is more favorable than $\mathcal{H}_j$, the we say $i \succ j$. Let i_0 denote the 'largest' state, i.e., $i_0 \succ j$ for all j, then we accept $\mathcal{H}_{i_0}$. Thus, in the following, we only need to determine the preference between $\mathcal{H}_i$ vs. $\mathcal{H}_j$. This is done as follows. For pre-chosen $A_{i,j} > 0$, and $B_{i,j} > 0$, we keep the accumulation of the process $S_n^{i,j}$ until it is bigger than $A_{i,j}$ or less than $-B_{i,j}$. Then we stop sampling and make a decision. If $S_n^{i,j} \ge A_{i,j}$ we say that $\mathcal{H}_i$ is more favorable or $i \succ j$; similarly if $S_n^{i,j} \le -B_{i,j}$, we say that $\mathcal{H}_j$ is more favorable or $j \succ i$.

Next we discuss how to choose $A_{i,j}, B_{i,j}$. It is clear that the values of $A_{i,j}$ and $B_{i,j}$ depend on the error probabilities. Normally, smaller error probabilities require larger $A_{i,j}$ and $B_{i,j}$. In [6], an estimate of such $A_{i,j}$ and $B_{i,j}$ are given in terms of ε such that the error probability is less than ε. However, the estimation in [6] is a purely theoretical approach and is quite conservative for our numerical simulation purpose.

In order to make more accurate estimates, we introduce a diffusion process to approximate the $S_n^{i,j}$ process and to estimate the associated $A_{i,j}$ and $B_{i,j}$. Then based on the diffusion approximation, we derive ordinary differential equations for the mean decision time and error probabilities. Solving these ODEs, we can obtain the appropriate $A_{i,j}$ and $B_{i,j}$, and, therefore, the mean decision time.

For each $i = 1, 2, \ldots, l+1$ and $k = 0, 1, 2 \ldots$, let

$$w_k^i = \alpha_i u_k + v_{k+1} - v_k.$$

Then for each i, $\{\sigma_i^{-1} w_k^i\}_{k=0}^{\infty}$ is a sequence of $N(0,1)$ random variables. Moreover, the sequence $\{w_{m+k(r+1)}^i\}_{k=0}^{\infty}$, is independent for each $m = 0, 1, \ldots, r$.

We assume that $\mathcal{H}_i$ holds and test it against an alternative $\mathcal{H}_j$. Let $K = 0$ for notational brevity. Then for any fixed i, j

$$S_n^{i,j} = c_1^{i,j} n + c_2^{i,j} \sum_{k=0}^{n-1} (w_k^i)^2.$$

For $k = 0,1,2,...$, let $\xi_k = (w_k^i)^2 - \sigma_i^2$. Then $\{\xi_k\}$ is a sequence of random variables with mean zero and finite variance. Moreover, $\{\xi_{m+k(r+1)}\}_{k=0}^{\infty}$ is iid for each $m = 0,1,2,...r$. Therefore, $\{\xi_k\}$ is a mixing sequence (cf. [2]).

For each i, k, let

$$\bar{\rho}_i(k) = \mathrm{cov}(w_0^i, w_k^i) = E(w_0^i w_k^i).$$

Then

$$\bar{\rho}_i(k) = \alpha_i^2 \rho_u(k) + 2\rho_v(k) - \rho_v(k-1) - \rho_v(k+1).$$

Let

$$\Sigma_i^2 := 2(\sigma_i^4 + 2(\bar{\rho}_i^2(1) + \cdots + \bar{\rho}_i^2(r)),$$

and for $0 \le t \le T$, $n = [t/\varepsilon]$, we define

$$W_i^i(n) = \sqrt{\varepsilon}\Sigma_i^{-1}(\xi_1 + \xi_2 + ... + \xi_n).$$

Then by [9, Theorem 7.1], there exists a standard brownian motion $W_0^i(t)$ such that

$$W_i^i(n) - \sqrt{\varepsilon}W_0^i(t/\varepsilon) = O(t^{1/2-\lambda}\varepsilon^{\lambda}) \ a.s. \tag{3.2}$$

on an enlarged probability space, where $\lambda = 1/300$. Note that $\sqrt{\varepsilon}W_0(t/\varepsilon)$ is also a standard Brownian motion on the t time scale.

Let $\zeta_i^{i,j}(t)$ denote a diffusion process

$$\zeta_i^{i,j}(t) = \mu_i^{i,j}t + \sqrt{2\varepsilon}\gamma_i^{i,j}(\sqrt{\varepsilon}W_0(t/\varepsilon)),$$

where $\mu_i^{i,j} = c_1^{i,j} + c_2^{i,j}\sigma_i^2$, $\gamma_i^{i,j} = c_2^{i,j}\sqrt{\sigma_i^4 + 2(\bar{\rho}_i^2(1) + \cdots + \bar{\rho}_i^2(r))}$. Then according to (3.2), $\zeta_i^{i,j}(t)$ is an approximation to $\varepsilon S_m^{i,j}$ under $\mathcal{H}_i$. We will use this diffusion approximation to obtain probabilities of Type I and Type II errors and mean decision times in terms of $A_{i,j}$ and $B_{i,j}$. These results are in good agreement with the results of numerical simulations reported in §5. More accurate approximations could, in principle, be obtained using large deviations results about the sequence $S_n^{i,j}$ (see Remark 3.2 below). However, explicit formulas like (3.3)-(3.6) are then no longer available.

Let $T_i^{i,j}$ be the first exit time of $\zeta_i^{i,j}(t)$ from $[-\varepsilon B_{i,j}, \varepsilon A_{i,j}]$, i.e.,

$$\mathcal{H}_i: \ T_i^{i,j} = \inf\{t, : \zeta_i^{i,j}(t) \ge \varepsilon A_{i,j} \text{ or } \zeta_i^{i,j}(t) \le -\varepsilon B_{i,j}\}.$$

Let $p_i^{i,j}(z) = P_{\zeta_i^{i,j}(0)=z}\{\zeta_i^{i,j}(T_i^{i,j}) = -\varepsilon B_{i,j} | \mathcal{H}_i\}$. Then $p_i^{i,j}(0)$ is the Type I error probability (later be taken to be $O(\varepsilon)$).

We now use the Dynkin formula to obtain (cf. [7, pp. 130])

$$\begin{cases} \varepsilon(\gamma_i^{i,j})^2(p_i^{i,j})'' + \mu_i^{i,j}(p_i^{i,j})' = 0 \\ p_i^{i,j}(-\varepsilon B_{i,j}) = 1, \ p_i^{i,j}(\varepsilon A_{i,j}) = 0. \end{cases}$$

Suppose $\eta_i^{i,j} = \mu_i^{i,j}/(\gamma_i^{i,j})^2$. Then the solution to the above ODE is given by

$$p_i^{i,j}(z) = (e^{-\eta_i^{i,j}z/\varepsilon} - e^{-\eta_i^{i,j}A_{i,j}})/(e^{\eta_i^{i,j}B_{i,j}} - e^{-\eta_i^{i,j}A_{i,j}}).$$

The Type I error probability is thus equal to

$$p_i^{i,j}(0) = (1 - e^{-\eta_i^{i,j}A_{i,j}})/(e^{\eta_i^{i,j}B_{i,j}} - e^{-\eta_i^{i,j}A_{i,j}}). \tag{3.3}$$

Similarly we can obtain the Type II error probability

$$p_j^{i,j}(0) = (1 - e^{\eta_j^{i,j} B_{i,j}})/(e^{-\eta_j^{i,j} A_{i,j}} - e^{\eta_j^{i,j} B_{i,j}}). \tag{3.4}$$

Next we consider the estimate of the expected decision time. Assume that $\mathcal{H}_i$ holds. We define $G_i^{i,j}(z) = E_{\zeta_i^{i,j}(0)=z} T_i^{i,j}$. Then $G_i^{i,j}(0)$ is the mean decision time. Moreover, by the Dynkin formula, $G_i^{i,j}$ satisfies the following ODE

$$\begin{cases} \varepsilon(\gamma_i^{i,j})^2 (G_i^{i,j})'' + \mu_i^{i,j}(G_i^{i,j})' + 1 = 0 \\ G_i^{i,j}(-\varepsilon B_{i,j}) = G_i^{i,j}(\varepsilon A_{i,j}) = 0. \end{cases}$$

Solving this equation, we obtain

$$G_i^{i,j}(z) = \frac{\varepsilon(A_{i,j} + B_{i,j})e^{-\eta_i^{i,j} A_{i,j}}}{\mu_i^{i,j}(e^{\eta_i^{i,j} B_{i,j}} - e^{-\eta_i^{i,j} A_{i,j}})} - \frac{\varepsilon(A_{i,j} + B_{i,j})e^{-\eta_i^{i,j} z/\varepsilon}}{\mu_i^{i,j}(e^{\eta_i^{i,j} B_{i,j}} - e^{-\eta_i^{i,j} A_{i,j}})} + \frac{\varepsilon A_{i,j} z}{\mu_i^{i,j}}.$$

Therefore, the expected decision time

$$ET_i^{i,j} = G_i^{i,j}(0) = \varepsilon(A_{i,j}/\mu_i^{i,j} - p_i^{i,j}(0)(A_{i,j} + B_{i,j})/\mu_i^{i,j}). \tag{3.5}$$

Similarly, we can define mean decision time T_j^* under $\mathcal{H}_j$ and obtain

$$\mathcal{H}_j : ET_j^{i,j} = \varepsilon(-B_{i,j}/\mu_j^{i,j} - p_j^{i,j}(0)(A_{i,j} + B_{i,j})/\mu_j^{i,j}). \tag{3.6}$$

Therefore, for any fixed i and j, we first postulate the value of $p_i^{i,j}(0), p_j^{i,j}(0)$. Then we solve (3.3) and (3.4) to obtain $A_{i,j}$ and $B_{i,j}$. Using the value of $A_{i,j}$ and $B_{i,j}$ and (3.5) and (3.6), we can get the estimate for the mean decision times. Elementary considerations show that $\mu_i^{i,j} > 0$ and $\mu_j^{i,j} < 0$, and therefore (3.3), (3.4) determine $A_{i,j} > 0$ and $B_{i,j} > 0$.

Example 3.1. For instance, if $l = 2$ and $\alpha_1 = -2, \alpha_2 = 1$, then $\eta_1^{1,2} > 0$, $\eta_2^{1,2} < 0$. Therefore, if we take $p_1^{1,2}(0) = p_2^{1,2}(0) = \varepsilon$, then, we get approximately $A_{1,2} = (\eta_2^{1,2})^{-1} \log \varepsilon$, $-B_{1,2} = (\eta_1^{1,2})^{-1} \log \varepsilon$. The expected decision times $T_1^{1,2}$ and $T_2^{1,2}$ are thus given by (3.5) and (3.6) which turn out to be $O(\varepsilon|\log \varepsilon|)$.

Remark 3.1. Let $p_i^{i,j}(0) = p_j^{i,j}(0) = \beta$. (In the Example 3.1 we took $\beta = \varepsilon$.) For small β, we have approximately $A_{i,j} = (\eta_j^{i,j})^{-1} \log \beta$ and $B_{i,j} = (\eta_i^{i,j})^{-1} \log \beta$, where $A_{i,j}$ and $B_{i,j}$ were determined using the diffusion approximation to $\varepsilon S_n^{i,j}$. Large deviations estimates for $S_n^{i,j}$ give more accurate choices of $A_{i,j}$ and $B_{i,j}$, but they are less explicit.

Remark 3.2. One may easily extend the QVT to the case where x_k, y_k are of the same dimension $\bar{n} > 1$. In this case, α_i is a $\bar{n} \times \bar{n}$ matrix, and I_i is a $\bar{n}$-dimensional polytope. The detectability condition in [3, §8] becomes for $h(x)$ piecewise linear, $g(x) =$ identity:

$$\mathrm{trace}(\alpha_i \alpha_i^*) \neq \mathrm{trace}(\alpha_j \alpha_j^*),\ i \neq j,$$

where $\alpha_i = \frac{\partial}{\partial x} H(x)$ for $x \in I_i$. The test statistic process S_n becomes

$$S_n^{i,j} = c_1^{i,j} n + c_2^{i,j} \sum_{k=K}^{K+n-1} |y_{k+1} - y_k|^2/\varepsilon$$

where

$$c_1^{i,j} = \log \frac{E|\alpha_j u_1 + v_2 - v_1|^2}{E|\alpha_i u_1 + v_2 - v_1|^2}$$

$$\text{and } c_2^{i,j} = \frac{1}{2}\left(\frac{1}{E|\alpha_j u_1 + v_2 - v_1|^2} - \frac{1}{E|\alpha_i u_1 + v_2 - v_1|^2}\right).$$

It should be noted that the QVT described above requires large computation if l is a large number. For example, if $l = 5$, then we need to compute 10 testing sequences $S_n^{i,j}$ for $(i > j)$ to make a decision. Now we give for the 1-dimensional case a modified version of QVT which will help to reduce the computation required in the previous version.

Suppose that $x_k \in I_i$ for $k \in D_{K,M}$ and x_k exits the interval $I_i = (x_{i-1}^*, x_i^*)$ at time $K+M$. Then either $x_{K+M} \sim x_{i-1}^*$ or x_i^*. Now we suppose $x_{K+M} \sim x_i^*$. Then during next time period, x_k will stay either in I_i or in I_{i+1} most probably. Therefore, we only have to calculate $S_n^{i+1,i}$ to examine the hypotheses $\mathcal{H}_i$ and $\mathcal{H}_{i+1}$. In the second part of this section, we will apply this modified version of QVT in the model with general f, g, and h.

QVT with general $f(x)$, $h(x)$ and $g(x)$

We now consider the testing problem with general $f(x)$, $h(x)$ and $g(x)$ that satisfy A1)-A4).

For any fixed $i = 1, 2, ..., l$, we first test $\mathcal{H}_i$ against $\mathcal{H}_{i+1}$. Recall the 'detectability condition' in A4) and let $x_i \to x_i^*$ and $x_{i+1} \to x_i^*$ in such a way that $h(x_i) = h(x_{i+1})$. Then (2.4) yields

$$|\alpha_i \phi'(x_i^*) g(x_i^*)|^2 \neq |\alpha_{i+1} \phi'(x_i^*) g(x_i^*)|^2. \tag{3.7}$$

This is equivalent to $\alpha_i^2 \neq \alpha_{i+1}^2$.

Let $\bar{\alpha}_i = \alpha_i \phi'(x_i^*) g(x_i^*)$, $\bar{\alpha}_{i+1} = \alpha_{i+1} \phi'(x_i^*) g(x_i^*)$. We now construct $(\bar{x}_k, \bar{y}_k)$ that approximates (x_k, y_k). Let $(\bar{x}_k, \bar{y}_k)$ be defined by

$$\left\{ \begin{array}{l} \bar{x}_{k+1} = \bar{x}_k + \sqrt{\varepsilon} u_k \\ \bar{y}_k = \left\{ \begin{array}{ll} \bar{\alpha}_i \bar{x}_k + \sqrt{\varepsilon} v_k & \text{if } \bar{x}_k < x_i^* \\ \bar{\alpha}_{i+1} \bar{x}_k + \sqrt{\varepsilon} v_k & \text{if } \bar{x}_k \geq x_i^* \end{array} \right. . \end{array} \right. \tag{3.8}$$

It is easy to see that the system (3.8) is 'detectable' due to (3.7). Let $\bar{\sigma}_j^2 = \bar{\alpha}_j^2 + 2$ for $j = i$ or $i+1$. Then similarly as in (3.1), we define a test statistic for (3.8):

$$\bar{S}_n = \bar{S}_n^{i,i+1} = \bar{c}_1 n + \bar{c}_2 \sum_{k=0}^{n-1} \bar{\Delta}_k^2 / \varepsilon \tag{3.9}$$

where $\bar{\Delta}_k = \bar{y}_{k+1} - \bar{y}_k$ and

$$\bar{c}_1 = \bar{c}_1^{i,i+1} = \log(\bar{\sigma}_{i+1}/\bar{\sigma}_i), \quad \bar{c}_2 = \bar{c}_2^{i,i+1} = (1/\bar{\sigma}_{i+1}^2 - 1/\bar{\sigma}_i^2)/2.$$

We also define a test statistic process for (1.2):

$$S_n = \bar{c}_1 n + \bar{c}_2 \sum_{k=0}^{n-1} \Delta_k^2 / \varepsilon \tag{3.10}$$

where $\Delta_k = y_{k+1} - y_k$. Then we show that $\bar{S}_\cdot^{i,j}$ is an approximation to $S_\cdot^{i,j}$. Note that the initial value x_0 is assumed to be given in (1.2). If x_0 is not equal to any of x_i^*, $i = 0, 1, \cdots, l+1$, then there exists i_0 such that $x_0 \in I_{i_0}$. Since we only consider the case that x_k stays in an interval I_i, $i = 0, 1, \cdots, l+1$, for a certain period of time, then $x_k \in I_{i_0}$ for all $k \in D_{0,M}$. Namely, $\mathcal{H}_{i_0}$ holds. Nothing is to be tested. Therefore, in the following, we assume x_k starts at x_i^*, $i = 0, 1, \cdots, l+1$.

Theorem 3.1. *Suppose $\mathcal{H}_i$ holds for any fixed $i = 1, 2, ..., l$ and $x_0 = x_i^*$. Let n^* be a stopping time such that $En^* = O(|\log \varepsilon|)$. Then, there exists $\varepsilon_0 > 0$ such that for $0 < \varepsilon \leq \varepsilon_0$,*

$$P(|S_{n^*} - \bar{S}_{n^*}| \geq \sqrt[8]{\varepsilon}) = O(\sqrt[8]{\varepsilon} |\log \varepsilon|).$$

Remark 3.3. Let n^* denote the decision time. Then in [6] it is shown that En^* is of order $|\log \varepsilon|$. Then the above theorem tells us that the probability that $S_\cdot$ and $\bar{S}_\cdot$ make different decision is small.

Proof. Let $\psi_k = E|x_k - x_i^*|^2$, then $\psi_0 = 0$ since x_k starts at x_i^*. By A1), we have

$$\psi_{k+1} \le (1 + C_1\varepsilon)\psi_k + C_1\varepsilon,\ 0 \le k \le [T/\varepsilon]$$

for some constant $C_1 > 0$. This implies $\psi_k \le (1 + C_1\varepsilon)^k - 1, k = 0, 1, 2, ..., [T/\varepsilon]$.

By direct computation,

$$\begin{aligned} S_n - \bar{S}_n = & \ \bar{c}_2 \textstyle\sum_{k=0}^{n-1}(\Delta_k^2 - \bar{\Delta}_k^2)/\varepsilon \\ = & \ \bar{c}_2 \textstyle\sum_{k=0}^{n-1}[(\alpha_i\phi'(x_k)g(x_k)u_k + v_{k+1} - v_k)^2 \\ & -(\alpha_i\phi'(x_i^*)g(x_i^*)u_k + v_{k+1} - v_k)^2] + O(\sqrt{\varepsilon n}) \\ = & \ \bar{c}_2 \textstyle\sum_{k=0}^{n-1}[\alpha_i(\phi'(x_k)g(x_k) - \phi'(x_i^*)g(x_i^*))]\Phi_k + O(\sqrt{\varepsilon n}) \end{aligned}$$

where $\Phi_k = u_k[\alpha_i(\phi'(x_k)g(x_k) + \phi'(x_i^*)g(x_i^*))u_k + 2(v_{k+1} - v_k)]$. It is easy to see that $E|\Phi_k|^2$ is uniformly bounded for $k \le [T/\varepsilon]$. Thus there exist constants C_2 and C_3 such that

$$\begin{aligned} E|S_n - \bar{S}_n| \le & \ C_2 \textstyle\sum_{k=0}^{n-1}(E|\phi'(x_k)g(x_k) - \phi'(x_i^*)g(x_i^*)|^2)^{\frac{1}{2}} + O(\sqrt{\varepsilon n}) \\ \le & \ C_3 C_g \textstyle\sum_{k=0}^{n-1}(E|x_k - x_i^*|^2)^{\frac{1}{2}} + O(\sqrt{\varepsilon n}) \\ \le & \ C_3 C_g \textstyle\sum_{k=0}^{n-1}[(1 + C_1\varepsilon)^k - 1]^{\frac{1}{2}} + O(\sqrt{\varepsilon n}) \\ \le & \ C_3 C_g \sqrt{n}[((1 + C_1\varepsilon)^n - 1 - C_1\varepsilon n)/\varepsilon]^{\frac{1}{2}} + O(\sqrt{\varepsilon n}). \end{aligned}$$

Observe that there exists $\varepsilon_0 > 0$ such that for $0 < \varepsilon \le \varepsilon_0$ and for $n \le \sqrt[8]{\varepsilon}$,

$$0 \le \frac{(1 + C_1\varepsilon)^n - 1 - C_1\varepsilon n}{C_1\varepsilon^2 n^2} \le 1.$$

This yields

$$E|S_n - \bar{S}_n| = O(\sqrt{\varepsilon} n^{\frac{3}{2}}). \tag{3.11}$$

Let $N_\varepsilon = \sqrt[-8]{\varepsilon}$. We then have

$$\begin{aligned} P(|S_{n^*} - \bar{S}_{n^*}| \ge \sqrt[8]{\varepsilon}) \le & \ P(|S_{n^*} - \bar{S}_{n^*}| \ge \sqrt[8]{\varepsilon}, n^* \le N_\varepsilon) + P(n^* \ge N_\varepsilon) \\ \le & \ \sqrt[-8]{\varepsilon} E|S_{n^*} - \bar{S}_{n^*}|\chi_{\{n^* \le N_\varepsilon\}} + N_\varepsilon^{-1} E n^* \\ \le & \ \sqrt[-8]{\varepsilon} E|S_{n^*} - \bar{S}_{n^*}|\chi_{\{n^* \le N_\varepsilon\}} + O(\sqrt[8]{\varepsilon}|\log\varepsilon|). \end{aligned} \tag{3.12}$$

Note that by (3.11),

$$\begin{aligned} & \sqrt[-8]{\varepsilon} E|S_{n^*} - \bar{S}_{n^*}|\chi_{\{n^* \le N_\varepsilon\}} \\ \le & \ \sqrt[-8]{\varepsilon} E(|S_1 - \bar{S}_1|\chi_{\{n^*=1\}} + \cdots + |S_{N_\varepsilon} - \bar{S}_{N_\varepsilon}|\chi_{\{n^*=N_\varepsilon\}}) \\ \le & \ \sqrt[-8]{\varepsilon} E(|S_1 - \bar{S}_1| + \cdots + |S_{N_\varepsilon} - \bar{S}_{N_\varepsilon}|) \\ = & \ O(\varepsilon^{\frac{1}{2} - \frac{1}{8}}(1^{\frac{3}{2}} + 2^{\frac{3}{2}} + \cdots + N_\varepsilon^{\frac{3}{2}})) \\ = & \ O(\sqrt[8]{\varepsilon}). \end{aligned}$$

Therefore,

$$P(|S_{n^*} - \bar{S}_{n^*}| \ge \sqrt[8]{\varepsilon}) \le O(\sqrt[8]{\varepsilon}) + O(|\log\varepsilon|\sqrt[8]{\varepsilon}) \le O(|\log\varepsilon|\sqrt[8]{\varepsilon}). \ \square$$

By this theorem, we may choose the confidence interval $[-B_{i,j}, A_{i,j}]$ as in (3.3) and (3.4) with $\bar{\alpha}_i$ and $\bar{\alpha}_{i+1}$ in the places of α_i and α_{i+1}, respectively. Then we only stop sampling when $S_n^{i,j}$ exits from $[-B_{i,j}, A_{i,j}]$ and say that $\mathcal{H}_i$ is more favorable if $S_n^{i,j} \ge A_{i,j}$ and $\mathcal{H}_2$ is more favorable if $S_n^{i,j} \le -B_{i,j}$.

4 Asymptotically optimal filters

Once a decision among the hypotheses $\mathcal{H}_i$ is made, an extended Kalman filter can be used to estimate the conditional mean $\hat{x}_k$. Let $(x_k^i, \varepsilon Q_k^i)$ be the extended Kalman filter (EKF). Then the EKF equations are:

$$\begin{cases} x_{k+1}^i = f_\varepsilon(x_k^i) + h_i'(f_\varepsilon(x_k^i))Q_{k+1}^i[y_{k+1} - h_i(f_\varepsilon(x_k^i))] \\ Q_{k+1}^i = [[f_\varepsilon'(x_k^i)]^2 Q_k^i + g(x_k^i)^2]/[(h_i'(f_\varepsilon(x_k^i))^2[(f_\varepsilon'(x_k^i))^2 Q_k^i + g(x_k^i)^2] + 1] \end{cases} \tag{4.1}$$

where $f_\varepsilon(x) = x + \varepsilon f(x)$. If $f(x)$ and $h(x)$ are linear and $g(x)$ is equal to a constant, then (4.1) is the standard Kalman filter equation. For (4.1) we can take the initial conditions $x_0^i = E(x_0)$, $Q_0^i = \text{var}(x_0)$. However, the influence of the initial data for x_0^i, Q_0^i are quickly lost for small ε, as seen in the continuous time case from [4, §5] or [3, §6].

We use $\tilde{x}_k$ to denote an approximation to $\hat{x}_k$. Let n^* be the decision time. Then, we define

$$\tilde{x}_k = x_k^i \text{ for } k \in D_{K+n^*, M-n^*} \text{ if } \mathcal{H}_i \text{ holds.}$$

where M is the time when the critical point crossing test fails (this critical point crossing test will be discussed in the next section). Corresponding estimates for the continuous time model [3, Theorem 7.5] indicate that the following should hold. Suppose $\mathcal{H}_i$ holds and that $\varepsilon(K_1 - K) \geq a > 0$. Then, with probability nearly one (cf. [6])

$$x_k^i - \hat{x}_k = O(\varepsilon^p), \text{ for } k \in D_{K_1, M-K_1+K},$$

for any $p < 1$.

For those $k \in D_{K,n^*}$, we may define $\tilde{x}_k$ to be anything, for example, we can define $\tilde{x}_k = (l+1)^{-1}\sum_i x_k^i$. No estimation for $\tilde{x}_k - \hat{x}_k$, $k \in D_{K,n^*}$ could be given on $k \in D_{K,n^*}$. Fortunately, the typical N is only an order of $|\log \varepsilon|$, provided the error probabilities are ε. So the actual error in a continuous time model is equal to

$$[T/\varepsilon]^{-1} E \sum_{K \leq k < K+n^*} |\tilde{x}_k - \hat{x}_k| = O(\varepsilon |\log \varepsilon|)$$

due to the fact that $E|\tilde{x}_k - \hat{x}_k|$ is uniformly bounded.

5 Numerical simulations

In this section, we first discuss conditions on y_k under which the state x_k crosses from I_i to I_j $(i \neq j)$.

Test for critical point crossings

We consider the following events

$$\begin{aligned} B_i &= \{x_k \in I_i, \text{ for } k \in D_{K,M}\}, \; i = 1, 2, ..., l+1 \\ C_i &= \{|y_k - h(x_i^*)| \geq c_i > 0, \text{ for } k \in D_{K,M}\}, \; i = 1, 2, ...l. \end{aligned} \tag{5.1}$$

By using elementary probabilistic arguments, one can prove as in [3, Prop 3.1] that:
Proposition 5.1. *For $i = 1, 2, ...l$, given $c_i > 0$, there exist $\varepsilon_0 > 0$, $k_0 > 0$ such that*

$$P((B_i \bigcup B_{i+1})^c | C_i) \leq e^{-k_0/\varepsilon}, \; 0 < \varepsilon < \varepsilon_0. \tag{5.2}$$

The estimate in this proposition is extremely conservative. It is based on estimating

$$P(\max_{k \in D_{K,M}} |u_k| > \varepsilon^{-1/2}\gamma), \; P(\max_{k \in D_{K,M}} |v_k| > \varepsilon^{-1/2}\gamma)$$

for some $\gamma > 0$ depending on c_i. However, critical point crossings of x_k are most likely to occur when $h(x_k) - h(x_i^*)$ is near the cutoff c_i.

The critical crossing test is described as follows: For given $c_1, ..., c_l > 0$, we only admit critical point crossing when $|y_k - h(x_i^*)| < c_i$ for $i = 1, ..., l$.

In our numerical simulations, we took

$$\begin{aligned} c_i &= \lambda\sqrt{\varepsilon}\sigma_i^0 \\ &= \lambda\sqrt{\varepsilon}\sqrt{\alpha_i^2\alpha_{i+1}^2\rho_u(0) + \alpha_i^2\rho_v(0) + \alpha_{i+1}^2\rho_v(0) - 2\alpha_i\alpha_{i+1}\rho_v(1)}/|\alpha_i - \alpha_{i+1}|. \end{aligned} \tag{5.3}$$

The choice of c_i involves a tradeoff between maintaining a large proportion of time steps k with $|y_k - h(x_i^*)| \geq c_i$ and reducing the probability that a critical point crossing of x_k is not detected. The form of (5.3) is suggested by elementary formulas for probabilities that x_k and x_{k+1} have opposite signs while $|y_k - h(x_i^*)| \geq c_i$ and $|y_{k+1} - h(x_i^*)| \geq c_i$. For example, if the following event A occurs

$$A = \{x_{k+1} \in I_i,\ x_k \in I_{i+1},\ |y_{k+1} - h(x_i^*)| \geq c_i,\ |y_k - h(x_i^*)| \geq c_i\} \tag{5.4}$$

and if $f(x) = 0, g(x) = 1$ and $h(x)$ is piecewise linear in (5.2), then

$$\varepsilon^{-1/2}|\alpha_i - \alpha_{i+1}|c_i \geq |\alpha_i\alpha_{i+1}||u_k + v_{k+1}/\alpha_i + v_k/\alpha_{i+1}|$$

The random variable $\alpha_i\alpha_{i+1}(u_k + v_{k+1}/\alpha_i + v_k/\alpha_{i+1})$ is normal $N(0, (\sigma_i^0)^2)$ where σ_i^0 is given in (5.3). For instance, λ is chosen to be 1.65 for .05 error probability.

Numerical experiments

Test procedure: Let $[K^{j_0}, K^{j_0} + M^{j_0}]$, $j_0 = 1, 2, ...$ denote successive discrete time intervals of maximum length in which $|y_k - h(x_i^*)| \geq c_i$ for $i = 1, 2, ..., l$. Thus

$$\begin{aligned} &|y_k - h(x_i^*)| \geq c_i,\ k \in D_{K^{j_0},M^{j_0}},\ K^{j_0} + M^{j_0} < K^{j_0+1}, \\ &\exists i \text{ such that } |y_k - h(x_i^*)| < c_i \text{ if } k \notin \cup_{j_0}[K^{j_0}, K^{j_0} + M^{j_0}]. \end{aligned} \tag{5.5}$$

Once we observe K^{j_0}, a quadratic variations test is used to decide among the alternatives

$$B_i = \{x_k \in I_i, \text{ for } k \in D_{K^{j_0},M^{j_0}}\} \tag{5.6}$$

A decision is taken at step $K^{j_0} + N^{j_0}$, provided $N^{j_0} \leq M^{j_0}$, where $K^{j_0} + N^{j_0}$ is the first time n such that $S_n^{i,j}$ exits from $[-B_{i,j}, A_{i,j}]$ for all $i > j$. There is no decision about the location of x_k if $N^{j_0} > M^{j_0}$ or if $|y_k - h(x_i^*)| < c_i$ for some i. Then based on the hypothesis test description given in §3, we carry out tests $\mathcal{H}_i$ against $\mathcal{H}_j$ for all $i > j$. Then after the end of all the tests, we accept $\mathcal{H}_{i_0}$ provided $i_0 \succ i$ for all i.

In the numerical experiments we took final time $T = 10$ on the continuous time scale. For step size $\varepsilon = .01$, this corresponds to a total number of steps $N_0 := 10\varepsilon^{-1} = 1000$ in all cases reported in Tables 1, 2 and 3 below. Also we took the initial value $x_0 = 0$ and the error probabilities $p_i^{i,j}(0) = .05$.

We use N_i^* to denote the time for reaching a decision in the discrete-time model, that is $N_i^* = \varepsilon^{-1}\max\{T_i^{i,j}, j \neq i\}$ where $T_i^{i,j}$ is the stopping time for the diffusion approximation process (see §3).

We consider three cases. For each of these we conducted 100 numerical simulations, and recorded the following: number of steps k with a decision, number of correct and incorrect sequential decisions among $\mathcal{H}_i$ and the number of steps to reach a decision. The latter is denoted by N_i in case $\mathcal{H}_i$ was chosen. Note also that N_i is the sample average. Let SD denote the percentage of steps with a decision about the location of x_k among the 100 simulations. In other words, SD equals $100N_0^{-1}$ times the number steps k

for which $|y_k - h(x_i^*)| \geq c_i$ and $N^{j_0} \leq M^{j_0}$. Let CD denote the percentage of correct decisions. Note that the numerical experiments were designed with $p_i^{i,j}(0) = .05$. Hence, CD should be near 95%. In all of the following three cases, we take $f(x) = 0$ since the effect of $f(x)$ is less important if ε is small enough (cf. [5]).

Case I. Correlated noise.

We take $l = 1$, $g(x) = 1$, $x_1^* = 0$ and

$$h(x) = \begin{cases} -2x & \text{if } x < x_1^* \\ x & \text{if } x \geq x_1^* \end{cases}$$

We also take the system noise in (1.2)

$$u_k = a_1 u_k^0 + a_2 u_{k+1}^0, \quad v_k = a_1 v_k^0 + a_2 v_{k+1}^0,$$

where u_k^0, v_k^0 are iid gaussian random variables. Let $\bar{\sigma} = a_1 + a_2$ which is the quadratic variation of the system (1.1) Then for different choice of a_1 and a_2 we obtain the following result.

Table 1

a_1	a_2	$\bar{\sigma}^2$	c_1	B	A	SD%	CD%	EN_1^*	N_1	EN_2^*	N_2
.71	.71	2	.18	2.8	6.9	86.0	96.8	7.8	11.7	40.9	37.5
.20	.98	1.4	.17	2.5	6.7	84.4	98.4	11.0	13.9	39.9	37.3

The numerical experiment in this case indicates that the correlated noise in our model does not make much difference from the uncorrelated noise model (compared with the results in [5]). Table 1 shows that the larger the $\bar{\sigma}$ the larger the SD. Now let us illustrate this point as follows. We consider the case that $u_k = a_1 u_k^0 + a_2 u_{k+1}^0$ $(a_1^2 + a_2^2 = 1)$ and v_k is iid. Then $c_1^{i,j}$, $c_2^{i,j}$, and $\mu_i^{i,j}$ are fixed while

$$(\gamma_i^{i,j})^2 = (c_2^{i,j})^2(\sigma_i^4 + (\alpha_i^2(\bar{\sigma}_u^2 - 1) + 4)^2/2).$$

Thus increase $\bar{\sigma}_u$ will reduce the error probability.

Case II. Three monotonicity intervals of $h(x)$.

We take $l = 2$, $g(x) = 1$ and $h(x)$ to be piecewise linear continuous function with

$$h'(x) = \begin{cases} 1 & \text{if } x < x_1^* \\ -4 & \text{if } x_1^* \leq x < x_2^* \\ 2 & \text{if } x_2^* \leq x. \end{cases}$$

In this case we assume the system noise u_k, v_k to be iid gaussian. Thus the confidence intervals are

$$[-B_{1,2}, A_{1,2}] = [-1.33, 7.31], \ [-B_{2,3}, A_{2,3}] = [-6.69, 1.63], \ [-B_{1,3}, A_{1,3}] = [-2.37, 4.68].$$

In the numerical experiments, we vary the location of x_1^* and keep $x_2^* = 0$.

Table 2

x_1^*	x_2^*	c_1	c_2	SD%	CD%	EN_1^*	N_1	EN_2^*	N_2	EN_3^*	N_3
- .5	0	.19	.25	52.1	92.8	44.8	41.0	2.7	4.9	32.9	43.6
- 1	0	.19	.25	70.6	95.8	44.8	46.9	2.7	6.7	32.9	48.2
- 2	0	.19	.25	80.9	96.8	44.8	45.7	2.7	7.1	32.9	48.3

Note that the closer the x_1^* and x_2^*, the more frequently the x_k passes through the critical points x_i^*, thus the smaller the M^{j_0} in (6.1). As a result, the SD drops down.

Case III. Non-constant $g(x)$.

For general $h(x)$ as in assumption A2) one can always transform the system by letting $\bar{x}_k = \phi(x_k)$ to the following one:

$$\bar{x}_{k+1} = \bar{x}_k + \epsilon F(\bar{x}_k) + \sqrt{\epsilon} G(\bar{x}_k) u_k$$
$$y_k = H(\bar{x}_k) + \sqrt{\epsilon} v_k,$$

where $F(x) = (\phi' f + \frac{1}{2}\phi'' g^2)(\phi^{-1}(x))$ and $G(x) = (\phi' g)(\phi^{-1}(x))$ with $H(x)$ piecewise linear. Thus we may simplify our model by taking $\phi(x) = x$.

We take $l = 1$, $g(x) = \sqrt{1+x^2}$, $x_1^* = 0$ and

$$h(x) = \begin{cases} \alpha_1 x & \text{if } x < x_1^* \\ x & \text{if } x_1^* \leq x. \end{cases}$$

Our numerical simulation in this case is given in the following table.

Table 3

α_1	α_2	c_1	B	A	SD%	CD%	EN_1^*	N_1	EN_2^*	N_2
- 2	1	.16	2.4	4.7	66.9	92.1	44.8	51.9	13.1	32.2
- 4	1	.18	1.3	7.3	79.7	94.4	14.4	16.3	0.7	4.4

We may also modify the above version of QVT by modifying the coefficients $c_1^{i,j}$ and $c_2^{i,j}$ in the $S_n^{i,j}$ process. This modification is as follows. Let

$$\sigma_i^2(k) = g^2(x_k^i) + 2, \ i = 1, 2, ..., l+1$$

where x_k^i is the output of the EKF from (4.1).

Instead of making $c_1^{i,j} c_2^{i,j}$ constants, we may take

$$c_1^{i,j}(k) = \log(\sigma_j(k)/\sigma_i(k)), \ c_2^{i,j}(k) = (1/\sigma_j^2(k) - 1/\sigma_i^2(k))/2.$$

Then the test statistic is

$$S_n^{i,j} = \sum_{k=K}^{K+n-1} (c_1^{i,j}(k) + c_2^{i,j}(k)\Delta_k^2/\epsilon).$$

By using this $S_n^{i,j}$, one can get a result which is little bit better.

Table 4

α_1	α_2	c_1	B	A	SD%	CD%	N_1	N_2
- 2	1	.16	2.4	4.7	67.2	92.3	44.2	33.0
- 4	1	.18	1.3	7.3	79.9	94.1	15.2	4.7

Remark 5.1. One might notice that once the observation $|y_k - h(x_i^*)| \geq c_i$, it is more stable after the first few steps after K^{j_0}. Therefore, in the actual simulations, we wait for a while (say six steps). Then we start to accumulate the $S_n^{i,j}$ process.

References

[1] B.D.O. Anderson and J.B. Moore, *Optimal Filtering*, Prentice Hall, (1979).

[2] P. Billingsley, *Convergence of Probability Measures*, John Wiley & Sons, New York, (1968).

[3] W. Fleming and E. Pardoux, Piecewise monotone filtering with small observation noise, *SIAM J. on Control and Optimiz.*, Vol. 27, No. 5, pp. 1156-1181, (1989).

[4] W. Fleming, D. Ji and E. Pardoux, Piecewise linear filtering with small observation noise, *Proc. 8th Int. Conf. on Analysis and Optimization of Systems, (Antibes Conf. 1988), Lect. N. in Control and Info. Sci.*, No. 111, pp 725-739, Springer (1988).

[5] W.H. Fleming, D. Ji, P. Salame, and Q. Zhang, Piecewise monotone filtering in discrete time with small observation noise, to appear in *IEEE Trans. Auto. Contr.*, (1991).

[6] W.H. Fleming and Q. Zhang, Nonlinear filtering with small observation noise: piecewise monotone observations, *Proceedings of the Conference in Honor of Moshe Zakai*, E. Merzbach, A. Shwartz, and E. Mayer-Wolf ed., Academic Press, New York, pp. 153-168, (1991).

[7] W.H. Fleming and R.W. Rishel, *Deterministic and Stochastic Optimal Control*, Springer-Verlag, New York, (1975).

[8] R.S. Liptser and A.N. Shiryayev, *Statistics of Random Processes*, Vol. I, Springer-Verlag, New York, (1977).

[9] W. Philipp and W. Stout, Almost sure invariance principles for partial sums of weakly dependent random variables, *Memoirs of the American Mathematical Society*, Vol. 2, No. 161, (1975).

[10] J. Picard, Filtrage de diffusions vectorielles faiblement bruitées, *Proc. 7th Int. Conf. on Analysis and Optimization of Systems (Antibes 1986), Lect. N. in Control and Inform. Sci. 83*, Springer, (1986).

PARTICLE APPROXIMATION FOR FIRST ORDER STOCHASTIC PARTIAL DIFFERENTIAL EQUATIONS*

Patrick FLORCHINGER†
Université de Metz
Département de Mathématiques
URA CNRS 399
Ile du Saulcy
F–57045 METZ Cédex

François LE GLAND
INRIA Sophia–Antipolis
Route des Lucioles
F–06565 VALBONNE Cédex

Abstract

A class of degenerate second order stochastic PDE is considered, for which a representation result in terms of stochastic characteristics has been proved by Krylov–Rozovskii [2] and Kunita [3,4]. An example of a stochastic PDE in this class has been exhibited in Florchinger–LeGland [1] as the result of a Trotter–like product formula for the Zakai equation of diffusion processes observed in correlated noise. Particle approximations are introduced for this class of stochastic PDE, and error estimates are provided which extend the results of Raviart [6] on first order deterministic PDE.

1 Introduction

Consider the following stochastic differential equation

$$dX_t = b(X_t)\,dt + \sigma(X_t)\,[dW_t - e(X_t)\,dt]\ , \tag{1.1}$$

where $\{W_t\,,\,t \geq 0\}$ is a d–dimensional standard Wiener process, and the associated stochastic flow of diffeomorphisms $\{\xi_{s,t}(\cdot)\,,\,0 \leq s \leq t\}$, and define

$$\Xi_{0,t}(x) \triangleq \exp\left\{\int_0^t e^*(\xi_{0,s}(x))\,dW_s - \tfrac{1}{2}\int_0^t |e(\xi_{0,s}(x))|^2\,ds + \int_0^t c(\xi_{0,s}(x))\,ds\right\}\ .$$

*Research partially supported by USACCE under Contract DAJA45–90–C–0008.
†also : INRIA Lorraine, CESCOM, Technopole de Metz 2000, 4 rue Marconi, F–57070 METZ.

Introduce the following partial differential operators

$$L \triangleq \tfrac{1}{2} \sum_{i,j=1}^{m} a^{i,j} \frac{\partial^2}{\partial x_i \partial x_j} + \sum_{i=1}^{m} b^i \frac{\partial}{\partial x_i} + c\,,$$

$$B_k \triangleq e_k + \sum_{i=1}^{m} \sigma_k^i \frac{\partial}{\partial x_i}\,, \qquad 1 \leq k \leq d\,,$$

with $a = \sigma\,\sigma^*$, and the stochastic PDE

$$dq_t = L^* q_t\, dt + \sum_{k=1}^{d} B_k^* q_t\, dW_t^k\,. \tag{1.2}$$

Because of the relation $a = \sigma\,\sigma^*$ between coefficients of higher order partial derivatives in operators L and B_k, equation (1.2) is a degenerate second order stochastic PDE or equivalently, after transformation into Stratonovich form, a first order stochastic PDE. Existence and representation results have been obtained by Kunita [4] for (generally nonlinear) first order stochastic PDE, based on the notion of stochastic characteristics.

In a previous work [1], the Zakai equation for the nonlinear filtering of diffusion processes observed in correlated noise has been considered. A decomposition of the Zakai equation has been introduced, exhibiting a degenerate second order stochastic PDE similar to (1.2) in the *correction step*. In addition, a time discretization scheme has been proposed for this degenerate second order stochastic PDE, with rate of convergence of order $\sqrt{\delta}$, where δ is the time step.

The purpose of this paper is to provide a discretization scheme of the degenerate second order stochastic PDE (1.2) with respect to the space variable $x \in \mathbf{R}^m$. This approximation relies on the representation of the solution in terms of stochastic characteristics, and approximation of the initial condition by a convex linear combination of Dirac masses. This kind of aproximation is called a ***particle approximation***, see Raviart [6].

More specifically, for any probability measure $\mu(dx)$ on $\mathbf{R}^m$, define the transformed measure $Q_t\,\mu(dx)$ by

$$\langle Q_t\,\mu, \phi\rangle = \int \phi(\xi_{0,t}(x))\, \Xi_{0,t}(x)\, \mu(dx)\,, \tag{1.3}$$

for any test function ϕ, or equivalently

$$Q_t\,\mu(A) = \int_{\xi_{0,t}^{-1}(A)} \Xi_{0,t}(x)\, \mu(dx)\,.$$

Note that, if ϕ is regular enough, then the Itô formula gives

$$d[\,\phi(\xi_{0,t}(x))\, \Xi_{0,t}(x)\,] = L\phi(\xi_{0,t}(x)) \cdot \Xi_{0,t}(x)\, dt + \sum_{k=1}^{d} B_k \phi(\xi_{0,t}(x)) \cdot \Xi_{0,t}(x)\, dW_t^k\,.$$

Therefore $\mu_t(dx) = Q_t\,\mu(dx)$ solves equation (1.2) in weak form, i.e.

$$d\mu_t = L^* \mu_t\, dt + \sum_{k=1}^{d} B_k^* \mu_t\, dW_t^k\,, \qquad \mu_0 = \mu\,. \tag{1.4}$$

Consider next the following two different assumptions on the original measure $\mu_0(dx)$:

□ Assume that the original measure $\mu(dx)$ has a density $q(x)$ with respect to the Lebesgue measure on $\mathbf{R}^m$, i.e. $\mu(dx) = q(x)\,dx$. Then, the transformed measure $Q_t\,\mu(dx)$ has itself a density $q_t(x)$ which satisfies

$$q_t(\xi_{0,t}(x)) \cdot J_{0,t}(x) = \Xi_{0,t}(x) \cdot q(x) \, ,$$

or in integrated form

$$\int_A q_t(x)\,dx = \int_{\xi_{0,t}^{-1}(A)} \Xi_{0,t}(x) \cdot q(x)\,dx \, .$$

Here, $J_{0,t}(\cdot)$ is the Jacobian (i.e. the determinant of the Jacobian matrix) of the stochastic flow $\xi_{0,t}(\cdot)$. In addition, the density $q_t(x)$ solves the degenerate second order stochastic PDE

$$dq_t = L^* q_t\,dt + \sum_{k=1}^{d} B_k^* q_t\,dW_t^k \, , \qquad q_0 = q \, . \tag{1.5}$$

□ Assume that the original measure $\mu(dx)$ is a convex linear combination of Dirac masses, also called *particles*

$$\mu(dx) = \sum_{i \in I} a^i\,\delta_{x^i}(dx) \, ,$$

where $\{a^i\,,\,i \in I\}$ are the particle weights, and $\{x^i\,,\,i \in I\}$ are the particle locations. Then, the transformed measure $Q_t\,\mu(dx)$ has a similar representation

$$Q_t \mu(dx) = \sum_{i \in I} a_t^i\,\delta_{x_t^i}(dx) \, ,$$

where the particles have been transported by the flow i.e. $x_t^i = \xi_{0,t}(x^i)$, and the weights have been updated according to $a_t^i = a^i\,\Xi_{0,t}(x^i)$.

The idea behind particle approximation for equation (1.2) is the following :

- given an initial condition $\mu_0(dx)$ with density $q_0(x)$, find an approximation $\mu_0^h(dx)$ in terms of a linear convex combination of Dirac masses,
- use the exact solution of equation (1.4) with the approximation $\mu_0^h(dx)$ as initial condition, as an approximation for the solution of the original equation (1.5), and get an error estimate if possible.

This can be illustrated by the following diagram

$$\begin{array}{ccc}
q_0(x)\,dx = \mu_0(dx) & \longrightarrow & \mu_0^h(dx) \\
\Big\downarrow Q_t & & \Big\downarrow Q_t \\
q_t(x)\,dx = \mu_t(dx) & \longrightarrow & \mu_t^h(dx)
\end{array}$$

The remainder of this section is devoted to recalling standard results concerning stochastic flows of diffeomorphisms and stochastic PDE.

Proposition 1.1 *Let $n \geq 0$ be fixed. Assume that*

- *b, σ and e have bounded derivatives up to order $(n+1)$,*
- *c has bounded derivatives up to order n.*

Then $\xi_{s,t}(\cdot)$ is a C^n-diffeomorphism in $\mathbf{R}^m$. In addition, the following estimates hold for all $p \geq 1$

$$\sup_{x \in \mathbf{R}^m} \mathbf{E}\left[|D^\alpha \xi_{s,t}(x)|^p\right] < \infty , \qquad 1 \leq |\alpha| \leq n ,$$

$$\sup_{x \in \mathbf{R}^m} \mathbf{E}\left[|D^\alpha \Xi_{s,t}(x)|^p\right] < \infty , \qquad 0 \leq |\alpha| \leq n .$$

Restricting to compact sets of $\mathbf{R}^m$, it is possible to invert the supremum and the mathematical expectation in the estimates above, see the Corollary 4.6.7 of Kunita [5]

Proposition 1.2 *Under the assumptions of the Proposition 1.1, there exists a constant $C > 0$, such that for any compact set $B \subset \mathbf{R}^m$ and $\varepsilon > 0$ the following uniform estimates hold for all $p \geq 1$*

$$\mathbf{E}\left[\sup_{x \in B} |D^\alpha \xi_{s,t}(x)|^p\right] \leq C\ [1 + \delta^{p-\varepsilon}] , \qquad 1 \leq |\alpha| \leq n ,$$

$$\mathbf{E}\left[\sup_{x \in B} |D^\alpha \Xi_{s,t}(x)|^p\right] \leq C\ [1 + \delta^{p-\varepsilon}] , \qquad 0 \leq |\alpha| \leq n ,$$

where $\delta = \delta(B)$ denotes the diameter of B.

For all $n \geq 0$, $p \geq 1$, let $W^{n,p} \equiv W^{n,p}(\mathbf{R}^m)$ denote the space of real-valued Lebesgue-measurable functions on $\mathbf{R}^m$ whose generalized derivatives up to order n are integrable in p-mean, and define the corresponding norm $\|\cdot\|_{n,p}$ and semi-norm $|\cdot|_{n,p}$ by

$$\|u\|_{n,p}^p \triangleq \sum_{0 \leq |\alpha| \leq n} \int |D^\alpha u(x)|^p \, dx \qquad \text{and} \qquad |u|_{n,p}^p \triangleq \sum_{|\alpha| = n} \int |D^\alpha u(x)|^p \, dx ,$$

respectively.

Consider the following degenerate second order stochastic PDE

$$dq_t = L^* q_t \, dt + \sum_{k=1}^d B_k^* q_t \, dW_t^k , \qquad q_0 = q . \tag{1.6}$$

Although no coercivity hypothesis is satisfied, the following existence, uniqueness and regularity result is proved in Krylov–Rozovskii [2].

Theorem 1.3 *Let $n \geq 1$ be fixed. Assume that*

- *a has bounded derivatives up to order* $\max(n, 2)$,
- *b, σ, c and e have bounded derivatives up to order n,*
- *the initial condition satisfies $q_0 \in W^{n,p}$.*

Then equation (1.6) has a unique solution $q \in M^p(0,T\,; W^{n,p})$. In addition

$$q \in L^p(\Omega\,; C_w([0,T]\,; W^{n,p})),$$

and the following estimate holds

$$\mathbf{E}[\sup_{0 \leq t \leq T} \|q_t\|_{n,p}^p] \leq \|q_0\|_{n,p}^p \, e^{CT} .$$

2 Quadrature–based particle approximation

With the quadrature formula (A.1)

$$\int g(x)\,dx \sim \sum_{i \in I} \omega^i \, g(x^i) ,$$

is associated the following particle approximation for the initial density $q_0(x)$

$$q_0(x)\,dx = \mu_0(dx) \sim \mu_0^h(dx) = \sum_{i \in I} \omega^i \, q_0(x^i) \, \delta_{x^i}(dx) . \tag{2.1}$$

This induces the following particle approximation for the solution $q_t(x)$ of equation (1.6)

$$q_t(x)\,dx = \mu_t(dx) \sim \mu_t^h(dx) = \sum_{i \in I} \omega^i \, \Xi_{0,t}(x^i)\, q_0(x^i)\, \delta_{\xi_{0,t}(x^i)}(dx) .$$

The following error estimate holds in Sobolev space with negative exponent, which extends the result of Raviart to the case of first order stochastic PDE.

Theorem 2.1 *Let $n \geq m$ be fixed. Assume that*

- *b, σ, c and e have bounded derivatives up to order $(n+1)$,*
- *the initial condition satisfies $q_0 \in W^{n,p}$.*

Then there exists a constant $C > 0$ independent of h, such that

$$\mathbf{E}\|\mu_t - \mu_t^h\|_{-n,p} \leq C\, h^n \, \|q_0\|_{n,p} .$$

PROOF. Let $\phi \in W^{n,p'}$ be an arbitrary test function. Since

$$\langle \mu_t, \phi \rangle = \int \phi(\xi_{0,t}(x))\, \Xi_{0,t}(x)\, q_0(x)\, dx , \qquad \langle \mu_t^h, \phi \rangle = \sum_{i \in I} \omega^i \, \phi(\xi_{0,t}(x^i))\, \Xi_{0,t}(x^i)\, q_0(x^i) ,$$

it follows from Theorem A.2 that

$$|\langle \mu_t, \phi \rangle - \langle \mu_t^h, \phi \rangle| \leq C\, h^n\, |g|_{n,1} \ ,$$

with $g = \phi \circ \xi_{0,t} \cdot \Xi_{0,t}\, q_0$, provided $g \in W^{n,1}$, $n \geq m$.
Under the assumptions on the coefficients, $\phi \circ \xi_{0,t} \in W^{n,p'}$ and $\Xi_{0,t} \cdot q_0 \in W^{n,p}$, for conjugate p and p'. Moreover, the generalized Leibniz formula yields

$$|g|_{n,1} \leq \sum_{(\alpha,\beta) \in I_n} \int |\chi_{\alpha,\beta}(x)\; D^\alpha \phi(\xi_{0,t}(x))\; D^\beta q_0(x)|\, dx \ ,$$

where I_n denotes the set of pairs (α, β) of multi-indices such that $|\alpha| + |\beta| \leq n$, and $\chi_{\alpha,\beta}(\cdot)$ are random fields involving the derivatives of $\xi_{0,t}(\cdot)$ and $\Xi_{0,t}(\cdot)$ up to order n. Using back and forth the changes of variable induced by the differomorphisms $\xi_{0,t}(\cdot)$ and $\xi_{0,t}^{-1}(\cdot)$, and the Hölder inequality, gives

$$\begin{aligned} |g|_{n,1} &\leq \sum_{(\alpha,\beta) \in I_n} \int |\chi_{\alpha,\beta}(\xi_{0,t}^{-1}(x))\; D^\alpha \phi(x)\; D^\beta q_0(\xi_{0,t}^{-1}(x))|\; [J_{0,t}(\xi_{0,t}^{-1}(x))]^{-1}\, dx \\ &\leq \sum_{(\alpha,\beta) \in I_n} \left\{ \int |D^\alpha \phi(x)|^{p'}\, dx \right\}^{1/p'} \left\{ \int |\chi_{\alpha,\beta}(\xi_{0,t}^{-1}(x))\; D^\beta q_0(\xi_{0,t}^{-1}(x))|^p \right. \\ &\qquad\qquad \left. [J_{0,t}(\xi_{0,t}^{-1}(x))]^{-p}\, dx \right\}^{1/p} \\ &\leq \|\phi\|_{n,p'} \sum_{(\alpha,\beta) \in I_n} \left\{ \int |\chi_{\alpha,\beta}(x)\; D^\beta q_0(x)|^p\; [J_{0,t}(x)]^{-(p-1)}\, dx \right\}^{1/p} . \end{aligned}$$

Therefore

$$\frac{|\langle \mu_t, \phi \rangle - \langle \mu_t^h, \phi \rangle|}{\|\phi\|_{n,p'}} \leq C\, h^n \sum_{(\alpha,\beta) \in I_n} \left\{ \int |\chi_{\alpha,\beta}(x)\; D^\beta q_0(x)|^p\; [J_{0,t}(x)]^{-(p-1)}\, dx \right\}^{1/p} ,$$

and

$$\begin{aligned} \mathbf{E}\|\mu_t - \mu_t^h\|_{-n,p} \leq C\, h^n \sum_{(\alpha,\beta) \in I_n} &\left\{ \int \mathbf{E}\left\{ |\chi_{\alpha,\beta}(x)|^p\; [J_{0,t}(x)]^{-(p-1)} \right\} \right. \\ &\qquad \left. |D^\beta q_0(x)|^p\, dx \right\}^{1/p} . \end{aligned}$$

From estimates in Proposition 1.1, it holds

$$\sup_{x \in \mathbf{R}^m} \mathbf{E}\left\{ |\chi_{\alpha,\beta}(x)|^p\; [J_{0,t}(x)]^{-(p-1)} \right\} < \infty \ ,$$

so that

$$\mathbf{E}\|\mu_t - \mu_t^h\|_{-n,p} \leq C\, h^n\, \|q_0\|_{n,p} \ . \qquad \square$$

Regularization

Let $\zeta(x)$ be a continuous cut–off function defined on $\mathbf{R}^m$, which satisfies

$$(i) \qquad \int \zeta(x)\, dx = 1 \ ,$$

$$(ii) \qquad \int x^\alpha\, \zeta(x)\, dx = 0 \ , \qquad 1 \leq |\alpha| \leq k-1 \ ,$$

$$(iii) \qquad \int |x|^k\, |\zeta(x)|\, dx < \infty \ ,$$

for some $k \geq 2$. For any $\varepsilon > 0$, $\zeta_\varepsilon(x)$ is defined by the following scaling

$$\zeta_\varepsilon(x) \triangleq \frac{1}{\varepsilon^m} \zeta(\frac{x}{\varepsilon}) .$$

With the particle approximation

$$\mu_t^h(dx) = \sum_{i \in I} \omega^i \, \Xi_{0,t}(x^i) \, q_0(x^i) \, \delta_{x_t^i}(dx) ,$$

is associated the regularized measure

$$\mu_t^{h,\varepsilon}(dx) = \mu_t^h * \zeta_\varepsilon(dx) = q_t^{h,\varepsilon}(x) \, dx ,$$

where the density $q_t^{h,\varepsilon}(x)$ is given by

$$q_t^{h,\varepsilon}(x) = \sum_{i \in I} \omega^i \, \Xi_{0,t}(x^i) \, q_0(x^i) \, \zeta_\varepsilon(x - x_t^i) .$$

The main result of this section is the following theorem, which is an extension of the Theorem 4.2 in [6], to the case of first order stochastic PDE.

Theorem 2.2 *Let $n > m$ be fixed. Assume that*

- *the cut-off function ζ satisfies (i)-(iii) for some $k \geq 2$, and $\zeta \in W^{n,1}$,*
- *b, σ, c and e have bounded derivatives up to order $(\ell + 1)$,*
- *the initial condition satisfies $q_0 \in W^{\ell,p}$,*

where $\ell = \max(k, n)$.
Then, there exists a constant C independent of both h and ε, such that

$$\left\{ \mathbf{E} \| q_t - q_t^{h,\varepsilon} \|_{0,p}^p \right\}^{1/p} \leq C \, \left\{ \varepsilon^k \, \|q_0\|_{k,p} + (h/\varepsilon)^n \, \|q_0\|_{n,p} \right\} .$$

PROOF. Obviously

$$q_t - q_t^{h,\varepsilon} = [q_t - q_t * \zeta_\varepsilon] + [q_t * \zeta_\varepsilon - q_t^{h,\varepsilon}] .$$

First, it follows from Lemma 4.4 in [6] that

$$\|q_t - q_t * \zeta_\varepsilon\|_{0,p} \leq C \, \varepsilon^k \, |q_t|_{k,p}$$

provided $q_t \in W^{k,p}$. Under the assumptions, Theorem 1.3 gives

$$\left\{ \mathbf{E} \|q_t - q_t * \zeta_\varepsilon\|_{0,p}^p \right\}^{1/p} \leq C \, \varepsilon^k \, \left\{ \mathbf{E} |q_t|_{k,p}^p \right\}^{1/p} \leq C \, \varepsilon^k \, \|q_0\|_{k,p} .$$

On the other hand, using the change of variable induced by the diffeomorphism $\xi_{0,t}^{-1}(\cdot)$, it holds for all $x \in \mathbf{R}^m$

$$q_t * \zeta_\varepsilon(x) - q_t^{h,\varepsilon}(x) = \int \Xi_{0,t}(z) \, q_0(z) \, \zeta_\varepsilon(x - \xi_{0,t}(z)) \, dz$$

$$- \sum_{i \in I} \omega^i \, \Xi_{0,t}(x^i) \, q_0(x^i) \, \zeta_\varepsilon(x - \xi_{0,t}(x^i)) = E(g(x, \cdot))$$

with $g(x,\cdot) = \Xi_{0,t}\, q_0 \cdot \zeta_\varepsilon(x - \xi_{0,t})$. Therefore, it follows from Theorem A.1 that for all $x \in \mathbf{R}^m$

$$|q_t * \zeta_\varepsilon(x) - q_t^{h,\varepsilon}(x)| \leq C\, h^n\, |g(x,\cdot)|_{n,1}$$

provided $g(x,\cdot) \in W^{n,1}$, $n \geq m$. Moreover, the generalized Leibniz formula yields

$$|g(x,\cdot)|_{n,1} \leq \sum_{(\alpha,\beta)\in I_n} \int |\chi'_{\alpha,\beta}(z)\, D^\beta q_0(z) D^\alpha \zeta_\varepsilon(x - \xi_{0,t}(z))|\, dz\ ,$$

where I_n denotes the set of pairs (α,β) of multi-indices such that $|\alpha| + |\beta| \leq n$, and $\chi'_{\alpha,\beta}(\cdot)$ are random fields involving the derivatives of $\xi_{0,t}(\cdot)$ and $\Xi_{0,t}(\cdot)$ up to order n. From the technical lemma below, it follows that

$$\int |g(x,\cdot)|_{n,1}^p\, dx \leq C \sum_{(\alpha,\beta)\in I_n} \left\{\int |D^\alpha \zeta_\varepsilon(x)|\, dx\right\}^p \left\{\int |\chi'_{\alpha\beta}(x)\, D^\beta q_0(x)|^p \right.$$
$$\left. [J_{0,t}(x)]^{-(p-1)}\, dx\right\}\ .$$

Making use of

$$D^\alpha \zeta_\varepsilon(x) = \frac{1}{\varepsilon^{m+|\alpha|}}\, D^\alpha \zeta(\frac{x}{\varepsilon})\ ,$$

taking mathematical expectation on both sides, and raising to the power $1/p$ gives

$$\left\{\mathbf{E}\int |g(x,\cdot)|_{n,1}^p\, dx\right\}^{1/p} \leq C\, \frac{1}{\varepsilon^n}\, \|\zeta\|_{n,1} \sum_{(\alpha,\beta)\in I_n} \left\{\int \mathbf{E}\left\{|\chi'_{\alpha,\beta}(x)|^p\, [J_{0,t}(x)]^{-(p-1)}\right\}\right.$$
$$\left. |D^\beta q_0(x)|^p\, dx\right\}^{1/p}\ .$$

From estimates in Proposition 1.1, it holds

$$\sup_{x\in\mathbf{R}^m} \mathbf{E}\left\{|\chi'_{\alpha,\beta}(x)|^p\, [J_{0,t}(x)]^{-(p-1)}\right\} < \infty\ .$$

Therefore

$$\left\{\mathbf{E}\|q_t * \zeta_\varepsilon - q_t^{h,\varepsilon}\|_{0,p}^p\right\}^{1/p} \leq C\, h^n \left\{\mathbf{E}\int |g(x,\cdot)|_{n,1}^p\, dx\right\}^{1/p}$$
$$\leq C\, (h/\varepsilon)^n\, \|\zeta\|_{n,1}\, \|q_0\|_{n,p}\ . \qquad \square$$

Lemma 2.3 *Let $f \in L^p$ and $g \in L^1$, and define*

$$I(x) = \int f(z)\, g(x - \xi_{0,t}(z))\, dz\ .$$

Then $I \in L^p$ and in addition

$$\left\{\int |I(x)|^p\, dx\right\}^{1/p} \leq \left\{\int |f(x)|^p\, [J_{0,t}(x)]^{-(p-1)}\, dx\right\}^{1/p} \int |g(x)|\, dx\ .$$

PROOF. Using back and forth the changes of variable induced by the differomorphisms $\xi_{0,t}(\cdot)$ and $\xi_{0,t}^{-1}(\cdot)$, and the Lemma 4.3 in [6], gives

$$I(x) = \int f(\xi_{0,t}^{-1}(z)) \, [J_{0,t}(\xi_{0,t}^{-1}(z))]^{-1} \, g(x-z) \, dz \ ,$$

and

$$\begin{aligned}\left\{\int |I(x)|^p \, dx\right\}^{1/p} &\leq \left\{\int |f(\xi_{0,t}^{-1}(x))|^p \, [J_{0,t}(\xi_{0,t}^{-1}(x))]^{-p} \, dx\right\}^{1/p} \int |g(x)| \, dx \\ &\leq \left\{\int |f(x)|^p \, [J_{0,t}(x)]^{-(p-1)} \, dx\right\}^{1/p} \int |g(x)| \, dx \ . \qquad \Box\end{aligned}$$

3 Adapted particle approximation

Consider the particle approximation (A.3) for the initial condition $\mu_0(dx)$

$$\mu_0(dx) \sim \mu_0^h(dx) = \sum_{i \in I} a^i \, \delta_{x^i}(dx) \ ,$$

where the particle weights $\{a^i \, , \, i \in I\}$ and the particle locations $\{x^i \, , \, i \in I\}$ are defined in the following way

$$a^i \triangleq \mu_0(B^i) = \int_{B^i} \mu_0(dx) \ , \qquad x^i \triangleq \frac{1}{a^i} \int_{B^i} x \, \mu_0(dx) \ ,$$

depending on the measure $\mu_0(dx)$. This induces the following particle approximation for the solution $\mu_t(x)$ of equation (1.4)

$$\mu_t(dx) \sim \mu_t^h(dx) = \sum_{i \in I} a^i \, \Xi_{0,t}(x^i) \, \delta_{\xi_{0,t}(x^i)}(dx) \ .$$

Parallel to the Theorem 2.1 above, the following error estimate holds in Sobolev space with negative exponent.

Theorem 3.1 *Assume that*

- *b, σ, c and e have bounded derivatives up to order* 3,
- *for all $i \in I$, the set $B^i \subset \mathbf{R}^m$ is compact.*

Then there exists a constant $C > 0$, such that

$$\mathbf{E}\|\mu_t - \mu_t^h\|_{-2,1} \leq C \sum_{i \in I} \delta_i^2 \, a^i \ ,$$

where $a^i = \mu_0(B^i)$ and $\delta_i = \delta(B^i)$ denotes the diameter *of the set B^i.*

PROOF. Let $\phi \in W^{2,\infty}$ be an arbitrary test function. Since

$$\langle \mu_t, \phi \rangle = \int \phi(\xi_{0,t}(x)) \, \Xi_{0,t}(x) \, \mu_0(dx) \ , \qquad \langle \mu_t^h, \phi \rangle = \sum_{i \in I} a^i \, \phi(\xi_{0,t}(x^i)) \, \Xi_{0,t}(x^i) \ ,$$

it follows from estimate (A.6) that

$$|\langle \mu_t, \phi\rangle - \langle \mu_t^h, \phi\rangle| \leq \tfrac{1}{2} \sum_{i\in I} |g|_{2,\infty,\widehat{B}^i}\, \delta_i^2\, a^i ,$$

with $g = \phi \circ \xi_{0,t} \cdot \Xi_{0,t}$, where $\widehat{B}^i$ denotes the *convex hull* of B^i. The generalized Leibniz formula yields

$$\begin{aligned} |g|_{2,\infty,B} &\leq \sum_{|\alpha|\leq 2} \sup_{x\in B} |\chi_\alpha(x)\; D^\alpha \phi(\xi_{0,t}(x))| \\ &\leq \sum_{|\alpha|\leq 2} \left[\sup_{x\in B} |\chi_\alpha(x)|\right] \left[\sup_{x\in \mathbb{R}^m} |D^\alpha \phi(x)|\right] \\ &\leq \|\phi\|_{2,\infty} \sum_{|\alpha|\leq 2} \sup_{x\in B} |\chi_\alpha(x)| , \end{aligned}$$

where $\chi_\alpha(\cdot)$ are random fields involving the derivatives of $\xi_{0,t}(\cdot)$ and $\Xi_{0,t}(\cdot)$ up to order 2. Therefore

$$\frac{|\langle \mu_t, \phi\rangle - \langle \mu_t^h, \phi\rangle|}{\|\phi\|_{2,\infty}} \leq \tfrac{1}{2} \sum_{i\in I} \sum_{|\alpha|\leq 2} \sup_{x\in \widehat{B}^i} |\chi_\alpha(x)|\, \delta_i^2\, a^i ,$$

and

$$\mathbf{E}\|\mu_t - \mu_t^h\|_{-2,1} \leq \tfrac{1}{2} \sum_{i\in I} \sum_{|\alpha|\leq 2} \mathbf{E}\left[\sup_{x\in \widehat{B}^i} |\chi_\alpha(x)|\right] \delta_i^2\, a^i .$$

From estimates in Proposition 1.2, it holds

$$\mathbf{E}\left[\sup_{x\in \widehat{B}^i} |\chi_\alpha(x)|\right] \leq C \left[1 + \delta_i^{2-\epsilon}\right] ,$$

for some p, where $\delta_i = \delta(B^i)$ denotes the diameter of both B^i and its convex hull $\widehat{B}^i$, so that

$$\mathbf{E}\|\mu_t - \mu_t^h\|_{-2,1} \leq C \sum_{i\in I} \left[1 + \delta_i^{2-\epsilon}\right] \delta_i^2\, a^i . \qquad \square$$

References

[1] P. FLORCHINGER and F. LE GLAND, Time–discretization of the Zakai equation for diffusion processes observed in correlated noise, *Stochastics and Stochastics Reports* **35** (4) 233–256 (1991).

[2] N.V. KRYLOV and B.L. ROZOVSKII, Characteristics of degenerating second–order parabolic Itô equations, *J.Soviet Math.* **32** (4) 336–348 (1982).

[3] H. KUNITA, Stochastic partial differential equations connected with nonlinear filtering, in: *Nonlinear Filtering and Stochastic Control (Cortona–1981)* (eds. S.K.Mitter and A.Moro) 100–169, Springer–Verlag (LNM–972) (1982).

[4] H. KUNITA, First order partial differential equations, in: *Stochastic Analysis (Katata and Kyoto–1982)* (ed. K.Itô) 249–269, North–Holland (1984).

[5] H. KUNITA, *Stochastic Flows and Stochastic Differential Equations*, Cambridge University Press (1990).

[6] P.A. RAVIART, An analysis of particle methods, in: *Numerical Methods in Fluid Dynamics (Como–1983)* (ed. F.Brezzi) 243–324, Springer-Verlag (LNM–1127) (1985).

A Particle approximation of functions

Consider the following quadrature formula on $\mathbb{R}^m$

$$\int g(x)\,dx \sim \sum_{i\in I} \omega^i\, g(x^i)\,, \tag{A.1}$$

where $\{x^i\,,\, i \in I\}$ is a coordinate grid of size $h > 0$, $I = \mathbb{Z}^m$ and $\omega^i = h^m$ is the Lebesgue measure of the m–dimensional cube B^i with center x^i and edge size h. For all $g \in C(\mathbb{R}^m)$, the quadrature error associated with the quadrature formula (A.1) is defined by

$$E_i(g) \triangleq \int_{B^i} g(x)\,dx - \omega^i\, g(x^i)\,, \qquad E(g) \triangleq \sum_{i\in I} E_i(g)\,.$$

The following estimate is proved in Raviart [6]

Theorem A.1 *There is a constant $C > 0$ independent of h such that*

$$|E(g)| \leq C\, h^n\, |g|_{n,1}\,,$$

for all $g \in W^{n,1}$, $n \geq m$.

Let $\mu(dx)$ be a probability measure on $\mathbb{R}^m$ having a continuous density $q(x)$ with respect to the Lebesgue measure, i.e. $\mu(dx) = q(x)\,dx$. With the quadrature formula (A.1) is associated the following particle approximation for the density $q(x)$

$$q(x)\,dx = \mu(dx) \sim \mu^h(dx) = \sum_{i\in I} \omega^i\, q(x^i)\, \delta_{x^i}(dx)\,, \tag{A.2}$$

so that, for any test function ϕ

$$\langle \mu, \phi\rangle = \int \phi(x)\, q(x)\,dx\,, \qquad \langle \mu^h, \phi\rangle = \sum_{i\in I} \omega^i\, \phi(x^i)\, q(x^i)\,.$$

The following result is proved in Raviart [6]

Theorem A.2 *There is a constant $C > 0$ independent of h such that*

$$\|\mu - \mu^h\|_{-n,p} \leq C\, h^n\, \|q\|_{n,p}\,,$$

for all $q \in W^{n,p}$, $n \geq m$.

PROOF. From Theorem A.1, it holds

$$|\langle\mu,\phi\rangle - \langle\mu^h,\phi\rangle| = |E(g)| \leq C\, h^n\, |g|_{n,1} ,$$

with $g = \phi\cdot q$, provided $g \in W^{n,1}$, $n \geq m$. The generalized Leibniz formula and the Hölder inequality yield

$$|g|_{n,1} \leq C\, \|\phi\|_{n,p'}\, \|q\|_{n,p} ,$$

for conjugate p and p', and therefore

$$\|\mu - \mu^h\|_{-n,p} = \sup_{\phi\in W^{n,p'}} \frac{|\langle\mu,\phi\rangle - \langle\mu^h,\phi\rangle|}{\|\phi\|_{n,p'}} \leq C\, h^n\, \|q\|_{n,p} . \qquad \square$$

Another possible approximation is to consider a partition $\{B^i\, , i \in I\}$ of $\mathbf{R}^m$, and to define the following particle approximation for the probability measure $\mu(dx)$

$$\mu(dx) \sim \mu^h(dx) = \sum_{i\in I} a^i\, \delta_{x^i}(dx) , \tag{A.3}$$

where the particle weights $\{a^i\, , i \in I\}$ and the particle locations $\{x^i\, , i \in I\}$ are defined in the following way

$$a^i \triangleq \mu(B^i) = \int_{B^i} \mu(dx) , \qquad x^i \triangleq \frac{1}{a^i} \int_{B^i} x\, \mu(dx) , \tag{A.4}$$

depending on the measure $\mu(dx)$ so that, for any test function ϕ

$$\langle\mu,\phi\rangle = \int \phi(x)\, \mu(dx) , \qquad \langle\mu^h,\phi\rangle = \sum_{i\in I} a^i\, \phi(x^i) .$$

For all $\phi \in C(\mathbf{R}^m)$, the quadrature error associated with the formula (A.3), is defined by

$$E'_i(\phi) \triangleq \int_{B^i} \phi(x)\, \mu(dx) - a^i\, \phi(x^i) , \qquad E'(\phi) \triangleq \sum_{i\in I} E'_i(\phi) .$$

Parallel to the Theorem A.2 above, the following result holds

Theorem A.3 *For any partition* $\{B^i\, , i \in I\}$

$$\|\mu - \mu^h\|_{-2,1} \leq \tfrac{1}{2} \sum_{i\in I} \delta_i^2\, a^i , \tag{A.5}$$

where $a^i = \mu(B^i)$ *and* $\delta_i = \delta(B^i)$ *denotes the diameter of the set* B^i.

PROOF. Let $\phi \in W^{2,\infty}$ be an arbitrary test-function. Using Taylor expansion around the point $x = x^i$ yields

$$\begin{aligned}\phi(x) &= \phi(x^i) + (x - x^i)^*\, D\phi(x^i) \\ &\quad + (x - x^i)^* \left\{\int_0^1 (1-u)\, D^2\phi[ux + (1-u)x^i]\, du\right\} (x - x^i) ,\end{aligned}$$

and the definition (A.4) gives

$$E_i'(\phi) = \int_{B^i} (x - x^i)^* \left\{ \int_0^1 (1-u)\, D^2\phi[ux + (1-u)x^i]\, du \right\} (x - x^i)\, dx \,.$$

Therefore

$$|E_i'(\phi)| \leq \tfrac{1}{2}\, |\phi|_{2,\infty,\widehat{B}^i} \int_{B^i} \|x - x^i\|^2 \, \mu(dx) \leq \tfrac{1}{2}\, |\phi|_{2,\infty,\widehat{B}^i}\, \delta_i^2\, a^i \,,$$

where $\widehat{B}^i$ denotes the *convex hull* of B^i. Then

$$|\langle \mu, \phi \rangle - \langle \mu^h, \phi \rangle| = |E'(\phi)| \leq \tfrac{1}{2} \sum_{i \in I} |\phi|_{2,\infty,\widehat{B}^i}\, \delta_i^2\, a^i \,, \tag{A.6}$$

and

$$\|\mu - \mu^h\|_{-2,1} = \sup_{\phi \in W^{2,\infty}} \frac{|\langle \mu, \phi \rangle - \langle \mu^h, \phi \rangle|}{\|\phi\|_{2,\infty}} \leq \tfrac{1}{2} \sum_{i \in I} \delta_i^2\, a^i \,. \qquad \Box$$

Remark A.4 If the partition $\{B_i\,,\, i \in I\}$ is *given*, with $\delta_i \leq C\, h$ for all $i \in I$, then

$$\|\mu - \mu^h\|_{-2,1} \leq C\, h^2 \,.$$

On the other hand, if the partition $\{B_i\,,\, i \in I\}$ has to be chosen so as to make the quadrature error as small as possible, then estimate (A.5) can be used to derive the following criterion

$$\delta_i^2\, a^i = c \qquad \text{for all } i \in I \,.$$

This criterion based on *equidistribution* of the local quadrature error, has the following interesting property

- a set with a large mass, will be split into some smaller subsets,
- conversely, neighbouring sets with small masses, will be packed together into one single set.

An Infinite-dimensional LP Solution to Control of a Continuous, Monotone Process

Arthur C. Heinricher*
Department of Mathematics
University of Kentucky
Lexington, Kentucky 40506-0027

Richard H. Stockbridge*
Department of Statistics
University of Kentucky
Lexington, Kentucky 40506-0027

1 Introduction.

Consider the following optimal control problem: A process increases randomly in time until it reaches a specified level at which time it is restarted at zero and the cycle is repeated. A reward is earned while the process is running, but a cost is incurred when the process is restarted at zero. The controller is faced with the following conflict: A high drift rate improves the reward but also shortens the time before the restart cost is incurred. The performance of a strategy is measured by the long-term average reward.

Motivation for this work comes from optimal control and replacement problems for systems subject to random deterioration and failure. This has been an active area in applied stochastic analysis for many years. A short list of references includes [1, 2, 4, 15], and the recent survey article [16].

In [7], we use a dynamic programming approach to solve the optimal control problem when the objective is to maximize the reward accumulated up to the first failure; there is no option to replace and no replacement cost. The replacement cost is introduced in [8] and the long-term average problem is reduced to a family of "single-cycle" control problems. Once again, the solution is obtained by dynamic programming techniques.

In this paper, we take a different point of view. The state process is characterized as a solution to a controlled version of Kurtz's [9, 10] *constrained martingale problem*. This leads directly to the stationary distributions for the process and an infinite-dimensional linear program for the optimal value. This linear programming approach was introduced by Manne in [11] and has been extended by several authors [5, 6, 12, 14, 17].

We describe the constrained martingale problem in §2 and the linear programming formulation in §3. The invariant measures for the replaced process are identified in §4 and the solution to the LP problem is described in §5. A policy improvement algorithm is exhibited in §6 for computing the solution.

1.1 Motivation for Martingale Problem Model.

The state of the system is modeled by the process

$$y_t = \max\{x_s;\ 0 \le s \le t\}, \tag{1}$$

where $(x_t;\ t \ge 0)$ is a diffusion

$$dx_t = f(u_t)dt + \sigma(u_t)dw_t\,. \tag{2}$$

These equations describe the evolution of the state process *between replacements*. Here, $u = (u_t;\ t \ge 0)$ is a control process and $w = (w_t;\ t \ge 0)$ is a standard Brownian motion.

Take a fixed state $\Delta > 0$ and reinitialize the process at $(x, y) = (0, 0)$ each time the running max reaches level Δ. A more detailed formulation is provided in [8, section 2]; the present paper does not require *a priori* that the control process $(u_t;\ t \ge 0)$ be reinitialized along with the state.

The process $((x_t, y_t);\ t \ge 0)$ without replacement is studied in [7] and [3]. The generator consists of the usual generator of a one-dimensional diffusion (in the first component)

$$A\phi(x, y, u) = \frac{1}{2}\sigma^2(u)\phi_{xx}(x, y) + f(u)\phi_x(x, y) \tag{3}$$

*The authors' research was supported by the National Science Foundation grant DMS-9006674.

and the fact that $(y_t;\, t \geq 0)$ increases only when $y_t = x_t$ is manifest in a boundary condition on the domain of A:

$$\phi_y(z,z) = 0 \quad \text{for all } z\,. \tag{4}$$

The processes (x_t, y_t) with instantaneous replacement to $(0,0)$ when $y_t = \Delta$ can be identified as the solution of a constrained martingale problem. To see this, consider the process $\phi(x_t, y_t)$, for $\phi \in C^2$. By Ito's formula for semimartingales (see [13, Theorem 33]),

$$\begin{aligned}
\phi(x_t,y_t) &= \phi(x_0,y_0) + \int_0^t \phi_x(x_{s-},y_{s-})dx_s + \int_0^t \phi_y(x_{s-},y_{s-})dy_s + \int_0^t \frac{1}{2}\phi_{xx}(x_{s-},y_{s-})d[x,x]_s^c \\
&\quad + \sum_{0\leq s\leq t} \{\phi(x_s,y_s) - \phi(x_{s-},y_{s-}) - \phi_x(x_{s-},y_{s-})\Delta x_s - \phi_y(x_{s-},y_{s-})\Delta y_s\} \\
&= \phi(x_0,y_0) + \int_0^t \left[f(u_s)\phi_x(x_s,y_s) + \frac{1}{2}\sigma(u_s)\phi_{xx}(x_s,y_s)\right] ds \\
&\quad + \int_0^t \phi_y(x_{s-},y_{s-})dy_s^c + \int_0^t \phi_x(x_s,y_s)\sigma(u_s)dw_s \\
&\quad + \sum_{0\leq s\leq t} \{\phi(x_s,y_s) - \phi(x_{s-},y_{s-})\}\,.
\end{aligned} \tag{5}$$

Define the operators A by (3) and B by

$$B\phi(x,y) = \phi(0,0) - \phi(x,y)\,. \tag{6}$$

If we define a counting measure $\lambda(t)$ by

$$\lambda(t) = \sum_{0\leq s\leq t} \delta_{(0,0)}(x_s,y_s)\,,$$

then, using the boundary condition (4), equation (5) says that

$$\phi(x_t,y_t) - \phi(x_0,y_0) - \int_0^t A\phi(x_s,y_s,u_s)ds - \int_0^t B\phi(x_{s-},y_{s-})d\lambda(s) \tag{7}$$

is a martingale for every function $\phi \in C^2$.

With this observation in mind, we define the dynamics of the system as a solution to a controlled version of a constrained martingale problem.

2 Formulation of the Model.

For a space S, $C_c^\infty(S)$ denotes the space of infinitely-differentiable functions on S having compact support, $D_S[0,\infty)$ denotes the space of functions from $[0,\infty)$ into S which are right continuous and have left limits and $\mathcal{P}(S)$ denotes the space of probability measures on S.

Decision Criterion. Let h be continuous on $[0,\Delta] \times U$ and let R denote the restart cost at the origin. The objective is to maximize the long-term average reward given by

$$\liminf_{t\to\infty} \frac{1}{t} E\left[\int_0^t h(Y_s,u_s)ds - \int_0^t R\, d\lambda(s)\right].$$

Dynamics. Let $E_0 = \{(x,y): -\infty \leq x \leq y, 0 < y < \Delta\}$, $E_1\, \{(x,y): -\infty \leq x \leq y, y = 0,\Delta\}$ and $E = E_0 \cup E_1$. Let $\mathcal{D} = \{\phi \in C_c^\infty(E): \phi_y(z,z) = 0,\ 0 \leq z \leq \Delta\}$ and define A and B on $\mathcal{D}$ by (3) and (6), respectively. We assume the controls take values in a compact set $U \subset \mathbb{R}$ and that the coefficients satisfy the following conditions:

(A1) f and σ are continuous functions on U;

(A2) there exists a constant α such that $0 < \alpha \leq f(u)$ $(u \in U)$.

Definition 2.1. *An $E \times U$-valued process (X,Y,u) taking values in $D_{E\times U}[0,\infty)$ is a solution of the (controlled) constrained martingale problem for (A, E_0, B, E_1) if there exist a filtration $\{\mathcal{F}_t\}$ and nondecreasing process λ such that* *(a)* (X, Y, u) *is* $\{\mathcal{F}_t\}$*-progressively measurable,*

(b) λ *is a counting process satisfying* $\int_0^t \chi_{E_1}(X_s, Y_s)d\lambda(s) = \lambda(t)$, *and*

(c) for each $\phi \in \mathcal{D}$, $\phi(X_t, Y_t) - \phi(X_0, Y_0) - \int_0^t A\phi(X_s, Y_s, u_s)ds - \int_0^t B\phi(X_{s-}, Y_{s-})d\lambda(s)$ *is an* $\{\mathcal{F}_t\}$*-martingale.*

Remark 2.1. *Existence of a solution to the constrained martingale problem for* (A, E_0, B, E_1) *is established by [9, Section 2] when the control process is constant,* $u_t \equiv u$, $t \geq 0$. *Existence of a solution with the optimal control is a result of Theorem 3.1.*

3 Linear Programming Formulation

Let (X, Y, u) be a solution of the constrained martingale problem for (A, E_0, B, E_1). Define the occupation measures $\{\mu_t^0\}_{t\geq 0}$ on $E \times U$ by

$$\mu_t^0(\Gamma) = \frac{1}{t}E\left[\int_0^t \chi_\Gamma(X_s, Y_s, u_s)\right] ds$$

for all Borel sets $\Gamma \subset E \times U$. Note that for each t, μ_t^0 is a probability measure. The collection of measures $\{\mu_t^0\}$ is tight since $E \times U$ is compact and therefore $\{\mu_t^0\}$ is relatively compact. Thus there exist weak limits, as $t \to \infty$, which are probability measures on $E \times U$.

Let $\{t_k\}$ be a sequence, $t_k \to \infty$, such that there exists a measure μ^0 with $\mu_{t_k}^0 \Rightarrow \mu^0$ and

$$\frac{1}{t_k}E\left[\int_0^{t_k} h(Y_s, u_s)ds - \int_0^{t_k} R d\lambda(s)\right] \to \liminf_{t\to\infty} \frac{1}{t}E\left[\int_0^t h(Y_s, u_s)ds - \int_0^t R\, d\lambda(s)\right]$$

as $t_k \to \infty$.

Now consider the measures $\{\mu_t^1\}_{t\geq 0}$ on E_1 given by

$$\mu_t^1(H) = \frac{1}{t}E\left[\int_0^t \chi_H(X_{s-}, Y_{s-})d\lambda(s)\right]$$

for Borel sets $H \subset E_1$. Since $\lambda(t)$ is a counting measure of the process at $(0,0)$ and $(X_{s-}, Y_{s-}) = (\Delta, \Delta)$ whenever $\lambda(t)$ increases, μ_t^1 concentrates its mass at (Δ, Δ) for each t. Thus $\mu_{t_k}^1(\Delta, \Delta) = E[\lambda(t_k)]/t_k$ is a sequence of positive real numbers. This sequence has a finite upper bound because the drift $f(u)$ and diffusion $\sigma(u)$ are bounded above, and hence the expected passage time from $(0,0)$ to (Δ, Δ) is bounded away from 0. Therefore, there exists a subsequence of t_k (without loss of generality, the entire sequence) for which $\mu_{t_k}^1$ converges to some finite measure μ^1 placing full mass at (Δ, Δ).

It now follows that

$$\begin{aligned}
&\liminf_{t\to\infty} \frac{1}{t}E\left[\int_0^t h(Y_s, u_s)ds - \int_0^t R\, d\lambda(s)\right] \\
&= \lim_{t_k\to\infty} \frac{1}{t_k}E\left[\int_0^{t_k} h(Y_s, u_s)ds - \int_0^{t_k} R\, d\lambda(s)\right] \\
&= \lim_{t_k\to\infty} \int h(y, u)d\mu_{t_k}^0 - \int R\, d\mu_{t_k}^1 \\
&= \int h(y, u)d\mu^0 - \int R\, d\mu^1.
\end{aligned}$$

In addition, since (X, Y, u) is a solution of the constrained martingale problem for (A, E_0, B, E_1), for each $\phi \in \mathcal{D}$,

$$\begin{aligned}
0 &= \frac{1}{t}E\left[\phi(X_t, Y_t) - \phi(X_0, Y_0) - \int_0^t A\phi(X_s, Y_s, u_s)ds - \int_0^t B\phi(X_{s-}, Y_{s-})d\lambda(s)\right] \\
&= \frac{1}{t}E[\phi(X_t, Y_t) - \phi(X_0, Y_0)] - \int A\phi(x, y, u)d\mu_t^0 - \int B\phi(x, y)d\mu_t^1.
\end{aligned}$$

Evaluating along the subsequence $\{t_k\}$ and letting $k \to \infty$, we obtain

$$\int A\phi(x, y, u)d\mu^0 + \int B\phi(x, y)d\mu^1 = 0 \text{ for each } \phi \in \mathcal{D}. \tag{8}$$

The same argument shows that any weak limits μ^0 and μ^1 (as $t \to \infty$) satisfy (8). Let

$$\widehat{\mathcal{P}} := \left\{(\mu^0,\mu^1) : \mu^0 \in \mathcal{P}(E \times U),\ \mu^1 \text{ is a finite measure on } E^1 \text{ and } (\mu^0,\mu^1) \text{ satisfy (8)}\right\}.$$

The above comments show that one way in which to evaluate the long-term average reward is to compute

$$\int h(y,u)d\mu^0 - \int R\,d\mu^1 \tag{9}$$

for the appropriate weak limits μ^0 and μ^1, with $\mu^0 \in \mathcal{P}(E \times U)$. The following theorem shows that, in fact, any measures $(\mu^0,\mu^1) \in \widehat{\mathcal{P}}$ correspond to a stationary solution of the constrained martingale problem for (A, E_0, B, E_1) and hence the long-term average reward is given by (9). This result, however, requires that we broaden the definition of solutions of the constrained martingale problem to allow *relaxed* controls.

Definition 3.1. *An $E\times\mathcal{P}(U)$-valued process (X,Y,Ψ) taking values in $D_{E\times\mathcal{P}(U)}[0,\infty)$ is a relaxed solution of the (controlled) constrained martingale problem for (A,E_0,B,E_1) if there exist a filtration $\{\mathcal{F}_t\}$ and nondecreasing process λ such that*

(a) (X,Y,Ψ) is $\{\mathcal{F}_t\}$-progressively measurable,

(b) λ is a counting process satisfying $\int_0^t \chi_{E_1}(X_s,Y_s)d\lambda(s) = \lambda(t)$, and

(c) for each $\phi \in \mathcal{D}$,

$$\phi(X_t,Y_t) - \phi(X_0,Y_0) - \int_0^t \int_U A\phi(X_s,Y_s,u)\Psi_s(du)ds - \int_0^t B\phi(X_{s-},Y_{s-})d\lambda(s)$$

is an $\{\mathcal{F}_t\}$-martingale.

Theorem 3.1. *Suppose $\mu^0 \in \mathcal{P}(E \times U)$ and μ^1 is a finite measure on E_1 which satisfy (8). Then there exists a stationary relaxed solution (X,Y,Ψ) of the constrained martingale problem for (A,E_0,B,E_1).*

PROOF: The proof is essentially an application of [10, Theorem 4.1].

Since every pair (μ^0,μ^1) satisfying (8) corresponds to a relaxed solution of the constrained martingale problem for (A,E_0,B,E_1) and the long-term average reward for the solution is given by (9), the long-term average problem is equivalent to an infinite-dimensional linear programming problem:

$$\text{Maximize} \int h(y,u)d\mu^0 - \int R\,d\mu^1 \tag{10}$$

subject to

$$\int A\phi(x,y,u)d\mu^0 + \int B\phi(x,y)d\mu^1 = 0 \quad \text{for each } \phi \in \mathcal{D}$$

and

$$(\mu^0,\mu^1) \in \widehat{\mathcal{P}}.$$

4 Identification of μ^0 and μ^1.

First consider the distribution μ^0 on $E \times U$. Let μ^0_E be the state marginal distribution $\mu^0_E(G) = \mu^0(G \times U)$ for all Borel sets $G \subset E$ and let ν be the regular conditional distribution satisfying

$$\mu^0(\Gamma) = \int_{E\times U} \chi_\Gamma(x,y,u)\nu(x,y,du)\mu^0_E(dx,dy)$$

for all Borel sets $\Gamma \subset E \times U$.

Assume for the moment that μ^0_E has a density $p(x,y)$; that is, μ^0 can be written as

$$\mu^0(\Gamma) = \int_{E\times U} \chi_\Gamma(x,y,u)\nu(x,y,du)p(x,y)\,dx\,dy. \tag{11}$$

Let $\bar{f}(x,y) = \int_U f(u)\,\nu(x,y,du)$ and $\bar{\sigma}^2(x,y) = \int_U \sigma^2(u)\nu(x,y,du)$. Using $\bar{f}$ and $\bar{\sigma}^2$ and formally integrating by parts, we have

$$\begin{aligned}\int A\phi\, d\mu^0 &= \int_0^\Delta \int_{-\infty}^y \int_U \left[f(u)\phi_x(x,y) + \frac{1}{2}\sigma^2(u)\phi_{xx}(x,y)\right] \nu(x,y,du)p(x,y)\,dx\,dy \\ &= \int_0^\Delta \int_{-\infty}^y \left[\bar{f}(x,y)p(x,y) - \frac{1}{2}\frac{\partial}{\partial x}\left(\bar{\sigma}^2(x,y)p(x,y)\right)\right] \phi_x(x,y)\,dx\,dy \\ &\quad + \int_0^\Delta \frac{1}{2}\bar{\sigma}^2(y,y)p(y,y)\phi_x(y,y)\,dy.\end{aligned}$$

Using the boundary condition (4), we obtain

$$\begin{aligned}&= \int_0^\Delta \int_{-\infty}^y \left[\bar{f}(x,y)p(x,y) - \frac{1}{2}\frac{\partial}{\partial x}\left(\bar{\sigma}^2(x,y)p(x,y)\right)\right] \phi_x(x,y)\,dx\,dy \\ &\quad + \int_0^\Delta \frac{1}{2}\bar{\sigma}^2(y,y)p(y,y)\left[\phi_x(y,y) + \phi_y(y,y)\right] dy.\end{aligned}$$

Since $\phi_x(y,y) + \phi_y(y,y) = \dfrac{d}{dy}(\phi(y,y))$, formal integration by parts again gives

$$\begin{aligned}&= \int_0^\Delta \int_{-\infty}^y \left[\bar{f}(x,y)p(x,y) - \frac{1}{2}\frac{\partial}{\partial x}\left(\bar{\sigma}^2(x,y)p(x,y)\right)\right] \phi_x(x,y)\,dx\,dy \\ &\quad - \int_0^\Delta \frac{1}{2}\frac{d}{dy}\left[\bar{\sigma}^2(y,y)p(y,y)\right] \phi(y,y)\,dy \\ &\quad \frac{1}{2}\bar{\sigma}^2(\Delta,\Delta)p(\Delta,\Delta)\phi(\Delta,\Delta) - \frac{1}{2}\bar{\sigma}^2(0,0)p(0,0)\phi(0,0).\end{aligned}$$

Since (8) must hold for all $\phi \in \mathcal{D}$, the density $p(x,y)$ should satisfy

$$\bar{f}(x,y)p(x,y) - \frac{1}{2}\frac{\partial}{\partial x}\left(\bar{\sigma}^2(x,y)p(x,y)\right) = 0 \tag{12}$$

with

$$\bar{\sigma}^2(y,y)p(y,y) \equiv \text{constant}. \tag{13}$$

Calling the constant $2K$, the solution is given by $p(x,y)$ of the form

$$p(x,y) = \frac{2K}{\bar{\sigma}^2(x,y)} \exp\left\{-2\int_x^y \frac{\bar{f}(t,y)}{\bar{\sigma}^2(t,y)}\,dt\right\} \qquad (x \le y, 0 < y < \Delta). \tag{14}$$

Thus (8) becomes

$$K[\phi(\Delta,\Delta) - \phi(0,0)] + \int_{E_1} [\phi(0,0) - \phi(x,y)]\,\mu^1(dx,dy) = 0$$

and it follows that the equation is satisfied when μ^1 is a point mass at (Δ,Δ) of size K. The normalizing constant K is:

$$K = \left[\int_0^\Delta \int_{-\infty}^y \frac{2}{\bar{\sigma}^2(x,y)} \exp\left\{-2\int_x^y \frac{\bar{f}(t,y)}{\bar{\sigma}^2(t,y)}dt\right\} dx\,dy\right]^{-1}. \tag{15}$$

Observe that the formal methods provide a density $p(x,y)$ which justifies the integration by parts. (Also note that the density $p(x,y)$ depends on the conditional distribution ν through $\bar{f}$ and $\bar{\sigma}^2$.) We now show that μ_E^0, in fact, has this density.

Proposition 4.1. *Let $(\mu^0,\mu^1) \in \hat{\mathcal{P}}$ and let μ_E^0 and ν be defined as above. Then μ_E^0 has the density given by (14) and (15).*

PROOF: Using the conditional distribution ν, define the probability measure $\tilde{\mu}_E^0$ on E having the density $p(x,y)$ of (14),(15) and define the probability measure $\tilde{\mu}^0$ on $E \times U$ by (11). Define the measure $\tilde{\mu}^1$ to have a point mass of size K at (Δ,Δ). We will show that $\mu_E^0 = \tilde{\mu}_E^0$.

Observe that μ^0 and μ^1 satisfy (8) by hypothesis and that $\tilde{\mu}^0$ and $\tilde{\mu}^1$ satisfy (8) by construction. Denoting the mass of μ^1 by K_0, we have, for all $\phi \in \mathcal{D}$,

$$\begin{aligned} \int A\phi(x,y,u)\,d\mu^0 &= -\int B\phi(x,y)\,d\mu^1 \\ &= -K_0\,[\phi(0,0)-\phi(\Delta,\Delta)] \\ &= -\frac{K_0}{K}K\,[\phi(0,0)-\phi(\Delta,\Delta)] \\ &= -\frac{K_0}{K}\int B\phi(x,y)\,d\tilde{\mu}^1 \\ &= \frac{K_0}{K}\int A\phi(x,y,u)\,d\tilde{\mu}^0. \end{aligned} \tag{16}$$

Since ν is the regular conditional distribution for both μ^0 and $\tilde{\mu}^0$, defining the operator $\tilde{A}$ by

$$\begin{aligned} \tilde{A}\phi(x,y) &= \int_U \left[f(u)\,\phi_x(x,y)+\frac{1}{2}\sigma^2(u)\,\phi_{xx}(x,y)\right]\nu(x,y,du) \\ &= \tilde{f}(x,y)\,\phi_x(x,y)+\frac{1}{2}\tilde{\sigma}^2(x,y)\,\phi_{xx}(x,y), \end{aligned}$$

(16) becomes

$$\begin{aligned} \int \tilde{A}\phi(x,y)\,d\mu_E^0 &= \frac{K_0}{K}\int \tilde{A}\phi(x,y)\,d\tilde{\mu}_E^0 \\ &= \frac{K_0}{K}\int \tilde{A}\phi(x,y)\,p(x,y)\,dx\,dy. \end{aligned} \tag{17}$$

Fix $a,b,c,d \in \mathbb{R}$ with $a<b$, $c<d$ and $a<d$. Define the function $I_G(x,y)$ to be the indicator of the set $G=\{(x,y): a\le x\le b, c\le y\le d\}$. The *idea* is to solve the ordinary differential equation

$$\tilde{A}\phi(x,y) = I_G(x,y) \tag{18}$$

for ϕ^* and then use this ϕ^* in (17) to show that

$$\mu_E^0(G) = \frac{K_0}{K}\tilde{\mu}_E^0(G).$$

However, the solutions to (18) have the form

$$\phi^*(x,y) = \int_{-\infty}^{x}\int_{-\infty}^{z}\frac{2}{\tilde{\sigma}^2(r,y)}I_G(r,y)\exp\left\{-2\int_r^z\frac{\tilde{f}(t,y)}{\tilde{\sigma}^2(t,y)}\,dt\right\}dr\,dz + C(y) \tag{19}$$

which are not elements of $\mathcal{D}$ (they are not $C_c^\infty(E)$), so we need to approximate.

Let $I_n(x,y)\in C_c^\infty(E)$ be a sequence of functions which converge pointwise to $I_G(x,y)$. Similarly, let f_n and σ_n^2 be sequences of functions in $C_c^\infty(E)$, with $\sigma_n^2>0$, which converge pointwise to $\tilde{f}$ and $\tilde{\sigma}^2$, respectively. Define ϕ_n by replacing $I_G(x,y)$, $\tilde{f}$ and $\tilde{\sigma}^2$ in (19) by $I_n(x,y)$, f_n and σ_n^2, respectively, and with $C(y)$ chosen (for each n) so that ϕ_n satisfies the boundary condition (4). Then $\phi_n\in\mathcal{D}$ and (17) is valid with ϕ_n. Since

$$\tilde{A}\phi_n(x,y) = \frac{\tilde{\sigma}^2(x,y)}{\sigma_n^2(x,y)}I_n(x,y)+\left[\tilde{f}(x,y)-\frac{\tilde{\sigma}^2(x,y)}{\sigma_n^2(x,y)}f_n(x,y)\right]\frac{\partial}{\partial x}\phi_n(x,y) \tag{20}$$

converges to $I_G(x,y)$ as $n\to\infty$, the bounded convergence theorem implies,

$$\begin{aligned} \int \tilde{A}\phi^*(x,y)\,d\mu_E^0 &= \lim_{n\to\infty}\int \tilde{A}\phi_n(x,y)\,d\mu_E^0 \\ &= \lim_{n\to\infty}\frac{K_0}{K}\int \tilde{A}\phi_n(x,y)\,d\tilde{\mu}_E^0 \\ &= \frac{K_0}{K}\int \tilde{A}\phi^*(x,y)\,d\tilde{\mu}_E^0. \end{aligned}$$

Therefore, $\mu_E^0(G)=\frac{K_0}{K}\tilde{\mu}_E^0(G)$ and since a, b, c and d are arbitrary, $\mu_E^0=\frac{K_0}{K}\tilde{\mu}_E^0$. Since both μ_E^0 and $\tilde{\mu}_E^0$ are probability measures, $K_0=K$ and the result is obtained. □

Remark 4.1.1. *The space E includes $-\infty$ as a possible value for the X process for compactness. The stationary density identified above assigns zero mass to $x = -\infty$ because of assumption (A2), which requires $f(u)$ strictly positive.*

Remark 4.1.2. *When $\nu(x, y, \cdot)$ does not actually depend on x, expressions (14) and (15) reduce considerably:*

$$p(x,y) = \frac{2K}{\bar{\sigma}^2(y)} \exp\left\{-2\frac{\bar{f}(y)}{\bar{\sigma}^2(y)}(y-x)\right\}$$

and

$$K = \left[\int_0^{\Delta} \frac{1}{\bar{f}(y)} dy\right]^{-1}$$

where $\bar{f}(y) = \int_U f(u)\nu(y, du)$ and $\bar{\sigma}^2(y) = \int_U \sigma^2(u)\nu(y, du)$. The stationary density for the running max alone, obtained by integrating out the x, agrees with [8, Theorem 3.6] and thus expressions (14) and (15) extend results in [8] to identify the stationary distribution of the (diffusion, running max) pair.

5 Solution of the LP.

Now that the elements of $\widehat{\mathcal{P}}$ have been identified, it is possible to solve the LP problem (10). From (14) and (15), it is clear that (μ^0, μ^1) are determined by the conditional distribution ν. So to solve the LP, it is sufficient to identify an optimal ν.

Let λ^* denote the optimal long-term average value. For each ν, let $p_\nu(x, y)$ and K_ν denote the quantities (14) and (15) obtained from ν. Then for every ν,

$$K_\nu \int_0^{\Delta} \int_{-\infty}^{y} \int_U h(y,u)\,\nu(x,y,du)\,\frac{1}{K_\nu}\,p_\nu(x,y)\,dx\,dy - K_\nu R \leq \lambda^*.$$

Due to (15), we have for every ν,

$$\int_0^{\Delta} \int_{-\infty}^{y} \int_U [h(y,u) - \lambda^*]\,\nu(x,y,du)\,\frac{1}{K_\nu}\,p_\nu(x,y)\,dx\,dy \leq R, \tag{21}$$

and an optimal ν will achieve equality. Thus, it is sufficient to identify a ν for which (21) is maximized. This is a (non-linear) *unconstrained* optimization problem.

Theorem 5.1. *For $0 \leq y \leq \Delta$, let $u_0(y)$ be the minimum element of the set of maximizers of the function $g(u) = \dfrac{h(y,u) - \lambda^*}{f(u)}$. Then the conditional distribution $\nu(x, y, \cdot) = \nu_0(y, \cdot) = \delta_{u_0(y)}(\cdot)$ maximizes (21) and is, therefore, an optimal conditional distribution for the long-term average problem. Observe that ν_0 does not depend on x and the long-term average value is*

$$\frac{\displaystyle\int_0^{\Delta} \frac{h(y,u_0(y))}{f(u_0(y))}\,dy - R}{\displaystyle\int_0^{\Delta} \frac{1}{f(u_0(y))}\,dy}. \tag{22}$$

PROOF: Consider (21) for an arbitrary ν.

$$\begin{aligned}
&\int_0^{\Delta} \int_{-\infty}^{y} \int_U [h(y,u) - \lambda^*]\,\nu(x,y,du)\,\frac{1}{K_\nu}\,p_\nu(x,y)\,dx\,dy \\
&= \int_0^{\Delta} \int_{-\infty}^{y} \int_U \frac{[h(y,u) - \lambda^*]}{f(u)} f(u)\nu(x,y,du) \cdot \frac{1}{K_\nu}\,p_\nu(x,y)dx\,dy.
\end{aligned}$$

Since $u_0(y)$ is chosen to maximize $\dfrac{h(y,u) - \lambda^*}{f(u)}$, this is

$$\leq \int_0^{\Delta} \int_{-\infty}^{y} \frac{[h(y,u_0(y)) - \lambda^*]}{f(u_0(y))} \frac{1}{K_\nu} \bar{f}(x,y)\,p_\nu(x,y)dx\,dy.$$

Using (12), integrating and then using (13), we get

$$= \int_0^{\Delta}\int_{-\infty}^{y} \frac{[h(y,u_0(y)) - \lambda^*]}{f(u_0(y))}\frac{1}{K_\nu}\frac{\partial}{\partial x}\left[\frac{1}{2}\bar{\sigma}^2(x,y)p_\nu(x,y)\right]dx\,dy$$
$$= \int_0^{\Delta}\frac{[h(y,u_0(y)) - \lambda^*]}{f(u_0(y))}\frac{1}{2K_\nu}\bar{\sigma}^2(y,y)p_\nu(y,y)dy$$
$$= \int_0^{\Delta}\frac{[h(y,u_0(y)) - \lambda^*]}{f(u_0(y))}dy\,.$$

In light of Remark 4.2, direct computations with ν_0 show that equality is achieved and that the long-term average value is (22). □

Remark 5.1. *Clearly, the choice of the minimum maximizer is for convenience; any maximizer may be chosen. Thus, there will not be a unique solution whenever $g(u)$ has more than one maximizer.*

6 Policy Improvement.

One difficulty with the solution of the previous section is that the value of λ^* is not known *a priori*. It is possible to overcome this by considering a family of optimization problems of the form (21) in which λ^* is replaced by the parameter λ.

Consider the problem

$$\text{Maximize} \int_0^{\Delta}\int_{-\infty}^{y}\int_U \frac{1}{K_\nu}[h(y,u) - \lambda]\,\nu(x,y,du)p_\nu(x,y)dx\,dy\,. \tag{23}$$

The proof of Theorem 5.1 applies to this problem as well and establishes that the conditional distribution ν which puts mass 1 at a maximizer of $\dfrac{h(y,u)-\lambda}{f(u)}$ also maximizes (23).

This structure can be used to provide a policy improvement algorithm. This algorithm (and proofs) are essentially the same as one given in [8, Section 4]. For completeness, we summarize the algorithm and provide an example.

The algorithm starts with $\lambda_0 = 0$ and identifies a sequence of conditional distributions $\{\nu_k\}$ $(k = 0,1,\ldots)$ and values $\{\lambda_k\}$ $(k = 0,1,\ldots)$ such that ν_k is an optimal distribution for (23) with parameter λ_k and λ_{k+1} is the long-term average value of the conditional distribution ν_k.

The Policy Improvement Algorithm:

Step 1: (Policy improvement step)
For $0 \le y \le \Delta$, define

$$u_k(y) = \min\left\{u \in U\colon\ \frac{h(y,u)-\lambda_k}{f(u)} = \max_{v\in U}\frac{h(y,v)-\lambda_k}{f(v)}\right\}$$

and let $\nu_k = \delta_{u_k(y)}$.

Step 2: (Value computation step)
For the conditional distribution ν_k, compute the long-term average value

$$\lambda_{k+1} = \frac{\displaystyle\int_0^{\Delta}\frac{h(y,u_k(y))}{f(u_k(y))}dy - R}{\displaystyle\int_0^{\Delta}\frac{1}{f(u_k(y))\,dy}}\,.$$

If $\lambda_{k+1} = \lambda_k$, then ν_k is an optimal distribution. If $\lambda_{k+1} > \lambda_k$, then return to Step 1 with the value λ_{k+1}.

Example. Suppose the operator A has drift coefficient $f(u) = e^u$ with $u \in U = [1,10]$, the reward rate is $h(y,u) = u - y$ and replacement occurs at level $\Delta = 10$ with a cost of $R = 0.25$. Note that replacement occurs when $h(y,u) \leq 0$ for all $u \in [1,10]$; when it is no longer possible to make a profit while running. The maximal value of $\dfrac{h(y,u)-\lambda}{f(u)}$ occurs at

$$u(y) = \begin{cases} 1 & \text{if } y+\lambda+1 < 0 \\ y+\lambda+1 & \text{if } 0 \leq y+\lambda+1 \leq 10 \\ 10 & \text{if } 10 \leq y+\lambda+1. \end{cases}$$

Thus, if the current value is λ, an optimal distribution is $\nu(y,\cdot) = \delta_{\{u(y)\}}(\cdot)$ and the next value can be easily computed by Step 2. The policy improvement algorithm produces the following successive values:

λ_0	0
λ_1	0.320367838
λ_2	0.384070986
λ_3	0.3861877815
λ_4	0.3861900246
λ_5	0.3861900246

Acknowledgement. The authors would like to thank Tom Kurtz for helpful discussions of the constrained martingale problem.

References

[1] R.S. Anderson, Replacement with nonconstant operating cost, *SIAM Journal on Control and Optimization*, **26**, 1076–1098, 1988.

[2] R.S. Anderson, *Long-run average maintenance problems*, preprint.

[3] E.N. Barron, *The Bellman equation for control of the running max of a diffusion and applications to look-back options*, preprint.

[4] E. Çinlar, Markov and Semi-Markov models for deterioration, *Reliability Theory and Models*, Academic Press, 1984.

[5] E.V. Denardo, On linear programming in a Markov decision problem, *Management Sci.*, **16**, 281–288, 1970.

[6] C. Derman, On sequential decisions and Markov Chains, *Management Sci.*, **9**, 16–24, 1962.

[7] A.C. Heinricher and R.H. Stockbridge, Optimal control of the running max, *SIAM J. Control Optim.*, **29**, 936–953, 1991.

[8] A.C. Heinricher and R.H. Stockbridge, *Long-term average control of a continuous, monotone process*, preprint.

[9] T.G. Kurtz, Martingale problems for constrained Markov processes, *Recent Advances in Stochastic Calculus*, J.S. Baras and V. Mirelli, eds., Springer-Verlag, New York, 1990. University of Maryland (to appear).

[10] T.G. Kurtz, A control formulation for constrained Markov processes, *Mathematics of Random Media*, Lectures in Applied Mathematics, **27**, 1991.

[11] A.S. Manne, Linear programming and sequential decisions, *Management Sci.*, **6**, 259–267, 1960.

[12] B.G. Pittel, A linear programming problem connected with optimal stationary control in a dynamic decision problem, *Theory Probab. Appl.*, **16**, 724–728, 1971.

[13] P. Protter, *Stochastic Integration and Differential Equations: A New Approach*, Springer-Verlag, 1990.

[14] R.H. Stockbridge, Time-average control of martingale problems: a linear programming formulation, *Ann. Probab.*, 18, 206–217, 1990.

[15] H.M. Taylor, Optimal replacement under additive damage and other failure models, *Naval Research Logistics Quarterly*, **22**, 1–18, 1975.

[16] C. Valdez-Flores and R.M. Feldman, A survey of preventive maintenance models for stochastically deteriorating single-unit systems, *Naval Research Logistics Quarterly*, **36**, 419–446, 1989.

[17] P. Wolfe and G.B. Dantzig, Linear programming in a Markov chain, *Oper. Res.*, **10**, 702–710, 1962.

An Optimal Control Depending on the Conditional Density of the Unobserved State[1]

by
K.L. Helmes and R.W. Rishel
Department of Mathematics
Lexington, Kentucky 40506

Abstract

The optimal control of a nonlinear partially observed stochastic control system is computed through using nonlinear filtering and linear quadratic techniques.

I. Introduction

From nearly the beginning of nonlinear filtering theory, the idea that the conditional probability distribution of the unobserved states should substitute as the state of the unobserved process of a partially observed control system has been used. Attempts to develop a dynamic programming theory using this idea were made by Kushner [8], Mortensen [9] and Beneš & Karatzas [1]. Severe technical complications were encountered in trying to develop this dynamic programming theory, and even today it is not in a satisfactory theoretical state. Use of the conditional probability distribution as the state is basic in the theories of Bensoussan [4] and Fleming [5]. General references are contained in [3].

Perhaps because of the technical problems, although the idea of using the conditional probability distribution as a state is over twenty-five years old, there are few if any genuine examples of partially observed optimal control laws depending on the conditional probability of the unobserved state, which have been computed explicitly. The objective of this paper is to give such an example.

Our example concerns a linear system in which additional noise, given by the output of a nonlinear system, is present. The case in which this additional noise was a finite-state Jump Markov process was discussed in [5]. We wish to express our appreciation to V.E. Beneš for suggesting extending the finite-state results of [5] to the present case, in which the nonlinear filtering equations are infinite-dimensional.

II. The Problem

Consider a partially observed control system whose observed process x is an n-dimensional stochastic process satisfying

$$dx_t = (Ax_t + Bu_t + F(y_t))dt + dW_t\,; \qquad x_0 = a \tag{1}$$

and whose unobserved process y is a q-dimensional stochastic process satisfying

$$dy_t = f(y_t)dt + dV_t. \tag{2}$$

In (1) and (2), W and V are independent n and q dimensional Wiener processes, A and B are $n \times n$ and $n \times m$ dimensional matrices. The functions $F(y)$ and $f(y)$ are n and q dimensional vector valued functions of the q-dimensional vector y, which are assumed to be bounded, twice continuously differentiable, with bounded first and second order partial derivatives. The initial value of y in (2) is a random vector y_0 which has the known probability density $P_0(y)$. This probability density is assumed to be twice continuously differentiable in y.

We can consider

$$z_t \triangleq \int_0^t F(y_s)ds + W_t \tag{3}$$

as the noise driving the system (1). It consists of the output

$$\int_0^t F(y_s)ds \tag{4}$$

of the nonlinear system (2) plus the Wiener process W_t.

[1]This work was partially supported by NSF grant DMS-9105649.

Let

$$\mathcal{F}_t^x \triangleq \sigma[x_s;\ 0 \le s \le t]$$
$$\mathcal{F}_t^z \triangleq \sigma[z_s;\ 0 \le s \le t] \tag{5}$$

be respectively the σ-fields generated by the past of x and the past of z up to time t. Let $\beta[0,T]$ denote the Borel field on $[0,T]$.

The class of admissible controls for the problem is defined to be the class of m-dimensional measurable stochastic processes u which satisfy the following two properties:

(i) For each control u there is a corresponding strong solution x of (1). By a strong solution of (1) we mean a solution which is a non-anticipative functional on the past of the driving noise of (1). That is, a solution x such that x_t can be represented as

$$x_t = \phi(t, (z_s;\ 0 \le s \le t)) \tag{6}$$

where ϕ is a $\beta[0,T] \times \mathcal{F}_T^z$-measurable functional so that for each t, $\phi(t,(z_s;\ 0 \le s \le t))$ is $\mathcal{F}_t^z$-measurable.

(ii) For each t, u_t is $\mathcal{F}_t^x$-measurable, where x is the solution of (1) corresponding to u. That is, the control u_t used at time t depends only on the measurements of its corresponding observed process made up to time t.

Consider a cost criterion, call it J^u, defined by

$$J^u \triangleq E\left[\int_0^T u_t' N u_t dt + x_T' Q x_T\right] \tag{7}$$

where N is a positive definite $m \times m$ matrix, Q a non negative definite $n \times n$ matrix and T is a fixed final time. We formulate the optimal control problem of minimizing J^u over the class of admissible controls.
Remark: Our assumptions on the class of admissible controls imply that for each t

$$\mathcal{F}_t^x = \mathcal{F}_t^z. \tag{8}$$

To see this, notice that property (i) implies

$$\mathcal{F}_t^x \subset \mathcal{F}_t^z. \tag{9}$$

Equation (1) implies

$$x_t - a - \int_0^t (Ax_s + Bu_s)ds = \int_0^t F(y_s)ds + W_t = z_t. \tag{10}$$

This and property (ii) imply

$$\mathcal{F}_t^z \subset \mathcal{F}_t^x. \tag{11}$$

Thus the two inequalities (9) and (11) imply (8).

III. Preliminary considerations

To state the results of the paper we will need some preliminary results. Notice from equation (2), and our assumptions, that y is a Markov process with a stationary transition probability density $P(t,y,z)$, that is

$$P[y_{t+s} \in A | y_s = y] = \int_A P(t,y,z)dz. \tag{12}$$

Let L_y be the partial differential operator given in terms of the coordinates of the vector y by

$$L_y = \sum_{i=1}^q f_i(y)\frac{\partial}{\partial y_i} + \frac{1}{2}\sum_{i,j=1}^q \frac{\partial^2}{\partial y_i \partial y_j} \tag{13}$$

Under our assumptions, the transition probability density $P(t,y,z)$ is the unique solution of the partial differential equation

$$\frac{\partial P}{\partial t} - L_y P = 0 \tag{14}$$

with initial condition

$$\lim_{t\to 0} P(t,y,z) = \delta(y-z) \tag{15}$$

where $\delta(x)$ is the δ-function.

From our remark that $\mathcal{F}_t^x = \mathcal{F}_t^z$, the conditional probability distribution of y_t given either $\mathcal{F}_t^x$ or $\mathcal{F}_t^z$ will be the same. Since neither z nor y depend on the control, this conditional distribution will be independent of the control. The assumptions on y_t and z_t imply, by Theorem 8.7 of Lipster and Shiryayev [6], that there will be a conditional density $\pi_t(y)$ of y_t given $\mathcal{F}_t^z$, which is the solution of

$$d\pi_t(y) = L_y^*\pi_t(y)dt + \pi_t(y)[F(y) - \int_{R_q} F(z)\pi_t(z)dz]d\nu_t \tag{16}$$

with initial condition

$$\pi_0(y) = P_0(y), \tag{17}$$

where ν_t is the innovations Brownian motion

$$d\nu_t \triangleq [F(y_t) - \int_{R_q} F(z)\pi_t(z)dz]dt + dW_t \tag{18}$$

and L_y^* is the operator

$$L_y^*\pi_t(y) \triangleq -\sum_{i=1}^{q}\frac{\partial}{\partial y_i}(f_i(y)\pi_t(y)) + \frac{1}{2}\sum_{i,j=1}^{q}\frac{\partial^2\pi_t(y)}{\partial y_i \partial y_j}. \tag{19}$$

Since the σ-field equality (8) holds $\pi_t(y)$ is also the conditional density of y_t given $\mathcal{F}_t^x$.

IV. Optimality Results

In terms of $\pi_t(y)$ and $P(t,y,z)$ define the stochastic process r by

$$r_t \triangleq e^{A(T-t)}x_t + \int_t^T \left[e^{A(T-s)}\int_{R_q}\int_{R_q} \pi_t(y)P(s-t,y,z)F(z)dzdy\right] ds \tag{20}$$

An argument can be given to show that, if one thinks of zero as a target one wants to hit at time T, and the zero control is used from t to T, then r_t is the expected miss at time T given that at time t x was at x_t.

Lemma 1 gives the stochastic differential equation satisfied by r.

Lemma 1. The stochastic process r satisfies

$$dr_t = B(t)u_t dt + \sigma(t,\pi_t(y))d\nu_t \tag{21}$$

where

$$B(t) \triangleq e^{A(T-t)}B, \tag{22}$$

$$\sigma(t,\pi_t(y)) \triangleq$$

$$e^{A(T-t)} + \int_t^T \left\{e^{A(T-s)}\int_{R_q}\int_{R_q}\pi_t(y)[F(y) - \int_{R_q}F(r)\pi_s(r)dr]P(s-t,y,z)F(z)dzdy\right\} ds, \tag{23}$$

and ν is the innovations process given by (18).

The proof of Lemma 1 follows from a calculation using Itô's stochastic differential rule. This is carried out in the appendix.

We are now in a position to state the main result of the paper.

Let $K(t)$ denote the solution of the matrix Riccati equation

$$\dot{K}(t) - K(t)B(t)N^{-1}B(t)'K(t) = 0 \tag{24}$$

with boundary condition

$$K(T) = Q. \tag{25}$$

Theorem 1. *For each admissible control u*

$$J^u \geq r_0' K(0) r_0 + E\left[\int_0^T \, trace\, \sigma(t, \pi_t(y))\sigma(t, \pi_t(y))' K(t) dt\right]. \tag{26}$$

For the control u^* defined by

$$u_t^* = -N^{-1} B(t)' K(t) r_t \tag{27}$$

equality holds in (26) and thus u^* is optimal.

The following Lemma, standard in linear quadratic theory, will be needed in the proof of Theorem 1. It can be proved by standard calculus arguments.

Lemma 2. If u and v are m-dimensional vectors, N is a positive definite $m \times m$ matrix, the minimum over u of

$$v'u + u'v + u'Nu \tag{28}$$

is attained when

$$u = -N^{-1} v \tag{29}$$

and is given by

$$-v'N^{-1}v. \tag{30}$$

Proof of Theorem 1. Consider the quantity $r_t' K(t) r_t$. Applying Itô's stochastic differential rule and (21) gives

$$\begin{aligned} dr_t' K(t) r_t &= r_t' K(t)[B(t) u_t dt + \sigma(t, \pi_t(y)) d\nu_t] \\ &+ [B(t) u_t dt + \sigma(t, \pi_t(y) d\nu]' K(t) r_t + r_t' \dot{K}(t) r_t dt \\ &+ \text{trace}\, \sigma(t, \pi_t(y))\sigma(t, \pi_t(y))' K(t) dt. \end{aligned} \tag{31}$$

Now add $u_t' N u_t dt$ to both sides of (31). Having done this we obtain a term of the form

$$r_t' K(t) B(t) u_t + u_t' B(t)' K(t) r_t + u_t' N u_t \tag{32}$$

multiplied by dt on the right side of (31). By Lemma 2, the minimum over u_t of (32) is

$$-r_t' K(t) B(t) N^{-1} B(t)' K(t) r_t \tag{33}$$

and is attained when

$$u_t = -N^{-1} B(t)' K(t) r_t \tag{34}$$

Since $K(t)$ is the solution of the matrix Riccati equation (24), the sum of (32) and $r_t' \dot{K}(t) r_t$ is nonnegative. Thus we obtain

$$\begin{aligned} d(r_t' K(t) r_t) + u_t' N u_t dt \;&\geq\; \text{trace}\, \sigma(t, \pi_t)\sigma(t, \pi_t(y))' K(t) dt \\ &+ 2 r_t K(t) \sigma(t, \pi_t(y)) d\nu_t \end{aligned} \tag{35}$$

Integrating (35) from 0 to T, using $K(T) = Q$, taking expected values, and using, since it's integrand is bounded, that the expected value of the stochastic integral is zero, we obtain

$$\begin{aligned} &E[r_T' Q r_T - r_0' K(0) r_0 + \textstyle\int_0^T u_t' N u_t dt] \\ &\geq E\left[\textstyle\int_0^T \text{trace}\, \sigma(t, \pi_t(y))\sigma(t, \pi_t(y))' K(t) dt\right]. \end{aligned} \tag{36}$$

Now from the definition (20) of r_t, r_0 is non random and $r_T = x_T$. Thus $-r_0' K(0) r_0$ may be taken out of the expected value sign. Now adding $r_0' K(0) r_0$ to both sides of (36) and using $r_T = x_T$, gives

$$\begin{aligned} &E[x_T' Q x_T + \textstyle\int_0^T u(t)' N u(t) dt] \\ &\geq r_0' K(0) r_0 + E\left[\textstyle\int_0^T \text{trace}\, \sigma(t, \pi_t(y))\sigma(t, \pi_t(y))' K(t) dt\right] \end{aligned}$$

which is (26).

Notice in the argument above, that if u_t is given by (27), then equality will hold in (35), (36) and (26). Thus we only need to show that (27) is an admissible control to show that it is optimal.

Let us first show that for the control (27) condition (i) holds, i.e., that there is a strong solution of (1). From (20) and (22) we see that the control (27) can be expressed as a linear combination of x_t and $\pi_t(y)$ by

$$u_t^* = g_t x_t + \int_{R_q} h(t,y)\pi_t(y)dy \tag{37}$$

where

$$g_t = -N^{-1}e^{A'(T-t)}BK(t)e^{A(T-t)} \tag{38}$$

and

$$h(t,y) = -N^{-1}e^{A'(T-t)}BK(t)\int_t^T \left[e^{A(T-s)}\int_{R_q} P(s-t,y,z)F(z)dz\right] ds. \tag{39}$$

For the control (37), the equation (1) becomes

$$dx_t = (Ax_t + Bg_t x_t + B\int_{R_q} h(t,y)\pi_t(y)dy + F(y_t))dt + dW_t \tag{40}$$

with initial condition

$$x_0 = a. \tag{41}$$

Combining results from [6] Theorem 8.7 and [7] Theorem 2.1, it can be shown that (16), (18) with initial condition (17) have a unique strong solution. That is, a solution such that $\pi_t(y)$ can be expressed by a $\beta[0,T] \times \mathcal{F}_T^z$-measurable function ψ, which is $\mathcal{F}_t^z$-measurable for each t as

$$\pi_t(y) = \psi(t,(z_s, 0 \le s \le t))(y). \tag{42}$$

Let $\chi(s,t)$ denote the transition matrix of the linear system

$$dx_t = (A + Bg_t)x_t dt. \tag{43}$$

Then we see that the solution x_t of (40) can be expressed as

$$x_t = \chi(0,t)a + \int_0^t \chi(s,t)\left[B\int_{R_q} h(s,y))\psi(s,(z_r;\ 0 \le r \le s))(y)dy\right] ds$$
$$+ \int_0^t \chi(s,t)dz_s. \tag{44}$$

Thus the strong solution of (1) corresponding to (27) is explicitly exhibited by (44).

Next let us show that u_t^* given by (27) satisfies condition (ii) by being $\mathcal{F}_t^x$ measurable for each t. From Theorem 1.1, Formula 1.10, of Kunita [7], it follows that $\pi_t(y)$ is also the unique solution of

$$\pi_t(y) = \int_{R_q} P_0(z)P(t,z,y)dz$$
$$+\int_0^t \left[\int_{R_q}\pi_s(z)P(t-s,z,y)F(z)dz - \int_{R_q}\pi_s(z)P(t-s,z,y)dz\int_{R_q}F(z)\pi_s(z)dz\right] d\nu_s \tag{45}$$

where ν is the innovation process (18). Writing (1) in terms of the innovations process using (18) and (37) we have

$$dx_t = \left[Ax_t + Bg_t x_t + B\int_{R_q} h(t,y)\pi_t(y)dy + \int_{R_q} F(y)\pi_t(y)dy\right] dt + d\nu_t. \tag{46}$$

Solving for $d\nu_t$ in (46) and substituting it in (45) we obtain the equation (47), driven by the process x, for $\pi_t(y)$

$$\pi_t(y) = \int_{R_q} P_0(z)P(t,z,y)dz$$

$$-\int_0^t \left\{ \left[\int_{R_q} \pi_s(z)P(t-s,z,y)F(z)dz - \int_{R_q} \pi_s(z)F(t-s,z,y)dz \int_{R_q} P(z)\pi_s(z)dz \right] \right.$$

$$\left. \left[B\int_{R_q} [h(s,y)+F(y)]\pi_s(y)dy \right] \right\} ds \quad (47)$$

$$+\int_0^t \left\{ \left[\int_{R_q} \pi_s(z)P(t-s,z,y)F(z)dz - \int_{R_q} \pi_s(z)P(t-s,z,y)dz \int_{R_q} F(z)\pi_s(z)dz \right] \right.$$

$$\left. [dx_s - (Ax_s + Bg_s x_s)] \right\} ds$$

Now we claim that arguments similar to but more complicated than those given in Theorem 2.1 of Kunita [7] show that (47) has a unique solution $\pi_t(y)$ which is $\mathcal{F}_t^x$-measurable for each fixed t. Thus (37) which equals (27) satisfies the measurability condition (ii).

V. Concluding Remarks

It appears from examining (37), (38) and (39), which give the form of the optimal control, that this control consists of two parts; the term $g_t x_t$ which tries to compensate for the deviation of x_t from zero, and the term $\int_{R_q} h(t,y)\pi_t(y)dy$ which estimates and tries to compensate for the nonlinear part of the noise. The latter portion of the control is truly infinite-dimensional, in that it depends on the probability densities $\pi_t(y)$.

To implement this control, the filtering equations (12), (14) would have to be solved in real time to obtain $\pi_t(y)$. Of course, since these are infinite-dimensional equations, this is not possible and only some finite dimensional approximation of $\pi_t(y)$ could be used to give an approximation of the optimal control in any real system.

Appendix

Proof of Lemma 1. Using the Itô stochastic differential rule, the interchange of integration and stochastic differentiation, and using (14), (15) and (16) we see that

$$d\left\{ \int_t^T \left[e^{A(T-s)} \int_{R_q}\int_{R_q} \pi_t(y)P(s-t,y,z)F(z)dzdy \right] ds \right\} \quad (48)$$

$$= \left\{ -e^{A(T-t)} \int_{R_q} \pi_t(z)F(z)dz \right\} dt$$

$$+ \left\{ \int_t^T \left[e^{A(T-s)} \int_{R_q}\int_{R_q} L_y^* \pi_t(y)P(s-t,y,z)F(z)dzdy \right] ds \right\} dt$$

$$+ \left\{ \int_t^T \left[e^{A(T-s)} \int_{R_q}\int_{R_q} \pi_t(y)[F(y) - \int_{R_q} F(r)\pi_t(r)dr]P(s-t,y,z)F(z)dzdy \right] ds \right\} d\nu_t$$

$$+ \left\{ \int_t^T \left[e^{A(T-s)} \int_{R_q}\int_{R_q} \pi_t(y)\tfrac{\partial}{\partial t}(P(s-t,y,z)F(z)dzdy \right] ds \right\} dt.$$

Now since from (14)

$$\frac{\partial}{\partial t}P(s-t,y,z) = -L_y P(s-t,y,z) \quad (49)$$

and L_y and L_y^* are adjoints of each other, the 2nd and 4th terms on the right side of (48) cancel each other. Using (48) with these terms cancelled, (2) and (1), we see that

$$dr_t = -Ae^{A(T-t)}x_t dt + e^{A(T-t)}[(Ax_t + Bu_t + F(y_t))dt + dW_t]$$

$$- \left[e^{A(T-t)} \int_{R_q} \pi_t(z)F(z)dz \right] dt \quad (50)$$

$$+ \left\{ \int_t^T \left[e^{A(T-s)} \int_{R_q}\int_{R_q} \pi_t(y)[F(y) - \int_{R_q} F(r)\pi_t(r)dr]P(s-t,y,z)F(z)dzdy \right] ds \right\} d\nu_t.$$

Now using the definition of ν in (18), rearranging and cancelling in (50) gives (21), (22) and (23).

References

[1] V.E. Beneš, I. Karatzas, "On the Relation of Zakai's and Mortensen's Equation", SIAM J. Control and Optimization 21 (1983) 472-489.

[2] A. Bensoussan, "Optimal Control of Partially Observed Diffusions", in "Advances in Filtering and Optimal Stochastic Control", W.H. Fleming, L.A. Gorostiza (Eds.), Proceedings of the IFIP - WG 7/1 Working Conference, Springer Lecture Notes in Control and Information Sciences, Vol. 42 (1982).

[3] N. Christopeit, M. Kohlmann (Eds.), "Stochastic Differential Systems", Proceedings of the 2nd Bad Honnef Conference, Springer Lecture Notes in Control and Information Sciences, vol. 43 (1982).

[4] W.H. Fleming, "Nonlinear Semigroups for Controlled Partially Observed Diffusions", SIAM Journal on Control and Optimization, Vol. 20, (1982).

[5] K. Helmes, R.W. Rishel, "The Solution of a Partially Observed Stochastic Control Problem in Terms of Predicted Miss", to appear.

[6] R.S. Lipster, A.N. Shiryayev, Statistics of Random Processes I - General Theory, Springer-Verlag (1977).

[7] H. Kunita, "Asymptotic Behavior of the Nonlinear Filtering Errors of Markov Processes" Journal of Multivariate Analysis 1 (1971), pp. 365-391.

[8] H.J. Kushner, "On the Dynamical Equations of Conditional Probability Density Functions with Application to Optimal Stochastic Control Theory", J. Math. Anal. Appl. 8 (1964), pp. 332-344.

[9] R.E. Mortensen, "Stochastic Optimal Control with Noisy Observations", International J. Control 4 (1966), pp. 455-464.

PARTIALLY OBSERVED CONTROL OF MARKOV PROCESSES

OMAR HIJAB

Temple University

The aim of this paper is to outline the proof of the existence of optimal controls in the partially observed setting in the simplest case: controlling the drift of a Brownian motion under partial observations. The general case is worked out in [3].

The point of view we adopt is a combination of results of P. L. Lions on the control of the Zakai equation [4] and results of the author on the control of diffusions in finite dimensions [2].

§1. THE THEOREM

The state equation is

$$dx = u(t)dt + d\xi, x(0) \in \mathbf{R}, \tag{1.1}$$

and the observation equation is

$$dy = c(x)dt + d\eta, y(0) = 0 \in \mathbf{R}, \tag{1.2}$$

where $u = u(t, y)$ depends only on the observations y and $(\xi, \eta) \in \mathbf{R}^2$ is a Brownian motion.

Fix a compact convex $U \subset \mathbf{R}$ and let $M = M(\mathbf{R})$ denote the set of probability measures on $\mathbf{R}$. Let $\|\varphi\|_k$ denote the norm in $C_b^k(\mathbf{R})$, let $\mu(\varphi)$ denote the integral of φ against μ over $\mathbf{R}$, and let $\|\mu\|_{-k} = \sup\{|\mu(\varphi)| : \|\varphi\|_k \leq 1\}$ denote the dual norm, where $k \geq 0$ and $\mu \in M$.

A *control* is a progressively measurable map $u : [0, \infty) \times C([0, \infty); \mathbf{R}) \to U$, $u = u(t, y)$. Assume the signal c is bounded. It follows then by Girsanov's theorem that, for each fixed $m \in M$, there is a one-to-one correspondence between the set of controls and the set of *systems starting from* m i.e. the set of laws Q of processes (x, y) such that $(\xi, \eta) \in \mathbf{R}^2$ is a Brownian motion and the law of $x(0)$ is m.

Given u and m the corresponding cost is

$$v^u(m) = E^Q\left(\int_0^\infty e^{-\lambda t} f(x(t), u(t))dt\right); \tag{1.3}$$

1980 *Mathematics Subject Classification* (1985 *Revision*). 60G35,49A10,93E11,93E20.

Key words and phrases. Markov Process, Partially Observed Control, Measure-Valued Diffusions, Infinite-Dimensional Bellman Equation.

Supported in part by a National Science Foundation Grant .

the *value function* is

$$v(m) = \inf_{u} v^u(m). \tag{1.4}$$

Here the infimum is over all controls, $m \in M$ is arbitrary, and $\lambda > 0$ is the discount factor. The problem is to characterize, for each m, the control u *optimal at* m, i.e. the control satisfying $v^u(m) = v(m)$.

It turns out [3], [4] that $v : M \to \mathbf{R}$ is weakly continuous and bounded. To obtain further regularity of v we make some definitions.

Fix $\mu \in M$ and let ν be a signed measure on $\mathbf{R}$. We say ν is *tangent to M at μ* if $\mu + t\nu \in M$ for $|t|$ small. A functional $\Phi : M \to \mathbf{R}$ is *differentiable at μ* if there is a bounded function $\varphi : \mathbf{R} \to \mathbf{R}$ such that the limit

$$\left.\frac{d}{dt}\right|_{t=0} \Phi(\mu + t\nu)$$

exists and equals $\nu(\varphi)$ for all ν tangent to M at μ. In this case φ is denoted $D\Phi(\mu)$ and is unique up to an additive constant. We say Φ is *differentiable* if Φ is differentiable at μ for all $\mu \in M$.

In general Dv does not exist. Nevertheless [4] v solves, in the viscosity sense, an infinite-dimensional Bellman equation

$$-\hat{A}v + H(\mu, Dv) + \lambda v = 0, \mu \in M, \tag{1.5}$$

where $\hat{A}$ is a diffusion generator on M and

$$H(\mu, p) = \sup_{u \in U}(\mu(-A^u p) - L(\mu, u)); \tag{1.6}$$

here $\mu \in M$, $p \in C_b^2(\mathbf{R})$, $A^u = u\partial_x + \frac{1}{2}\partial_x^2$ is the state generator, and $L(\mu, u) = \mu(f(\cdot, u))$.

Assume $f \in C_b^2(\mathbf{R}^2)$ and $f_{uu}(x, u) > 0$ on $\mathbf{R} \times U$. Then $L_{uu} > 0$ on $M \times U$ and standard convexity reasoning shows (see for example the Appendix in [2]) that the supremum in (1.6) is attained at a unique point $\mathbf{u}(\mu, p) \in U$ where the map $(\mu, p) \mapsto \mathbf{u}(\mu, p)$ satisfies

$$|\mathbf{u}(\mu, p) - \mathbf{u}(\mu', p')| \leq C_r \|\mu - \mu'\|_{-1} + C_r \|p - p'\|_1, \tag{1.7}$$

$\mu, \mu' \in M$, $p, p' \in C_b^2(\mathbf{R})$, $\|p\|_2 \leq r$, $\|p'\|_2 \leq r$, for all $r > 0$.

For each $t \geq 0$ let $\mu_m^u(t)$ denote the (normalized) conditional probability distribution of $x(t)$ given $y(s), 0 \leq s \leq t$,

$$\mu_m^u(t)(\varphi) = E^Q(\varphi(x(t)) | y(s), 0 \leq s \leq t).$$

Theorem. *Assume $c \in C_b^2(\mathbf{R})$, $f \in C_b^2(\mathbf{R}^2)$, and $f_{uu} > 0$ on $\mathbf{R} \times U$; then there exists $\lambda_1 \geq 0$ and $C > 0$ such that, for $\lambda > \lambda_1$, v is differentiable and*

(1) $\|Dv(\mu)\|_2 \leq C$, $\mu \in M$,
(2) $|v(\mu) - v(\mu')| \leq C\|\mu - \mu'\|_{-2}$, $\mu, \mu' \in M$,
(3) $\|Dv(\mu) - Dv(\mu')\|_1 \leq C\|\mu - \mu'\|_{-1}$, $\mu, \mu' \in M$.

Moreover for each m and $\lambda > \lambda_1$ there exists exactly one control u_m^* optimal at m and a control u satisfies

$$u(t) = \mathbf{u}(\mu_m^u(t), Dv(\mu_m^u(t))), t \geq 0,$$

iff $u = u_m^*$.

The theorem remains valid for a wide class of systems: The state space may be taken to be $\mathbf{R}^d$, the convexity assumptions can be dropped, the drift coefficients in the state generator A^u can be controlled in a nonlinear fashion, and U can be an arbitrary complete separable metric space. What is necessary is the estimate (1.7). Moreover $L(\mu, u)$ may be any nonlinear functional on $M \times U$ satisfying estimates similar to (1), (2), (3). If the diffusion coefficients in A^u are controlled, then C_b^4 smoothness is required of c, f and the appropriate modification of the Theorem is valid [3].

§2. Outline of Proof

We remark that the estimates below are stated for the general case [3]. In the special case under consideration, it is easy to improve some of them substantially.

Let $\Phi : M \to \mathbf{R}$ be weakly continuous and differentiable and let μ, ν be in M. Then $\mu(t) = (1-t)\mu + t\nu \in M$ and $\mu(t) = \mu(s) + (t-s)(\nu - \mu)$ for $0 \leq s < t \leq 1$; this implies $\nu - \mu$ is tangent to M at $\mu(t)$ for $0 < t < 1$ and so $f(t) = \Phi(\mu(t))$ is differentiable on $0 < t < 1$ with $f'(t) = (\nu - \mu)(D\Phi(\mu(t)))$. Thus

$$\Phi(\nu) - \Phi(\mu) = \int_0^1 (\nu - \mu)(D\Phi(\mu(t)))dt. \tag{2.1}$$

Since $|\mu(\varphi)| \leq \|\varphi\|_2 \|\mu\|_{-2}$, it follows that Φ is Lipschitz on M relative to $\|\cdot\|_{-2}$ whenever $\|D\Phi(m)\|_2 \leq C$. This shows that (2) is implied by (1) in the Theorem. Note also that if $\mu_n \to \mu$ weakly then by Ascoli's theorem $\|\mu_n - \mu\|_{-2} \to 0$. This implies that $\Phi : M \to \mathbf{R}$ is weakly continuous whenever Φ is Lipschitz relative to $\|\cdot\|_{-2}$. This shows (2) in the Theorem implies v is weakly continuous. Note also that $\|\cdot\|_{-2} \leq \|\cdot\|_{-1} \leq \|\cdot\|_0 \leq 1$ on M.

We will need to work with weak sense controls following Fleming and Pardoux [1]. *A generalized control* is a filtered probability space $\alpha = (\Omega, \mathcal{Y}, \mathcal{Y}_t, P)$ equipped with $\mathcal{Y}_t$ progressively measurable processes (y, u) such that u is U-valued and y is a $\mathcal{Y}_t$ Brownian motion. A generalized control α is a (strict-sense) control if in addition u is a progressively measurable function of y.

Let x, $\mathcal{X}$, $\mathcal{X}_t$, $t \geq 0$, denote the canonical process, Borel σ-algebra, and canonical filtration respectively on $C([0,\infty); \mathbf{R})$. For each measurable $\beta : [0, \infty) \to U$ and $m \in M$ let E_m^β denote the expectation (on $\mathcal{X}$)

$$E_m^\beta(\Phi(x)) = \int_{\mathbf{R}} E\left(\Phi\left(a + \int_0^\cdot \beta(t)dt + \xi(\cdot)\right)\right) dm(a).$$

Given a generalized control α and $m \in M$ let $Q^0 = Q^0(\alpha, m)$ be the unique law on $\mathcal{X} \times \mathcal{Y}$ such that

(1) the marginal of Q^0 on $\mathcal{Y}$ is P and
(2) the conditional expectation of x given $\mathcal{Y}$, under Q^0, is E_m^u.

It follows then that y is an $\mathcal{X}_t \times \mathcal{Y}_t$ Brownian motion under Q^0 and hence

$$R(t) = \exp\left(\int_0^t c(x(s))dy(s) - \frac{1}{2}\int_0^t c(x(s))^2 ds\right), t \geq 0,$$

is well-defined.

The *generalized system* starting at m and corresponding to α is the law $Q = Q(\alpha, m)$ on $\mathcal{X} \times \mathcal{Y}$ satisfying $dQ/dQ^0 = R(t)$ on $\mathcal{X}_t \times \mathcal{Y}_t$ for all $t \geq 0$. Clearly the law of (x, y, u) under $Q = Q(\alpha, m)$ depends only on m and the law of (y, u) under P.

Let $v^\alpha(m)$ denote the right side of (1.3) where the expectation is against $Q = Q(\alpha, m)$ and let $v'(m) = \inf_\alpha v^\alpha(m)$. We say a generalized control is *optimal at* m if $v^\alpha(m) = v'(m)$.

For each measurable $\beta : [0, \infty) \to U$ and $(s, x) \in [0, \infty) \times \mathbf{R}$ let $E^\beta_{s,x}$ denote the expectation (on $\mathcal{X}$)

$$E^\beta_{s,x}(\Phi(x)) = E\left(\Phi\left(x + \int_s^{\cdot \vee s} \beta(t)dt + \xi(\cdot \vee s) - \xi(s)\right)\right).$$

Given a generalized control α and $(s, x) \in [0, \infty) \times \mathbf{R}$ let $Q^0 = Q^0(\alpha, s, x)$ be the unique law on $\mathcal{X} \times \mathcal{Y}$ such that

(1) the marginal of Q^0 on $\mathcal{Y}$ is P and
(2) the conditional expectation of x given $\mathcal{Y}$, under Q^0, is $E^u_{s,x}$.

It follows then that y is an $\mathcal{X}_t \times \mathcal{Y}_t$ Brownian motion under Q^0 and hence

$$R(t|s) = \exp\left(\int_s^t c(x(r))dy(r) - \frac{1}{2}\int_s^t c(x(r))^2 dr\right), t \geq s,$$

is well-defined. Let

$$T_{t,s}\varphi(x) = E^{Q^0}(\varphi(x(t))R(t|s)|\mathcal{Y}_t) = E^u_{s,x}(\varphi(x(t))R(t|s)), t \geq s.$$

Then $T_{t,s}$ is an operator-valued $\mathcal{Y}_t$-measurable random variable.

Differentiation under the expectation sign yields the following.

Lemma 1. *There exists $C > 0$ such that for all $t \geq s \geq 0$ there is a $\mathcal{Y}_t$-measurable random variable $M(t, s, x) \geq 0$ satisfying*

(1) $E^P(M(t, s, x)| \mathcal{Y}_s) \leq C(1 + (t - s))$ *for all* $x \in \mathbf{R}$,
(2) *for* $\varphi \in C^2_b(\mathbf{R})$, $x \in \mathbf{R}$,

$$|\partial^k_x T_{t,s}\varphi(x)| \leq M(t, s, x)\|\varphi\|_2, k = 0, 1, 2,$$

almost surely. □

Set

$$p(s) = p(s, \cdot) = E^P\left(\int_s^\infty e^{-\lambda(t-s)} T_{t,s} f(\cdot, u(t))dt \middle| \mathcal{Y}_s\right). \tag{2.2}$$

Then $p(s)$ is a bounded function on $\mathbf{R}$ for each $s \geq 0$, almost surely. An immediate consequence of the above Lemma is

Lemma 2. *There exists $C > 0$ such that $p(t)$ is in $C_b^2(\mathbf{R})$ and*

$$\|p(t)\|_2 \leq C\frac{\lambda+1}{\lambda^2}$$

for $t \geq 0$ and $\lambda > 0$, almost surely. □

Let $\mu(t) \in M$ denote the conditional distribution $\mu(t)(\varphi) = E^Q(\varphi(x(t))|\mathcal{Y}_t)$ of $x(t)$ given $\mathcal{Y}_t$; here $Q = Q(\alpha, m)$. This is defined for any α and agrees with $\mu_m^u(t)$ as defined in §1 when α is strict-sense.

Lemma 3. *Let $\lambda > 0$ and suppose α is optimal at m. Then*

$$u(t) = \mathbf{u}(\mu(t), p(t)), t \geq 0. \quad \square \tag{2.3}$$

This Lemma is obtained for $t = 0$ by an Euler-Lagrange argument and then extended to all $t > 0$ by the Markov property. This Lemma shows that $p(t) \in C_b^2(\mathbf{R})$ plays the role of the co-state (adjoint) process dual to $\mu(t) \in M$.

More explicitly, given a constant control $a \in U$ and $\epsilon > 0$ and an optimal (generalized) control $u(\cdot)$ define $u^\epsilon(t)$ by setting it equal to a for $t < \epsilon$ and equal to $u(t)$ elsewhere; let $v^\epsilon(m)$ denote the corresponding cost. Since $v^\epsilon(m) \geq v^0(m) = v^u(m)$, it follows that $(d/d\epsilon)v^\epsilon(m) \geq 0$ at $\epsilon = 0$. Performing the differentiation explicitly we obtain an inequality valid for all $a \in U$ which yields (2.3) at $t = 0$.

We now have a formula (2.2) yielding the dependence of $p(t)$ on $u(t)$ and a formula (2.3) yielding the dependence of $u(t)$ on $\mu(t)$, $p(t)$; to obtain a closed system we need to know the dependence of $\mu(t)$ on u. This is given by the Bayes formula as follows.

Lemma 4. *For $t \geq 0$, $\varphi \in C_b(\mathbf{R})$, $m \in M$, $\lambda > 0$, and any α,*

$$\mu(t)(\varphi) = \frac{E_m^u(\varphi(x(t))R(t))}{E_m^u(R(t))}, \tag{2.4}$$

almost surely $Q(\alpha, m)$. □

Following Fleming-Pardoux [1] one can impose a notion of convergence on the set of generalized controls α such that

(1) the set of generalized controls is compact,
(2) the function $(m, \alpha) \mapsto v^\alpha(m)$ is continuous,

where the weak topology is on M. This leads to

Lemma 5. *For each m and $\lambda > 0$ there is a generalized control optimal at m.* □

Since this aspect of the subject has been known since the early 1980's, we omit the definition of the appropriate notion of convergence but we note that to handle the nonlinear dependence of f on u one must (temporarily) work with a slightly larger class of generalized controls valued in $M(U)$ instead of U and modify Lemma 3 accordingly in order to obtain Lemmas 3 and 5 *as stated above.*

The crucial estimate is to compare two generalized controls, one optimal at m_1 and one optimal at m_2. However two generalized controls can be compared only if they are defined on the same probability space. Because of this, we use the Watanabe-Yamada technique to establish

Lemma 6. *Given two generalized controls $(\Omega_i, \mathcal{Y}_t^i, P_i, y_i, u_i)$, $i = 1, 2$, there exists a probability space $(\Omega^*, \mathcal{Y}_t^*, P^*)$ supporting processes (y^*, u_1^*, u_2^*) such that the law of (y_i, u_i) under P_i equals the law of (y^*, u_i^*) under P^*.* □

The Watanabe-Yamada technique was invented to establish pathwise or strong uniqueness for initial value problems. Here we use it to establish Lipschitz dependence of $(\mu(t), p(t))$ on the *boundary* conditions at $t = 0$ and $t = \infty$, $\mu(0) = m$ and $e^{-\lambda t}p(t)|_{t=\infty} = 0$.

Now consider two generalized controls optimal at m_1 and m_2 respectively. By Lemma 6 we can assume they are defined over the same filtration $\mathcal{Y}_t$ with the same process y and probability measure P. Let $\mu_i(t)$, $p_i(t)$, denote the corresponding objects, $i = 1, 2$.

For $\rho : [0, \infty) \to [0, \infty)$ set

$$K\rho(t) = K_{C,\lambda}\rho(t) = C\int_0^t e^{C(t-s)}\rho(s)ds + C\int_t^\infty e^{(C-\lambda)(s-t)}\rho(s)ds.$$

Set

$$\rho(t) = E^P\left(\|\mu_1(t) - \mu_2(t)\|_{-1}^2 + \|p_1(t) - p_2(t)\|_1^2\right).$$

Here E^P denotes expectation against the background probability common to both generalized controls. Then we have two closed systems (2.2), (2.3), (2.4), one for each generalized control. Straightforward but tedious estimates yield the crucial estimate:

Lemma 7. *There exists $C > 0$ such that for $\lambda > 0$*

$$\rho(t) \le Ce^{Ct}\|m_1 - m_2\|_{-1}^2 + K_{C,\lambda}\rho(t), t \ge 0. \quad \square$$

Now set $\lambda_1 = \max(6C, 1)$. Then for $\lambda > \lambda_1$ we have $\rho(t) \le C' \le C'e^{3Ct}$ for some C' and $K(e^{3Ct}) \le \delta e^{3Ct}$ with $\delta < 1$. Because of this iterating the inequality in Lemma 7 yields

Lemma 8. *For $\lambda > \lambda_1$,*

$$\rho(t) \le Ce^{3Ct}\|m_1 - m_2\|_{-1}^2, t \ge 0.$$

In particular

$$\|p_1(0) - p_2(0)\|_1 \le C\|m_1 - m_2\|_{-1}. \quad \square$$

Recalling the definition of $Q^0(\alpha, m)$, $Q(\alpha, m)$ it follows that

$$\begin{aligned} v^\alpha(m) &= E^P\left(\int_0^\infty e^{-\lambda t}E_m^u(f(x(t), u(t))R(t))dt\right) \\ &= \int_{\mathbf{R}} E^P\left(\int_0^\infty e^{-\lambda t}E_{0,x}^u(f(x(t), u(t))R(t))dt\right)dm(x) = m(p(0)) \end{aligned}$$

is affine in m. From this we have

Lemma 9. *For each α, m, and $\lambda > 0$, $Dv^\alpha(m) = p(0)$ and the map $(\alpha, m) \mapsto Dv^\alpha(m)$ is continuous in an appropriate sense.* □

If we set $F(m)$ equal to $Dv^\alpha(m)$ for any α optimal at m, Lemmas 8 and 9 show that $F(m)$ is well-defined and Lipschitz relative to $\|\cdot\|_{-1}$, when $\lambda > \lambda_1$.

Lemma 10. *For $\lambda > \lambda_1$, v' is differentiable on M and $Dv' = F$. Moreover if α is optimal at m, then*

$$u(t) = \mathbf{u}(\mu(t), Dv'(\mu(t))), t \geq 0. \quad \square \tag{2.5}$$

To establish this, fix $m \in M$ and ν tangent to M at m and let α_t be an optimal generalized control at $m + t\nu$. Then for $t > 0$

$$\frac{v'(m+t\nu) - v'(m)}{t} \leq \frac{v^{\alpha_0}(m+t\nu) - v^{\alpha_0}(m)}{t}.$$

Thus

$$\limsup_{t \downarrow 0} \frac{v'(m+t\nu) - v'(m)}{t} \leq \nu(Dv^{\alpha_0}(m)) = \nu(F(m)).$$

Also

$$\frac{v'(m+t\nu) - v'(m)}{t} \geq \frac{v^{\alpha_t}(m+t\nu) - v^{\alpha_t}(m)}{t} = \int_0^1 \nu(Dv^{\alpha_t}(m+st\nu))ds.$$

Passing to a subsequence $\alpha_t \to \alpha_0'$ we obtain, since $(\alpha, m) \mapsto v^\alpha(m)$ is continuous, α_0' is optimal at m and hence

$$\liminf_{t \downarrow 0} \frac{v'(m+t\nu) - v'(m)}{t} \geq \nu(Dv^{\alpha_0'}(m)) = \nu(F(m))$$

since $(\alpha, m) \mapsto \nu(Dv^\alpha(m))$ is continuous (Lemma 9). Thus

$$\lim_{t \downarrow 0} \frac{v'(m+t\nu) - v'(m)}{t} = \nu(F(m)).$$

Replacing ν by $-\nu$ shows that $Dv'(m) = F(m)$ for all m.

Since $Dv'(m) = p(0) = Dv^\alpha(m)$ we obtain (2.5) at $t = 0$ from Lemma 3. For $t > 0$ we use the Markov property to reduce to $t = 0$. This establishes Lemma 10.

Lemma 11. *Suppose $\lambda > \lambda_1$. Then every α optimal at m is necessarily induced by a strict-sense control. Moreover for each m there is exactly one strict-sense control u solving the feedback*

$$u(t) = \mathbf{u}(\mu_m^u(t), Dv(\mu_m^u(t))), t \geq 0, \tag{2.6}$$

and $v(m) = v'(m)$.

To establish this one first notes the feedback $F(m) = \mathbf{u}(m, Dv'(m))$ is Lipschitz relative to $\|\cdot\|_{-1}$ for $\lambda > \lambda_1$ by Lemma 10. One then solves the fixed point formula (2.5) by iteration to produce a strong solution. This can be done for any feedback F Lipschitz relative to $\|\cdot\|_{-1}$. Moreover the same iteration procedure shows that any weak solution of the fixed point formula (2.5) equals the strong solution. This then implies $v = v'$ and hence (2.5) implies (2.6). The proof of Lemma 11 is entirely pathwise i.e. no moments are involved. The details are in [3].

Combining Lemmas 2, 9, 10, 11 yields the differentiability of v together with the estimates $\|Dv(m)\|_2 \leq C$, $\|Dv(m) - Dv(m')\|_1 \leq C\|m - m'\|_{-1}$, for $\lambda > \lambda_1$. This completes the outline of the proof.

We conclude with some remarks concerning the roles of the Kushner-Stratonovitch and Zakai equations for the normalized and unnormalized conditional distributions respectively.

A natural technique to estimate quantities such as $\|\mu(t)\|_{-k}$ or $\|\mu_1(t) - \mu_2(t)\|_{-k}$ is by time differentiation, using either of these equations, and solving the resulting differential inequalities; however this method is not available here as $\|\cdot\|_{-k} : M \to \mathbf{R}$ is not differentiable, at least not in a convenient sense. Because of this difficulty one must, if one follows this approach, use a differentiable norm by interpreting $\|\cdot\|_{-k}$ as an L^2 Sobolev norm and reformulating the Theorem accordingly. This is done in [4] for a different though related result. Since we instead deal directly with the Bayes formula (2.4) and do not use the above technique, we avoid this problem entirely. Nevertheless the idea that in these types of problems one *must* use the $(\|\varphi\|_1, \|\mu\|_{-1})$ duality instead of the simpler $(\|\varphi\|_0, \|\mu\|_0)$ duality first appears in [4].

Another remark concerns the fundamental estimate in Lemma 7. Throughout we work with the ***normalized*** conditional distribution. If instead in defining $\rho(t)$ we work with the ***unnormalized*** conditional distribution, the estimate in Lemma 7 fails. We explain this phenomenom in terms of the differential equations of filtering; although ultimately we avoid them, the difficulty persists. This is because the Zakai equation is ***bilinear*** in μ, u and so the moments of the difference $\|\mu_1(t) - \mu_2(t)\|_{-1}$ cannot be estimated in terms of the difference of two controls $|u_1(t) - u_2(t)|$. In the normalized case the Kushner-Stratonovitch equation is also bilinear (in fact a little worse) but we have now the ***a priori*** estimate $\|\mu(t)\|_0 = 1$ ***almost surely*** whose effect is to make the coefficients bounded, resolving the difficulty. This difficulty persists even for bilinear equations in finite dimensions (e.g. when the state space is finitely many points instead of $\mathbf{R}$).

References

1. W. H. Fleming & E. Pardoux, *Existence of Optimal Control for Partially Observed Diffusions*, SIAM J. Control 20 (1982), 261-283.
2. O. Hijab, *Control of Degenerate Diffusions in* $\mathbf{R}^d$, Trans. Amer. Math. Soc. (October 1991).
3. O. Hijab, *Partially Observed Control of Markov Processes, IV*, To Appear.
4. P. L. Lions, *Viscosity Solutions of Fully Nonlinear Second-Order Equations and Optimal Stochastic Control in Infinite Dimensions, Part II: Optimal Control of Zakai's Equation*, CEREMADE preprint #8825 (1989).

Department of Mathematics, Temple University, Broad & Montgomery, Philadelphia, PA 19122

E-mail: hijab@euclid.math.temple.edu

Numerical Approximation for Nonlinear Filtering and Finite–Time Observers

Matthew R. James
Department of Mathematics
University of Kentucky
Lexington, KY 40506
USA

François LeGland
INRIA–Sophia Antipolis
2004, Route des Lucioles
F–06565 Valbonne Cedex
FRANCE

I Models of Partially Observed Systems

We consider partially observed systems of the form

$$\left\{ \begin{aligned} dX_t &= b(X_t)\,dt + \sigma(X_t)\,dW_t \\ dY_t &= h(X_t)\,dt + dV_t \end{aligned} \right. \tag{1}$$

where $\{W_t\,,\,t \geq 0\}$ and $\{V_t\,,\,t \geq 0\}$ are independent Wiener processes of appropriate dimensions, with covariance matrices I and R respectively. We are interested in the state estimation problem, under various hypotheses concerning $a = \sigma\,\sigma^*$ and R.

Consider first the two extreme cases : If $a \equiv 0$ and $R \equiv 0$, we are dealing with an *observer* problem for the deterministic system

$$\left\{ \begin{aligned} \dot{X}_t &= b(X_t) \\ z_t &= h(X_t) \end{aligned} \right. \tag{2}$$

At the other extreme, if R is non-singular, we are dealing with a *nonlinear filtering* problem for the diffusion process (1) . We can also easily handle the following intermediate case : If R is non-singular and $a \equiv 0$, we are again dealing with a *nonlinear filtering* problem but the state equation is now an ODE

$$\left\{ \begin{aligned} \dot{X}_t &= b(X_t) \\ dY_t &= h(X_t)\,dt + dV_t \end{aligned} \right.$$

Let us point out that the solution of the state estimation problem is radically different, depending on whether R is non–singular or identically zero. On the other hand, whether a is non–singular, singular or identically zero only affects the algorithms to be used.

Our purpose is to present, for each of the three main cases described above, a solution to the state estimation problem, and to suggest some numerical approximation procedures. The general idea is to study the asymptotics $R \to 0$. As a by–product, we expect to obtain some numerical algorithms for the nonlinear filtering problem, that are robust when the non–singular matrix R is *small*.

II Solutions to State Estimation Problems

We assume for simplicity that $b \in C_b^1(\mathbf{R}^m, \mathbf{R}^m)$ and $h \in C_b^1(\mathbf{R}^m, \mathbf{R}^p)$, unless otherwise stated.

Let us begin with the *nonlinear filtering* (NLF) problem.

□ When R is non-singular, the Bayesian approach to the state estimation problem is to compute the unnormalized conditional probability distribution $\mu_t(dx)$ of the state X_t, given the past observations $\mathcal{Y}_t = \sigma(Y_s\,, 0 \leq s \leq t)$. By definition

$$\langle \mu_t, f \rangle = \mathbf{E}^\dagger[f(X_t)\, Z_t \mid \mathcal{Y}_t]\,,$$

for any test function f, where

$$Z_t^s = \exp\left\{\int_s^t h^*(X_\tau) R^{-1}\, dY_\tau - \tfrac{1}{2}\int_s^t |h(X_\tau)|^2_{R^{-1}}\, d\tau\right\} \quad \text{and} \quad Z_t = Z_t^0\,,$$

and $P^\dagger$ is the *reference* probability measure. The probability distribution $\mu_t(dx)$ satisfies a stochastic PDE in weak sense. Usually, this p.d.f. has a density w.r.t. the Lebesgue measure, i.e. $\mu_t(dx) = p_t(x)\, dx$. A sufficient condition for this to hold, is that the probability distribution $\mu_0(dx)$ of the initial condition X_0 has already a density w.r.t. the Lebesgue measure, i.e. $\mu_0(dx) = p_0(x)\, dx$. We will assume that $p_0(x) > 0$ for all $x \in \mathbf{R}^m$. The unnormalized conditional density $p_t(x)$ is the unique solution of the Zakai equation

$$dp_t = L^* p_t\, dt + p_t h^* R^{-1}\, dY_t\,, \tag{3}$$

with initial condition $p_0(x)$, where L^* is the formal adjoint of the second-order partial differential operator

$$L = \tfrac{1}{2}\,\mathrm{tr}[\, a\, \frac{\partial^2}{\partial x^2}\,] + b \cdot \frac{\partial}{\partial x}\,,$$

associated with the SDE

$$dX_t = b(X_t)\, dt + \sigma(X_t)\, dW_t\,. \tag{4}$$

□ If in addition $a \equiv 0$, then the Zakai equation ((3)) becomes a first order stochastic PDE, for which a representation result is available in terms of the flow $\Phi_t(x)$ associated with the ODE

$$\dot{X}_t = b(X_t)\,. \tag{5}$$

Actually, define

$$J_{t-s}(x) = \det[\, \frac{\partial \Phi_{t-s}}{\partial x}(x)\,] = \exp\left\{\int_s^t \mathrm{div}\, b(\Phi_{\tau-s}(x))\, d\tau\right\}\,,$$

$$\Xi_{s,t}(x) = \exp\left\{\int_s^t h^*(\Phi_{\tau-s}(x)) R^{-1}\, dY_\tau - \tfrac{1}{2}\int_s^t |h(\Phi_{\tau-s}(x))|^2_{R^{-1}}\, d\tau\right\}\,.$$

In this case, the unique solution of the Zakai equation ((3)) satisfies

$$p_t(\Phi_{t-s}(x)) \cdot J_{t-s}(x) = \Xi_{s,t}(x) \cdot p_s(x)\,, \tag{6}$$

or equivalently, introducing the logarithmic transform $W_t(x) = -\log p_t(x)$

$$W_t(\Phi_{t-s}(x)) - \log J_{t-s}(x) = W_s(x) - \log \Xi_{s,t}(x)\,. \tag{7}$$

We turn now to the *observer* problem.

Let $\{x_t^*\,, 0 \leq t \leq T\}$ denote the *true* state trajectory producing the available observation trajectory $\{z_t\,, 0 \leq t \leq T\}$. The idea is to build an observer by considering the limit of a sequence of nonlinear filtering problems with noise covariances going to zero. Two different cases are possible

- Introduce small noises of similar intensities in both the state equation and the observation, i.e. set $a = \varepsilon I$ and $R = \varepsilon I$,

- Introduce a small noise in the observation only, i.e. set $a \equiv 0$ and $R = \varepsilon I$.

□ In the first case, it is proved in James [2] that

$$-\varepsilon \log p_t^\varepsilon(x) \xrightarrow{\varepsilon \downarrow 0} m_t'(x)$$

in probability uniformly on compact subsets of $x \in \mathbf{R}^m$, where up to an additive constant independent of x, $m_t'(x)$ is the unique solution of the Hamilton-Jacobi equation

$$\frac{\partial m_t'}{\partial t} + \tfrac{1}{2}\left|\frac{\partial m_t'}{\partial x}\right|^2 + b \cdot \frac{\partial m_t'}{\partial x} - V_t = 0 , \tag{8}$$

with initial condition $m_0'(x) = 0$, in the *viscosity* sense, where

$$V_t(x) = \tfrac{1}{2}|z_t - h(x)|^2 .$$

In addition, $m_t'(x)$ is the value function associated with the following control problem. Introduce first the action functional

$$I_t(\xi) = \tfrac{1}{2}\int_0^t |\dot{\xi}_s - b(\xi_s)|^2 \, ds$$

if $\xi \in C([0,T]; \mathbf{R}^m)$ is absolutely continuous, and $I_t(\xi) = +\infty$ otherwise. Define also

$$F_t(\xi) = \int_0^t V_s(\xi_s)\, ds = \tfrac{1}{2}\int_0^t |z_s - h(\xi_s)|^2 \, ds .$$

Then

$$m_t'(x) = \inf\left\{I_t(\xi) + F_t(\xi) \; : \; \xi_t = x\right\} .$$

Clearly $m_t'(x) \geq 0$ and $m_t'(x_t^*) = 0$ for the *true* state trajectory, and we define our observer as the set

$$\widehat{x}_t' = \operatorname*{argmin}_{x \in \mathbf{R}^m} \; m_t'(x) = \left\{x \in \mathbf{R}^m \; : \; m_t'(x) = 0\right\} . \tag{9}$$

Obviously $x_t^* \in \widehat{x}_t'$ for all $t \geq 0$. It is proved in James [3] that, provided the deterministic system ((2)) is *observable* on $[0,T]$ (i.e. the map $x_0 \mapsto \{z_s\,,\, 0 \leq s \leq t\}$ is injective), the set-valued observer ((9)) is actually a *finite-time observer* (FTO) on $[0,T]$ (meaning that $\widehat{x}_t'$ is defined in terms of a recursive system with the property that $\widehat{x}_t' = \{x_t^*\}$ for all $t \geq T$).

□ In the second case, it follows from equation ((7)) that

$$-\varepsilon \log p_t^\varepsilon(x) = \varepsilon W_t^\varepsilon(x) \xrightarrow{\varepsilon \downarrow 0} m_t(x) ,$$

in probability uniformly on compact subsets of $x \in \mathbf{R}^m$, where up to an additive constant independent of x, $m_t(x)$ is given by

$$m_t(\Phi_t(x)) = \int_0^t V_s(\Phi_s(x))\, ds \quad \text{or} \quad m_t(x) = \int_0^t V_s(\Phi_{t-s}^{-1}(x))\, ds ,$$

i.e. $m_t(x) = F_t(\xi^{t,x})$, where $\xi^{t,x}$ is the unique solution of the ODE ((5)) ending in x at time t. In addition, $m_t(x)$ is the unique solution of the linear first-order PDE

$$\frac{\partial m_t}{\partial t} + b \cdot \frac{\partial m_t}{\partial x} - V_t = 0 \;,$$

satisfying the initial condition $m_0 = 0$. Just as above, it is clear that $m_t(x) \geq 0$ and $m_t(x_t^*) = 0$ for the *true* state trajectory, and we define our observer as the set

$$\hat{x}_t = \operatorname*{argmin}_{x \in \mathbf{R}^m} \; m_t(x) = \left\{ x \in \mathbf{R}^m \; : \; m_t(x) = 0 \right\} \; . \tag{10}$$

Here again, it is obvious that $x_t^* \in \hat{x}_t$ for all $t \geq 0$, and in addition the set-valued observer defined by ((10)) is actually a FTO on $[0, T]$, provided the deterministic system ((2)) is *observable* on $[0, T]$.

Note that $m_t(x) = F_t(\xi^{t,x})$ where $I_t(\xi^{t,x}) = 0$ (i.e. $\xi^{t,x}$ solves the ODE ((5)) exactly) and $\xi_t^{t,x} = x$, whereas in the definition of $m_t'(x)$, a penalty $I_t(\xi)$ is put on those trajectories ξ that do not solve the ODE ((5)). This is a less severe requirement, and is reflected in the relation $m_t'(x) \leq m_t(x)$. Note however that $\hat{x}_t = \hat{x}_t'$. This is the set of those points that are *indistinguishable* from the true state x_t^*. In conclusion, the observer ((10)) is more *precise* than the observer ((9)), whereas the latter is expected to be more *robust* w.r.t. modeling errors.

III Numerical Approximation

In this section, we restrict ourselves to the situation where the state satisfies an ODE, in which case the solution to the NLF problem is given by ((6)), where R is non-singular, and the corresponding FTO is given by ((10)), where $R \equiv 0$.

Concerning the approximation of the NLF ((6)), we wish to compute an approximate normalized conditional density $p_k^{\Delta,\delta}(x)$ (where Δ and δ denote the time discretization step and the space discretization step respectively) with the following property

($\star$) as $\Delta, \delta \downarrow 0$

$$\mathbf{E} \int_{\mathbf{R}^m} |p_{[t/\Delta]}^{\Delta,\delta}(x) - c_t p_t(x)| \, dx \longrightarrow 0 \quad \text{for all } t \geq 0 \;,$$

where c_t is a normalization constant.

Concerning the approximation of the FTO ((10)), our approach is to build a family $\hat{x}_k^{\Delta,\delta}$ with the following property

($\star\star$) if the deterministic system ((2)) is observable on $[0, T]$, then as $\Delta, \delta \downarrow 0$

$$\operatorname{dist}(\hat{x}_{[t/\Delta]}^{\Delta,\delta}, \{x_t^*\}) \longrightarrow 0 \quad \text{for all } t \geq T \;.$$

A necessary and sufficient condition for ($\star\star$) to hold is $\operatorname{dist}(\hat{x}_{[t/\Delta]}^{\Delta,\delta}, \hat{x}_t) \to 0$ as $\Delta, \delta \downarrow 0$. The approximate observer $\hat{x}_k^{\Delta,\delta}$ will be defined in terms of an approximate value function $m_k^{\Delta,\delta}(x)$, i.e.

$$\hat{x}_k^{\Delta,\delta} = \left\{ x \in \mathbf{R}^m \; : \; m_k^{\Delta,\delta}(x) \leq c^{\Delta,\delta} \right\} \;,$$

and a sufficient condition for ($\star\star$) to hold is $c^{\Delta,\delta} \downarrow 0$ and $m_{[t/\Delta]}^{\Delta,\delta}(x) \to m_t(x)$ uniformly on compact subsets of $\mathbf{R}^m$, as $\Delta, \delta \downarrow 0$.

Time Discretization

Consider a uniform partition $0 = t_0 < \cdots < t_k < \cdots$ of the time interval $[0, \infty)$, with time step $\Delta = t_k - t_{k-1}$. The first step is to sample the available observation trajectory.

The nonlinear filtering problem. If noisy observations $\{Y_t \,,\, t \geq 0\}$ are available, we first build the following sequence of compressed observations

$$y_k^\Delta = \frac{1}{\Delta}[Y_{t_k} - Y_{t_{k-1}}] = \frac{1}{\Delta}\int_{t_{k-1}}^{t_k} h(X_s)\,ds + \frac{1}{\Delta}[V_{t_k} - V_{t_{k-1}}]$$

and we use the approximate model

$$\left\{\begin{aligned} \dot{X}_t &= b(X_t) \\ y_k^\Delta &= h(X_{t_k}) + v_k^\Delta \end{aligned}\right. \tag{11}$$

where $\{v_k^\Delta\,,\ k = 1,2,\cdots\}$ is a Gaussian white noise sequence with covariance matrix R/Δ.

The solution of the NLF problem for the approximate model ((11)) is given in terms of the *a priori* and *a posteriori* conditional probability densities defined by

$$p_{k-\frac{1}{2}}^\Delta(x)\,dx = P(X_{t_k} \in dx \mid \mathcal{Y}_{k-1}^\Delta) \quad \text{and} \quad p_k^\Delta(x)\,dx = P(X_{t_k} \in dx \mid \mathcal{Y}_k^\Delta)\,,$$

respectively, where $\mathcal{Y}_k^\Delta = \sigma(y_1^\Delta, \cdots, y_k^\Delta)$. The transition from $p_{k-1}^\Delta(x)$ to $p_k^\Delta(x)$ is divided into two steps

- *prediction step* : Transport by the flow gives $p_{k-\frac{1}{2}}^\Delta(x) = T_\Delta\ p_{k-1}^\Delta(x)$ where $\{T_t\,,\ t \geq 0\}$ is the semigroup associated with the linear first-order PDE

$$\frac{\partial p_t}{\partial t} = L^* p_t\,. \tag{12}$$

An explicit solution is available for this equation

$$p_{k-\frac{1}{2}}^\Delta(\Phi_\Delta(x)) \cdot J_\Delta(x) = p_{k-1}^\Delta(x)\,, \tag{13}$$

or equivalently

$$\int_A p_{k-\frac{1}{2}}^\Delta(x)\,dx = \int_{\Phi_\Delta^{-1}(A)} p_{k-1}^\Delta(x)\,dx\,, \tag{14}$$

for all Borel set $A \subset \mathbf{R}^m$.

- *correction step* : According to the Bayes formula

$$p_k^\Delta(x) = c_k \cdot \Psi_k^\Delta(x) \cdot p_{k-\frac{1}{2}}^\Delta(x)\,, \tag{15}$$

where

$$\Psi_k^\Delta(x) = \exp\left\{-\tfrac{1}{2}\Delta\ |y_k^\Delta - h(x)|_{R^{-1}}^2\right\}\,,$$

is the *likelihood function* for the estimation of X_{t_k} in the approximate model ((11)), based on the observation y_k^Δ alone, and c_k is a normalization constant.

Introducing the logarithmic transform $W_k^\Delta(x) = -\log p_k^\Delta(x)$, it follows from from ((13)) and ((15)) that

$$W_k^\Delta(x) - \log J_\Delta(\Phi_\Delta^{-1}(x)) = -\log c_k + W_{k-1}^\Delta(\Phi_\Delta^{-1}(x)) + \tfrac{1}{2}\Delta\ |y_k^\Delta - h(x)|_{R^{-1}}^2\,. \tag{16}$$

The observer problem. If perfect observations $\{z_t\,,\ t \geq 0\}$ are available, i.e. $R \equiv 0$, we can simply use $z_k = z_{t_k}$, and our model becomes

$$\left\{\begin{aligned} \dot{X}_t &= b(X_t) \\ z_k &= h(X_{t_k}) \end{aligned}\right. \tag{17}$$

Introducing the asymptotics $R = \varepsilon I$ in the NLF problem and sending ε to zero, it follows from equation ((16)) that

$$-\varepsilon \log p_k^{\Delta,\varepsilon}(x) = \varepsilon W_k^{\Delta,\varepsilon}(x) \xrightarrow{\varepsilon \downarrow 0} m_k^\Delta(x)\,,$$

in probability uniformly on compact subsets of $x \in \mathbf{R}^m$, where $m_k^\Delta(x)$ satisfies the following relation

$$m_k^\Delta(x) = m_{k-1}^\Delta(\Phi_\Delta^{-1}(x)) + \Delta\, V_k^\Delta(x) ,$$

where

$$V_k^\Delta(x) = \tfrac{1}{2}|z_k^\Delta - h(x)|^2 \quad \text{and} \quad z_k^\Delta = \frac{1}{\Delta}\int_{t_{k-1}}^{t_k} z_s\, ds = \frac{1}{\Delta}\int_{t_{k-1}}^{t_k} h(X_s)\, ds \neq z_k .$$

It is clear that $m_k^\Delta(x) \geq 0$. However, because the averaged observation z_k^Δ used in the definition of $m_k^\Delta(x)$ is different from the actual observation z_k, we have $V_k^\Delta(x_{t_k}^*) \neq 0$ in general for the *true* state trajectory. Therefore, we decide to use the actual observation z_k in the definition of $m_k^\Delta(x)$, instead of the averaged observation z_k^Δ, i.e.

$$m_k^\Delta(x) = m_{k-1}^\Delta(\Phi_\Delta^{-1}(x)) + \Delta\, V_k(x) , \tag{18}$$

where

$$V_k(x) = \tfrac{1}{2}|z_k - h(x)|^2 .$$

This relation can be divided into two steps

- *prediction step* : Transport by the flow gives $m_{k-\frac{1}{2}}^\Delta(x) = S_\Delta\, m_{k-1}^\Delta(x)$ where $\{S_t\,,\, t \geq 0\}$ is the semigroup associated with the linear first-order PDE

$$\frac{\partial m_t}{\partial t} + b \cdot \frac{\partial m_t}{\partial x} = 0 . \tag{19}$$

  An explicit solution is available for this equation

$$m_{k-\frac{1}{2}}^\Delta(\Phi_\Delta(x)) = m_{k-1}^\Delta(x) . \tag{20}$$

- *correction step* : The contribution of the new observation z_k to the approximate value function is given by

$$m_k^\Delta(x) = m_{k-\frac{1}{2}}^\Delta(x) + \Delta\, V_k(x) .$$

We note that

$$m_{k-\frac{1}{2}}^\Delta(x) = F_{k-1}^\Delta(\xi^{t_k,x}) \quad \text{and} \quad m_k^\Delta(x) = F_k^\Delta(\xi^{t_k,x}) ,$$

where $\xi^{t,x}$ is the unique solution of the ODE ((5)) ending in x at time t, and the functional $F_k^\Delta(\xi)$ satisfies for all $\xi \in C([0,T];\, \mathbf{R}^m)$

$$F_k^\Delta(\xi) = F_{k-1}^\Delta(\xi) + \Delta\, V_k(\xi_{t_k}) = \Delta\, \left\{V_1(\xi_{t_1}) + \cdots + V_k(\xi_{t_k})\right\} .$$

Now it is clear that $m_k^\Delta(x) \geq 0$ and $m_k^\Delta(x_{t_k}^*) = 0$ for the *true* state trajectory, and we define our observer as the set

$$\hat{x}_k^\Delta = \operatorname*{argmin}_{x \in \mathbf{R}^m} m_k^\Delta(x) = \left\{x \in \mathbf{R}^m \,:\, m_k^\Delta(x) = 0\right\} . \tag{21}$$

Obviously $x_{t_k}^* \in \hat{x}_k^\Delta$ for all k, and one can verify using the explicit formulas that $m_{[t/\Delta]}^\Delta(x) \to m_t(x)$ uniformly on compact subsets as $\Delta \downarrow 0$, with the consequence that property ($\star\star$) holds for this discrete-time approximation.

Model Approximation and PDE Discretization

To obtain computable algorithms, it is necessary to discretize the linear first-order PDE (12) or (19) involved in the prediction step. Generally speaking, two classes of methods can be used : in the *finite difference* approximation (FD) a fixed bounded grid is used, and partial differential operators are approximated by finite differences on grid points, whereas in the *flow-based* approximation (FLOW) the explicit representation (13) or (20) is used to move grid points (or alternatively cells) along the flow of the ODE (5) .

A Finite Difference

A finite difference nonlinear filter. To derive a finite difference algorithm, we must first constrain the nonlinear filtering problem to a bounded domain. Let $D \subset \mathbf{R}^m$ be a m–dimensional open cube. After Dupuis-Ishii [1], we constrain the ODE (5) to the convex set $\bar{D}$ as follows. For $x \in \partial D$, let $\nu(x) = \{ \nu \in \mathbf{R}^m : |\nu| = 1, \langle \nu, x - z \rangle \leq 0 \text{ for all } z \in \bar{D} \}$ denote the set of inward unit normals. For $x \in \bar{D}$, $v \in \mathbf{R}^m$, the projection $\pi(x,v)$ of the velocity vector v at x is given by v if $x \in D$, or $v + [\langle v, -\nu^*(x,v)\rangle \vee 0]\, \nu^*(x,v)$ if $x \in \partial D$, where $\nu^*(x,v)$ is an element of $\nu(x)$ which maximizes $\langle v, -\nu\rangle$, $\nu \in \nu(x)$. Define then $\tilde{b}(x) = \pi(x, b(x))$, $x \in \bar{D}$. By Theorem 5.1 of Dupuis-Ishii [1], there exists a unique absolutely continuous solution of the constrained ODE

$$\dot{\xi}_s = \tilde{b}(\xi_s) \quad \text{a.e. } 0 \leq s \leq t \tag{22}$$

satisfying $\xi_0 = x \in \bar{D}$.

A finite difference algorithm is obtained using a Markov chain scheme similar to those described in Kushner [5]. Let $\mathbf{R}^m_\delta$ denote a coordinate grid of size $\delta > 0$. We define a system of neighborhoods $N_\delta(x) = \{ z \in \mathbf{R}^m_\delta : z = x \text{ or } z = x \pm \delta e_i \text{ for some } i = 1, \ldots, m \}$ for $x \in \mathbf{R}^m_\delta$, where $e_i \in \mathbf{R}^m$ denotes the i-th unit vector. We define next $\bar{D}^\delta = \bar{D} \cap \mathbf{R}^m_\delta$, $D^\delta = \{ x \in \bar{D}^\delta : N_\delta(x) \subset \bar{D}^\delta \}$, and $\partial D^\delta = \bar{D}^\delta \setminus D^\delta$. We define the jump intensity matrix $L_\delta(x,z)$ of a pure jump Markov process $\{\overline{X}^\delta_t, t \geq 0\}$ taking values in $\bar{D}^\delta$ by

$$L_\delta(x,z) = \begin{cases} -|\tilde{b}(x)|_1/\delta & \text{if } z = x, \\ \tilde{b}^\pm_i(x)/\delta & \text{if } z = x \pm \delta e_i \text{ and } i = 1, \ldots, m \\ 0 & \text{if } z \notin N_\delta(x), \end{cases} \tag{23}$$

with the notation $|u|_1 = |u_1| + \cdots + |u_m|$ for any $u = (u_1, \cdots, u_m)$. If we use an *implicit* time discretization scheme, we obtain the finite difference equation

$$p^{\Delta,\delta}_k(x) - \Delta \sum_{z \in N_\delta(x)} L^*_\delta(x,z)\, p^{\Delta,\delta}_k(z) = c_k \cdot \Psi^\Delta_k(x) \cdot p^{\Delta,\delta}_{k-1}(x)\,, \tag{24}$$

for $x \in \bar{D}^\delta$ and $k = 1, \cdots$, where c_k is a normalization constant, and the initial condition $p^{\Delta,\delta}_0(x)$ is a suitable approximation of the density $p_0(x)$. This relation can be divided into two steps

- *prediction step* : Transport by the flow gives

$$[I - \Delta\, L^*_\delta]\, p^{\Delta,\delta}_{k-\frac{1}{2}}(x) = p^{\Delta,\delta}_{k-1}(x)\,.$$

- *correction step* : According to the Bayes formula

$$p^{\Delta,\delta}_k(x) = c_k \cdot \Psi^\Delta_k(x) \cdot p^{\Delta,\delta}_{k-\frac{1}{2}}(x)\,,$$

  where c_k is a normalization constant.

The following result is proved in Kushner [5] using weak convergence $\overline{X}^\delta_\cdot \Longrightarrow X_\cdot$ as $\delta \downarrow 0$.

Theorem 1 *Property* $(\star)$ *holds for the finite difference nonlinear filtering algorithm.*

A finite difference observer. To derive a finite difference algorithm, we still need to constrain the observer problem to a bounded domain. However, because we are going to approximate (18) , we must consider the ODE (5) as running backward in time, before we constrain it to the convex set $\bar{D}$. We use the same definition as above for the set $\nu(x)$ of inward unit normals. For $x \in \bar{D}$, $v \in \mathbf{R}^m$, the projection $\pi(x,v)$ of the velocity vector v at x is now given by v if $x \in D$, or $v + [\langle v, \nu^*(x,v)\rangle \vee 0]\, \nu^*(x,v)$ if $x \in \partial D$, where $\nu^*(x,v)$ is an element of $\nu(x)$ which maximizes $\langle v, \nu\rangle$, $\nu \in \nu(x)$. Define then $\tilde{b}(x) = -\pi(x, -b(x))$, $x \in \bar{D}$. By Theorem 5.1 of Dupuis–Ishii [1] again, there exists a unique absolutely continuous solution $\xi = \xi^{x,t}$ of the constrained ODE

$$\dot{\xi}_s = \tilde{b}(\xi_s) \quad \text{a.e. } 0 \leq s \leq t\,, \tag{25}$$

satisfying $\xi_t = x \in \bar{D}$.

Select $\beta \in C(\mathbf{R}^m)$ non-negative, $\beta \equiv 0$ on $D' \subset D$, with $D' \cap \partial D = \emptyset$, and $\beta > 0$ on ∂D. Now define the value function for $x \in \bar{D}$, $t \geq 0$ by

$$m_t(x) = \beta(\xi_0) + \int_0^t [V_s(\xi_s) + \beta(\xi_s)]\, ds\,, \tag{26}$$

where ξ is the solution of the constrained ODE (25). Then $m_t(x)$ is the unique viscosity solution of the Hamilton–Jacobi equation, see Lions [6]

$$\begin{cases} \dfrac{\partial m_t}{\partial t} + b \cdot \dfrac{\partial m_t}{\partial x} - V_t - \beta = 0 & \text{in } D \times (0,S]\,, \\[2ex] \qquad\qquad -\nu \cdot \dfrac{\partial m_t}{\partial x} = 0 & \text{on } \partial D \times (0,S]\,, \end{cases} \tag{27}$$

satisfying the initial condition $m_0(x) = \beta(x)$ for $x \in D$. In addition, m satisfies in the viscosity sense

$$\frac{\partial m_t}{\partial t} + \tilde{b} \cdot \frac{\partial m_t}{\partial x} - V_t - \beta = 0 \quad \text{on } \partial D \times (0,S]\,. \tag{28}$$

Define the corresponding observer as the set

$$\hat{x}_t = \operatorname*{argmin}_{x \in \mathbf{R}^m} m_t(x) = \left\{ x \in \mathbf{R}^m \,:\, m_t(x) = 0 \right\}\,. \tag{29}$$

Let $\mathcal{I} = \{x_0 \in D' \,:\, \Phi_s(x_0) \in D',\ 0 \leq s \leq S\}$. If $x_0^* \in \mathcal{I}$, then $x_t^* \in \hat{x}_t$ for all $0 \leq t \leq S$, and the observer (29) defines a FTO provided the deterministic system ((2)) is *observable* on $[0,T]$, see James [4].

We again use a Markov chain finite difference scheme. However, we discretize the boundary equation (28) rather than the boundary condition in (27). We use the same definition as above for the grid $\mathbf{R}^m_\delta$, the system of neighborhoods $N_\delta(x)$, and the subsets $\bar{D}^\delta$, D^δ and $\partial D^\delta = \bar{D}^\delta \setminus D^\delta$ of the grid $\mathbf{R}^m_\delta$. Assume that $v = \delta/\Delta$ is a fixed real number, indicating the "speed" of the algorithm, satisfying

$$v \geq \max_{x \in \bar{D}} |\tilde{b}(x)|_1\,. \tag{30}$$

We define the transition probabilities $\pi^{\Delta,\delta}(x,z) = P(\xi^{\Delta,\delta}_{k-1} = z \mid \xi^{\Delta,\delta}_k = x)$ for a *backward* Markov chain $\{\xi^{\Delta,\delta}_k,\ k = [S/\Delta], \ldots, 0\}$ by

$$\pi^{\Delta,\delta}(x,z) = \begin{cases} 1 - |\tilde{b}(x)|_1/v & \text{if } z = x, \\[2ex] \tilde{b}_i^{\mp}(x)/v & \text{if } z = x \pm \delta e_i \text{ and } i = 1, \ldots, m \\[2ex] 0 & \text{if } z \notin N_\delta(x)\,. \end{cases} \tag{31}$$

Note that $\mathbf{E}[\xi^{\Delta,\delta}_{k-1} - \xi^{\Delta,\delta}_k \mid \xi^{\Delta,\delta}_k = x] = -\Delta\, \tilde{b}(x)$.

If we replace $\Phi_\Delta^{-1}(x)$ in (18) by the state $\xi^{\Delta,\delta}_{k-1}$ of the backward Markov chain starting at $\xi^{\Delta,\delta}_k = x$ and take expectations, we obtain the finite difference equation

$$m_k^{\Delta,\delta}(x) = \sum_{z \in N_\delta(x)} \pi^{\Delta,\delta}(x,z)\, m_{k-1}^{\Delta,\delta}(z) + \Delta\, [V_k(x) + \beta(x)] \,, \tag{32}$$

for $x \in \bar{D}^\delta$ and $k = 1, \dots, [S/\Delta]$, with initial condition $m_0^{\Delta,\delta}(x) = \beta(x)$. This relation can be divided into two steps

- *prediction step* : Transport by the flow gives
$$m_{k-\frac{1}{2}}^{\Delta,\delta}(x) = \pi^{\Delta,\delta}\, m_{k-1}^{\Delta,\delta}(x) \,.$$

- *correction step* : The contribution of the new observation z_k to the approximate value function is given by
$$m_k^{\Delta,\delta}(x) = m_{k-\frac{1}{2}}^{\Delta,\delta}(x) + \Delta\, [V_k(x) + \beta(x)] \,.$$

The finite difference observer set is defined by

$$\hat{x}_k^{\Delta,\delta} = \operatorname*{argmin}_{x \in D^\delta} \; m_k^{\Delta,\delta}(x) \,. \tag{33}$$

Obviously, there is no reason for this approximate observer to satisfy the non-asymptotic consistency property : in general we can not guarantee that $x_{t_k}^* \in \hat{x}_k^{\Delta,\delta}$.

The following result is proved in James [4].

Theorem 2 *If $x_0^* \in \mathcal{I}$, then property $(\star\star)$ holds for the finite difference observer algorithm.*

Remark 3 It is also shown in [4] that under additional regularity assumptions $\mathrm{dist}(\hat{x}_{[t/\Delta]}^{\Delta,\delta}, x_t^*) = O(\sqrt{\delta})$ as $\delta \downarrow 0$.

Remark 4 The speed constraint (30) which appears in the finite difference observer algorithm is actually a stability condition for the *explicit* time–discretization scheme used in (32) . From the probabilistic point of view, it ensures that (31) defines transition probabilities. We do not need a similar constraint for the finite difference NLF algorithm, because we are using there an *implicit* time–discretization scheme.

B Flow–Based Approximation

Let us first describe the approximate model we are going to use.

We assume that at each time t_k, a partition $\{B_k^i \,,\, i \in I_k^\delta\}$ of the state space $\mathbf{R}^m$ is given, and we define the discrete I_k^δ–valued state n_k^δ by the relation

$$\left\{n_k^\delta = i\right\} = \left\{X_{t_k} \in B_k^i\right\} \,. \tag{34}$$

The idea behind our approximation is to suppose that, at each step of the algorithm, any information (e.g. probability distribution, likelihood function, value function) about the continuous state X_{t_k} is immediately compressed into some information about the discrete state n_k^δ. We can think of memory constraint as a justification for this compression mechanism. As a consequence, whenever information is needed about the continuous state X_{t_k}, it has to be deduced from the corresponding information about the discrete state n_k^δ, resulting in *compression error.*

Making explicit use of the flow associated with ((5)), we have

$$\left\{\xi_{t_k} \in B_k^i\right\} = \left\{\xi_{t_{k-1}} \in \Phi_\Delta^{-1}(B_k^i)\right\} = \bigcup_{j \in I_{k-1}^\delta[i]} \left\{\xi_{t_{k-1}} \in B_{k-1}^j \cap \Phi_\Delta^{-1}(B_k^i)\right\} \,, \tag{35}$$

where

$$I^\delta_{k-1}[i] = \left\{ j \in I^\delta_{k-1} \ : \ B^j_{k-1} \cap \Phi^{-1}_\Delta(B^i_k) \neq \emptyset \right\} ,$$

provided ξ solves the ODE ((5)). Notice that in general the set $I^\delta_{k-1}[i]$ has finite cardinality.

Various possible choices are available for the partitions, e.g.

- $I^\delta_k = I^\delta_{k-1} \equiv I^\delta$ and $B^i_k = \Phi_\Delta(B^i_{k-1})$ for all $i \in I^\delta$. In this case $n^\delta_k = n^\delta_{k-1}$, i.e. the discrete state process is constant over time, but the sets B^i_k can become very complicated after some steps.
- $I^\delta_k = I^\delta_{k-1} \equiv I^\delta$ and $B^i_k = B^i_{k-1}$ for all $i \in I^\delta$. In this case, the partition is constant over time, but updating the discrete state can be cumbersome.

Between these two extreme cases, a trade–off has to be found in order to reduce the computational burden of updating both the partition and the discrete probability distribution : B^i_k should both be "close" to $\Phi_\Delta(B^i_{k-1})$ and have a simple geometry.

A flow–based nonlinear filter. According to our approximation approach, we introduce the discrete *a priori* and *a posteriori* conditional probability distributions

$$\bar{\mu}^i_{k-\frac{1}{2}} = P(X_{t_k} \in B^i_k \mid \mathcal{Y}^\Delta_{k-1}) \quad \text{and} \quad \bar{\mu}^i_k = P(X_{t_k} \in B^i_k \mid \mathcal{Y}^\Delta_k) ,$$

respectively, where again $\mathcal{Y}^\Delta_k = \sigma(y^\Delta_1, \cdots, y^\Delta_k)$. Making use of (35) transport by the flow gives

$$\begin{aligned}
\bar{\mu}^i_{k-\frac{1}{2}} &= \sum_{j \in I^\delta_{k-1}[i]} P(X_{t_{k-1}} \in B^j_{k-1} \cap \Phi^{-1}_\Delta(B^i_k) \mid \mathcal{Y}^\Delta_{k-1}) \\
&= \sum_{j \in I^\delta_{k-1}[i]} \lambda(B^j_{k-1} \cap \Phi^{-1}_\Delta(B^i_k)) \cdot \frac{1}{\lambda(B^j_{k-1} \cap \Phi^{-1}_\Delta(B^i_k))} \int_{B^j_{k-1} \cap \Phi^{-1}_\Delta(B^i_k)} p^\Delta_{k-1}(x)\, dx \\
&\simeq \sum_{j \in I^\delta_{k-1}[i]} \lambda(B^j_{k-1} \cap \Phi^{-1}_\Delta(B^i_k)) \cdot \frac{1}{\lambda(B^j_{k-1})} \int_{B^j_{k-1}} p^\Delta_{k-1}(x)\, dx \\
&\simeq \sum_{j \in I^\delta_{k-1}[i]} \bar{\mu}^j_{k-1} \, \frac{\lambda(B^j_{k-1} \cap \Phi^{-1}_\Delta(B^i_k))}{\lambda(B^j_{k-1})} .
\end{aligned}$$

Next, according to the Bayes formula

$$\begin{aligned}
\bar{\mu}^i_k &= c_k \int_{B^i_k} \Psi^\Delta_k(x)\, p^\Delta_{k-\frac{1}{2}}(x)\, dx \\
&= c_k \int_{B^i_k} p^\Delta_{k-\frac{1}{2}}(x)\, dx \cdot \frac{\int_{B^i_k} \Psi^\Delta_k(x)\, p^\Delta_{k-\frac{1}{2}}(x)\, dx}{\int_{B^i_k} p^\Delta_{k-\frac{1}{2}}(x)\, dx} \\
&\simeq c_k \cdot \bar{\mu}^i_{k-\frac{1}{2}} \cdot \max_{x \in B^i_k} \Psi^\Delta_k(x) ,
\end{aligned}$$

where c_k is a normalization constant. This approximation can be justified in the small noise case, using the Laplace asymptotic formula.

To obtain a computable algorithm, we introduce new discrete probability distributions $\mu^i_{k-\frac{1}{2}}$ and μ^i_k, and the corresponding densities

$$p^{\Delta,\delta}_{k-\frac{1}{2}}(x) = \mu^i_{k-\frac{1}{2}} / \lambda(B^i_k) \qquad \text{and} \qquad p^{\Delta,\delta}_k(x) = \mu^i_k / \lambda(B^i_k) \qquad \text{iff } x \in B^i_k .$$

We then *define* the transition from $\{\mu_{k-1}^i\,,\, i \in I_{k-1}^\delta\}$ to $\{\mu_k^i\,,\, i \in I_k^\delta\}$, by the following two steps

- *prediction step* : Transport by the flow gives

$$\mu_{k-\frac{1}{2}}^i = \sum_{j\in I_{k-1}^\delta[i]} \mu_{k-1}^j \, \frac{\lambda(B_{k-1}^j \cap \Phi_\Delta^{-1}(B_k^i))}{\lambda(B_{k-1}^j)} \; .$$

- *correction step* : The contribution of the new observation y_k^Δ is given by

$$\mu_k^i = c_k \cdot R_k^i \cdot \mu_{k-\frac{1}{2}}^i \; , \tag{36}$$

where c_k is a normalization constant, and

$$R_k^i = \max_{x\in B_k^i} \Psi_k^\Delta(x) \; ,$$

is the *generalized likelihood function* for the estimation of n_k^δ based on the observation y_k^Δ alone.

Theorem 5 *In the case* $I_k^\delta = I_{k-1}^\delta \equiv I^\delta$, *let* $\{B_0^i\,,\, i \in I^\delta\}$ *denote a finite partition of a bounded domain* D *with* $\mathrm{diam}(B_0^i) \leq \delta$. *Then property* $(\star)$ *holds for this flow-based nonlinear filtering algorithm.*

A flow-based observer. According to our approximation approach, we introduce the *a priori* and *a posteriori* discrete value functions

$$\widetilde{m}_{k-\frac{1}{2}}^i = \inf\left\{F_{k-1}^\Delta(\xi^{t_k,x}) \;:\; x \in B_k^i\right\} \quad \text{and} \quad \widetilde{m}_k^i = \inf\left\{F_k^\Delta(\xi^{t_k,x}) \;:\; x \in B_k^i\right\} \; , \tag{37}$$

respectively. Making use of ((35)) transport by the flow gives

$$\widetilde{m}_{k-\frac{1}{2}}^i = \inf_{j\in I_{k-1}^\delta[i]} \inf\left\{F_{k-1}^\Delta(\xi^{t_{k-1},x}) \;:\; x \in B_{k-1}^j \cap \Phi_\Delta^{-1}(B_k^i)\right\} \geq \inf_{j\in I_{k-1}^\delta[i]} \widetilde{m}_{k-1}^j \; .$$

Next, by definition of the functional $F_k^\Delta(\xi)$,

$$\widetilde{m}_k^i = \inf\left\{F_{k-1}^\Delta(\xi^{t_k,x}) + \Delta\, V_k(x) \;:\; x \in B_k^i\right\} \geq \widetilde{m}_{k-\frac{1}{2}}^i + \Delta \inf_{x\in B_k^i} V_k(x) \; .$$

Thus the discrete value functions satisfy difference *inequalities*. Unfortunately, this does not give a recursive mechanism for computation. Instead, we introduce new discrete value functions $m_{k-\frac{1}{2}}^i$ and m_k^i, and the corresponding value functions

$$m_{k-\frac{1}{2}}^{\Delta,\delta}(x) = m_{k-\frac{1}{2}}^i \qquad \text{and} \qquad m_k^{\Delta,\delta}(x) = m_k^i \qquad \text{iff } x \in B_k^i \; .$$

We then *define* the transition from $\{m_{k-1}^i\,,\, i \in I_{k-1}^\delta\}$ to $\{m_k^i\,,\, i \in I_k^\delta\}$ by the following two steps

- *prediction step* : Transport by the flow gives

$$m_{k-\frac{1}{2}}^i = \inf_{j\in I_{k-1}^\delta[i]} m_{k-1}^j \; .$$

- *correction step :* The contribution of the new observation z_k is given by

$$m_k^i = m_{k-\frac{1}{2}}^i + \Delta \inf_{x\in B_k^i} V_k(x) \; .$$

By construction, it is clear that $m_k^{\Delta,\delta}(x) \geq 0$ and $m_k^{\Delta,\delta}(x_{t_k}^*) = 0$ for the *true* state trajectory, and we define our observer as the set

$$\hat{x}_k^{\Delta,\delta} = \operatorname*{argmin}_{x \in \mathbb{R}^m} m_k^{\Delta,\delta}(x) = \left\{ x \in \mathbb{R}^m \,:\, m_k^{\Delta,\delta}(x) = 0 \right\} , \tag{38}$$

or equivalently

$$\hat{x}_k^{\Delta,\delta} = \bigcup_{i \in \hat{I}_k^\delta} B_k^i \quad \text{with} \quad \hat{I}_k^\delta = \left\{ i \in I_k^\delta \,:\, m_k^i = 0 \right\} .$$

By an inductive comparison argument, it is easy to show that $m_k^{\Delta,\delta}(x) \leq m_k^{\Delta}(x)$, with the consequence that $\hat{x}_k^{\Delta} \subset \hat{x}_k^{\Delta,\delta}$. Therefore, $x_{t_k}^* \in \hat{x}_k^{\Delta,\delta}$.

Theorem 6 *In the case $I_k^\delta = I_{k-1}^\delta \equiv I^\delta$ and $B_k^i = \Phi_\Delta(B_{k-1}^i)$ for all $i \in I^\delta$, let $\{B_0^i,\, i \in I^\delta\}$ denote a finite partition of a bounded domain D with* $\operatorname{diam}(B_0^i) \leq \delta$. *If $x_0^* \in D$, then property $(\star\star)$ will hold for this flow-based observer algorithm.*

As noticed in James [3], the only thing that matters is the argmin set, not the value function itself. This remark can be used to design a simplified algorithm for the construction of the set-valued observer ((38)). We introduce the piecewise-constant logical value functions $\overline{m}_k^{\Delta,\delta}(x)$ taking values TRUE or FALSE, and defined iteratively by the following relations

$$\overline{m}_{k-\frac{1}{2}}^i = \bigvee_{j \in I_{k-1}^\delta[i]} \overline{m}_{k-1}^j ,$$

$$\overline{m}_k^i = \overline{m}_{k-\frac{1}{2}}^i \wedge \overline{V}_k^i ,$$

where

$$\overline{V}_k^i = \begin{cases} \text{TRUE} & \text{if } \inf_{x \in B_k^i} V_k(x) = 0 \\ \text{FALSE} & \text{otherwise} \end{cases}$$

It is clear that $\overline{m}_k^i$ = TRUE iff $m_k^i = 0$, so that an equivalent expression for the set-valued observer ((38)) is given by

$$\hat{x}_k^{\Delta,\delta} = \bigcup_{i \in \hat{I}_k^\delta} B_k^i \quad \text{with} \quad \hat{I}_k^\delta = \left\{ i \in I_k^\delta \,:\, \overline{m}_k^i = \text{TRUE} \right\} .$$

Corollary 7 *Under the assumptions of Theorem 6, property $(\star\star)$ will hold for the simplified algorithm.*

Remark 8 In the particular case where $I_k^\delta = I_{k-1}^\delta \equiv I^\delta$ and $B_k^i = \Phi_\Delta(B_{k-1}^i)$ for all $i \in I^\delta$, the algorithms exhibit a parallel structure explicitly. On the other hand, these algorithms assume that certain calculations can be made exactly. This is not always possible, in which case one would have e.g. to discretize the ODE (5) or use the following approximations

$$\frac{1}{\lambda(B_k^i)} \int_{B_k^i} p(x)\, dx \simeq p(x_k^i) \qquad \max_{x \in B_k^i} \Psi_k^{\Delta}(x) \simeq \Psi_k^{\Delta}(x_k^i) ,$$

$$\inf_{x \in B_k^i} m(x) \simeq m(x_k^i) \qquad \inf_{x \in B_k^i} V_k(x) \simeq V_k(x_k^i) ,$$

where x_k^i is some point in B_k^i.

IV Numerical Experiments

A A One Dimensional Example

We consider a one dimensional model with

$$b(x,t) = -0.2x + 0.8\cos(2.5t) \qquad h(x) = \mathrm{sgn}(x)\ .$$

Even though the observation function is *discontinuous*, the convergence results are still valid, see James [4]. The location of the trajectory is determined at the first time t^* it crosses the origin, so the system is observable.

Figure 1 (below) shows results for the simplified (logical) flow-based observer algorithm, with the choice $I_k^\delta = I_{k-1}^\delta \equiv I^\delta$ and $B_k^i = \Phi_\Delta(B_{k-1}^i)$ for all $i \in I^\delta$, $\Delta = 0.05$, $\delta = 0.02$, and noise-free observations. The estimate $\hat{x}_t$ is a one-dimensional set for times t before t^*, and zero-dimensional after this time.

Figure 2 illustrates the numerical results obtained from the finite difference nonlinear filter algorithm. Here, $\Delta = 0.05$, $\delta = 0.005$, $R = 10^{-4}$, and the observation path was noise-free. Notice the jumps in the conditional mean trajectory and the peaking of the conditional density function each time the origin is crossed. Numerical viscosity causes the density to spread between these times.

Figure 3 shows results for the finite difference observer algorithm, with $\delta = 0.02$, $\Delta = 0.0198$, $v = 1.01$, and noise-free observations. The plot of the value function clearly shows the valley containing the state trajectory.

Figure 4 shows results for the flow-based nonlinear filtering algorithm, with the choice $I_k^\delta = I_{k-1}^\delta \equiv I^\delta$ and $B_k^i = \Phi_\Delta(B_{k-1}^i)$ for all $i \in I^\delta$, $\Delta = 0.05$, $\delta = 0.02$, $R = 10^{-4}$, and noise-free observations. Marginals for the conditional density are shown for times before and after time t^*.

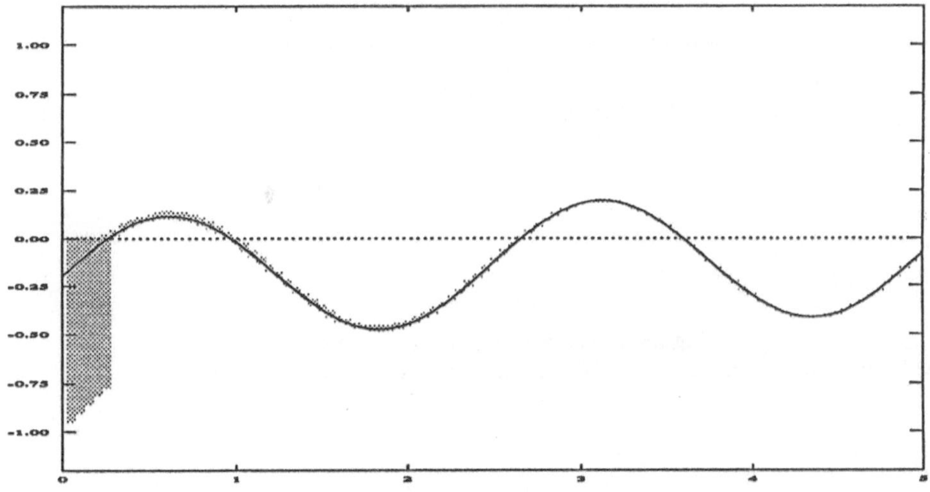

Figure 1. Flow-based observer, simplified algorithm.
State x_t and estimate $\hat{x}_t$ trajectories.

(a)

(b)

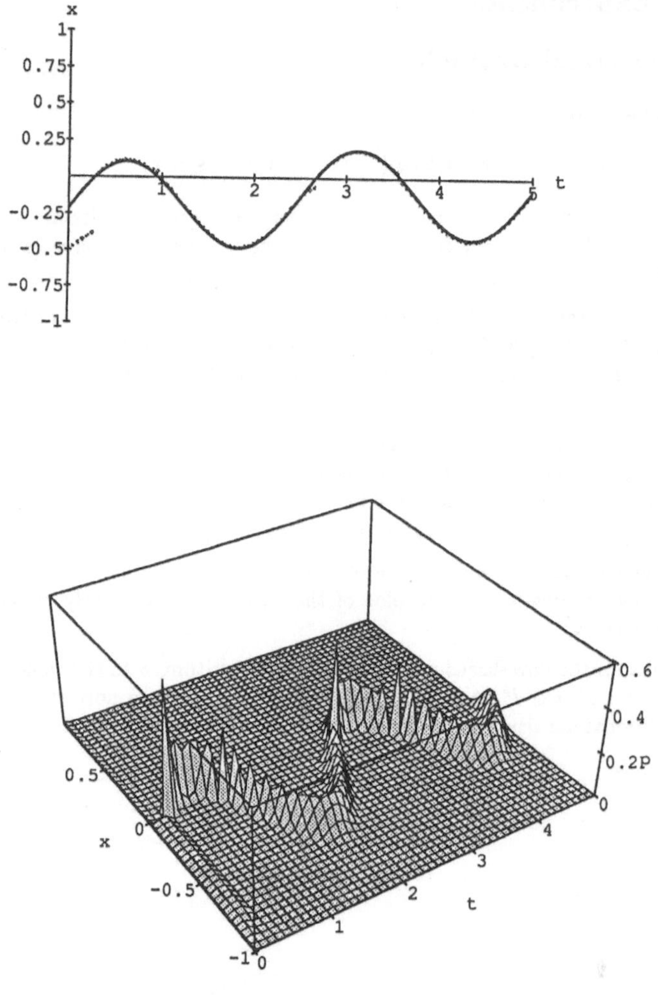

Figure 2. Finite difference nonlinear filter.
(a) State x_t and conditional mean $\mathbf{E}[x_t|\mathcal{Y}_t]$ trajectories; (b) Conditional density function.

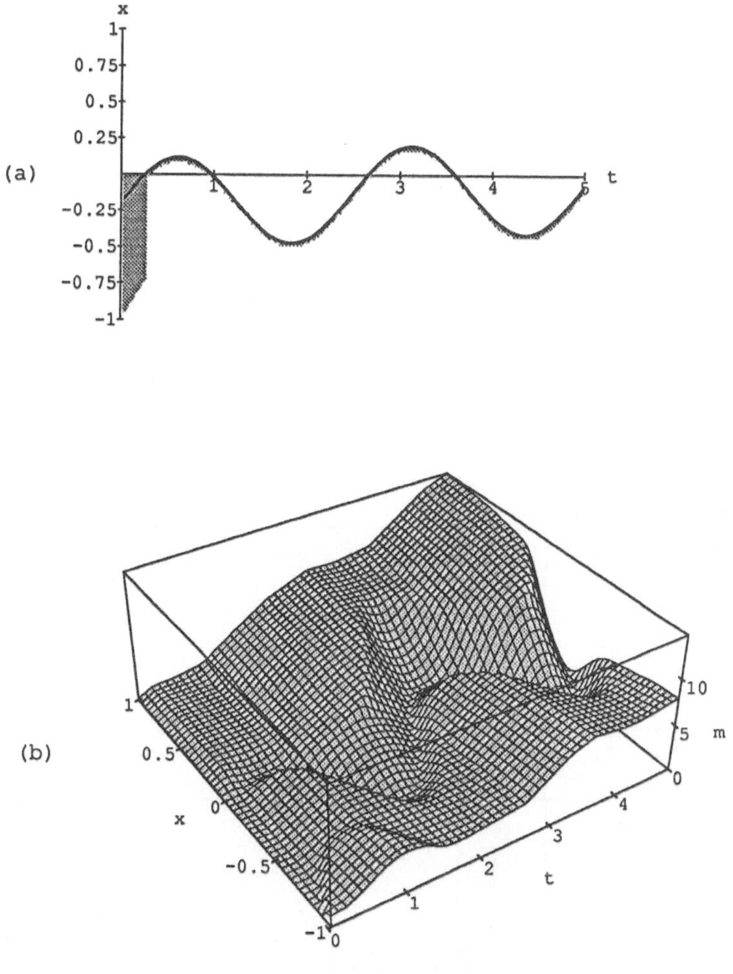

Figure 3. Finite difference observer.
(a) State x_t and estimate $\hat{x}_t$ trajectories; (b) Value function.

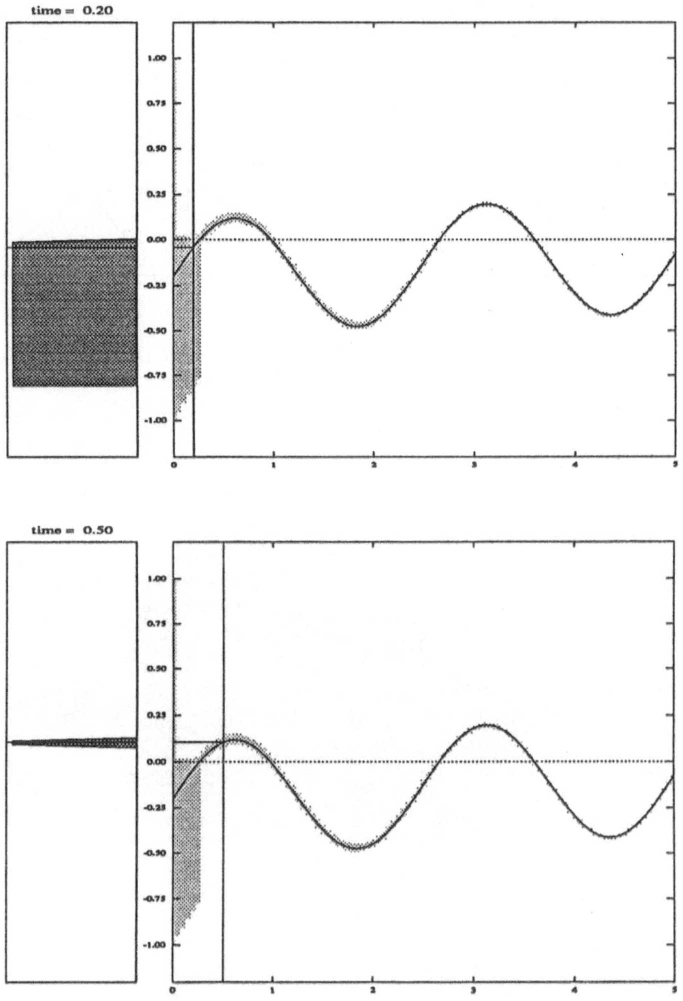

(a)

(b)

Figure 4. Flow-based nonlinear filter.
(a) State x_t trajectory, 90% confidence region, density at $t = 0.2$;
(b) State x_t trajectory, 90% confidence region, density at $t = 0.5$.

B A Four Dimensional Example

We consider here the problem of target motion analysis, which is to estimate the trajectory (position and velocity) of a target moving at constant speed along a straight line at the surface of the sea. We suppose that bearings-only measurements are available in discrete time, taken from a moving observation platform. If the observation platform itself moves at constant speed along a straight line, the problem is ***non-observable.*** However, as soon as the observation platform changes its course, the problem becomes ***observable.*** Assuming that the direction of motion of the target is known, which is true in the case of perfect observations, we can reduce the problem to three dimensions. The state vector is $X = (x, y, v)$ and the state equation

$$\dot{x}_t = v_t \qquad \dot{y}_t = 0 \qquad \dot{v}_t = 0 \, .$$

The observation function is

$$h(x, y, v, t) = \arctan[\, \frac{x - x_t^P}{y - y_t^P} \,] \, ,$$

where (x_t^P, y_t^P) is the (known) position of the observation platform at time t.

For this problem, the flow is known explicitly, and the flow-based algorithms (for both the nonlinear filtering and the observer case) are explicitly parallelizable. A variant of the flow-based NLF algorithm has been implemented at INRIA on a 16K Connection Machine from Thinking Machines Corporation. Numerical experiments have been carried out, using noisy observations with standard deviations ranging between one and five degrees. The goal is to find better maneuvers, and to investigate them off-line. The filter is not intended to be run in real-time on the ship.

Acknowledgment: Research supported by NSF Grant "USA-France (INRIA) Collaborative Research in Stochastic Control" NSF-INT-89-06965.

References

[1] P. DUPUIS and H. ISHII, On Lipschitz continuity of the solution mapping to the Skorokhod problem, with applications, *Stochastics and Stochastics Reports* **35** (1+2) 31–62 (1991).

[2] M.R. JAMES, *Asymptotic Nonlinear Filtering and Large Deviations with Application to observer Design*, Ph.D. Dissertation, University of Maryland (SRC Technical Report Ph.D. 88-1) (1988).

[3] M.R. JAMES, Finite time observer design by probabilistic–variational methods, to appear, *SIAM Journal on Control and Optimization* **29** (4) (1991).

[4] M.R. JAMES, A numerical method for finite time observers, preprint (1990).

[5] H.J. KUSHNER, *Probability Methods for Approximations in Stochastic Control and for Elliptic Equations*, Academic Press, New York, 1977.

[6] P.L. LIONS, Neumann type boundary conditions for Hamilton–Jacobi equations, *Duke Math. J.* **52** (3) 793–820 (1985).

A NUMERICAL METHOD FOR STOCHASTIC SINGULAR CONTROL PROBLEMS WITH NONADDITIVE CONTROLS

Harold J. Kushner[1]
Luiz F. Martins[2]

Division of Applied Mathematics
Brown University, Box F
Providence, R.I. 02912
April, 1991

Abstract

We consider stochastic systems with "singular" controls, where the control might be non-additive. It is shown that the Markov chain approximation method yields a convergent numerical method.

1 Introduction

In modelling physical processes, situations occur where several impulses or control actions occur in "quick succession"; one might say that there is a possibility of "multiple simultaneous" impulses. This might occur naturally. But it is commonly due to the nature of the scaling in developing an approximation to the original physical model, as in [1], [2], where events which might be separated by "small" time in the original model are squeezed together due to the scaling used to get a "limit" or simpler approximation. Essentially, an impulsive action is taken, the state changes accordingly, a follow-up action is taken "immediately after", and so on. A brief example is given in Section 2. The modelling of such problems and the numerical methods for the solution of associated optimal control problems require special care. It will be shown that the Markov chain approximation method [3], [4], [5] can be easily adapted to this problem. The basic systems model will be written as

$$x(t) = x + \int_0^t b(x(s))ds + \int_0^t \sigma(x(s))dw(s) + \int_0^t g(x(s^-)) \circ dJ(s) \tag{1.1}$$
$$+L(t) - U(t).$$

The integral

$$\int_0^t g(x(s^-)) \circ dJ(s) \tag{1.2}$$

is not necessarily a Stieltjes integral, and its proper definition is one of the main matters of concern. The appropriate definition of this term will follow naturally from the physical problems of interest and will be given in Section 2. The "non-additivity" of the effects of the controls is due to the $g(x)$ term in (1.1), since the effects of two impulses in $J(\cdot)$ in quick succession will not equal the effect of their sum being applied together. There is a filtration $\mathcal{F}_t$ such that $w(\cdot)$ is a standard $\mathcal{F}_t$-Wiener process, and $x(\cdot)$ and $J(\cdot)$ are $\mathcal{F}_t$-adapted processes.

[1]This work is supported in part by Grants AFOSR 89-0015, ARO-DAAL-03-86-0171, NSF-ECS 89-13351.
[2]Also in Mathematics Dept. of Universidade Feleral de Rio Grande do Sul. Supported in part by NSF-ECS 89-13351.

Let E_x^J denote the expectation given $x(0) = x$ and under control $J(\cdot)$. The cost functional will be

$$W(x, J) = E_x^J \int_0^\infty e^{-\beta t} k(x(t)) dt + E_x^J \int_0^\infty e^{-\beta t} |dJ|(t) \tag{1.3}$$

where $\beta > 0$ and $k(\cdot)$ is assumed to be continuous.

The system is one dimensional and $x(t)$ is confined to the interval $[0, B]$. $L(\cdot)$ and $U(\cdot)$ are the "reflection" terms at the boundaries 0 and B, resp. The multidimensional problem is treated in an essentially similar way and will be commented on at the end of the paper. $J(\cdot)$ is a "singular" control, a right continuous process which is of bounded variation on each bounded time interval, although we will need to keep track of the "pieces" of jumps in $J(\cdot)$ which are due to particular control actions.

The functions $b(\cdot), g(\cdot)$ and $\sigma(\cdot)$ are bounded and Lipschitz continuous. The Lipschitz continuity is used only to guarantee uniqueness and to avoid details which are peripheral to our main purpose. All that is actually needed is continuity and weak sense uniqueness of the solutions to (2.3) below. For simplicity, let $\sigma(x) > 0$ on $[0, B]$, although the convergence Theorem 5.1 can be shown to hold even if this condition fails. Define

$$V(x) = \inf_J W(x, J),$$

where the inf is over the admissible controls, which will be defined in the next section.

The Markov chain approximation method of [3], [4], [5] is a useful approach to numerical methods for stochastic and deterministic control problems. Reference [5] dealt with "singular control problems" (vector case), where the $g(\cdot)$ in (1.1) did not depend on x (which is the general situation in the current literature on the singular control problem). Nevertheless, the numerical methods and proofs in [5] can be readily adapted to the problem at hand, and heavy use will be made of [5] in the sequel.

Let $h > 0$ be a scalar approximation parameter. The basic approach involves the construction of a controlled finite state Markov chain $\{\xi_n^h, n < \infty\}$, whose "local behavior" resembles that of (1.1) in the sense given in Section 3. An appropriate cost function for the chain is defined, and the optimal value $V^h(x)$ computed. One needs to prove that $V^h(x) \to V(x)$. The proofs use the methods of weak convergence. The path spaces will always be $D^k[0, \infty)$, the space of CADLAG functions with the Skorohod topology [7], [8].

It turns out that, although the proper definition of (1.2) might be subtle, the Markov chain approximation picks out the correct one, in that the weak limits of suitable continuous time interpolations of the approximating chains have the representation (1.1) with the desired definition of (1.2).

2 Interpretation of the singular control term (1.2).

The forms of the model (1.1) which are of interest to us arise as scaled limits of a sequence of physical processes in either discrete or continuous time. The interpretation given to the integral (1.2) must be consistent with the "limits of the paths" of the physical processes. One example will be given to illustrate the approach as well as the issues to be dealt with.

We follow the ideas of [6] concerning limits of sequences of processes driven by approximations of Poisson jump processes, and of [2], which dealt with optimal control problems on similar models and actually motivated this work.

A "physical model". Let $\delta > 0$ and define the difference $\Delta w(n\delta) = w(n\delta + \delta) - w(n\delta)$. Let $C < \infty$ be given. Define the controlled process $\{X_n^\delta\}$ by $X_0^\delta = 0$ and

$$X^\delta_{n+1} = X^\delta_n + \delta b(X^\delta_n) + \sigma(X^\delta_n)\Delta w(n\delta) \tag{2.1}$$
$$+ g(X^\delta_n)\Delta J^\delta_n + \Delta L^\delta_n - \Delta U^\delta_n, |\Delta J^\delta_n| \le C$$

where ΔL^δ_n and ΔU^δ_n are the reflection terms which serve to keep X^δ_n in $[0, B]$. I.e.,

$$\Delta L^\delta_n = \max[-(X^\delta_n + \delta b(X^\delta_n) + \sigma(X^\delta_n)\Delta w(n\delta) + g(X^\delta_n)\Delta J^\delta_n), 0]$$

$$\Delta U^\delta_n = \max[(X^\delta_n + \delta b(X^\delta_n) + \sigma(X^\delta_n)\Delta w(n\delta) + g(X^\delta_n)\Delta J^\delta_n - B), 0]$$

The control term ΔJ^δ_n is $\mathcal{F}(n\delta)$-measurable. Let the cost be

$$W^\delta(x, J^\delta) = E^{J^\delta}_x \sum_{n=0}^{\infty} e^{-\beta n\delta}[k(X^\delta_n)\delta + |\Delta J^\delta_n|]. \tag{2.2}$$

We use (2.1) as a "canonical" model, since it allows as to illustrate the main issues most clearly. Identical results would be obtained if $\{\Delta W^\delta_n\}$ were replaced by a sequence $\{\xi^\delta_n\}$ of mutually independent and identically distributed random variables variance δ and $\sup_{n \le T/\delta} |\xi^\delta_n| \xrightarrow{\delta} 0$. If $\{\xi^\delta_n\}$ were a mixing sequence, with a fast enough mixing rate, then weak convergence methods of the type used in [9] can be used to get the results, but there will be the usual "correction term" in (1.1).

Our interest in (1.1), (1.3) is as a limit of suitably interpolated processes and costs such as (2.1), (2.2). It is (2.1) which is to be considered to be the physical or "applications" model. (1.1) is to be treated as a convenient and simplifying approximation for small $\delta > 0$. An example of an "advertising model" where this is of interest is in [2]. It is often easier to solve the numerical problem for an approximation to (1.1), than for (2.1) directly.

Define $J^\delta(t) = \sum_{n\delta < t} \Delta J^\delta_n$ and define $L^\delta(\cdot), U^\delta(\cdot)$ analogously. Define $X^\delta(t) = X^\delta_n$ for $t \in [n\delta, n\delta + \delta)$.

Example. The advertising example in [2] is roughly of the following form, where x denotes the market share of a given product and ΔJ^δ_n is the money spent on advertising in period n. Thus $B = 1$. We would have $b(x) < 0$ for $x \in (0, 1]$ and $b(0) = 0$ since in the absence of advertising the average market share would slip. The function $g(\cdot)$ is convex and satisfies $g(1) = 0, g(0) > 0$ but small. For specificity, suppose that $g(\cdot)$ is strictly increasing in a neighborhood of $x = 0$. Thus, as current market share increases, the advertising is more effective, up to a point. An incremental increase in market share would require more investment if that share is sufficiently large, since $g(\cdot)$ decreases near $x = 1$.

The upper bound C on ΔJ^δ_n reflects resource limitations and also helps guarantee that market share will not change unrealistically in a short period. The basic control problem concerns the investment decisions. If the initial market share is small, then the basic control decision involves a resolution of the conflict between immediate return over exploiting the (initial) increased effectiveness of advertising as market share increases.

We note that the optimal control problem is not to be solved analytically here. The numerical method gives an arbitrarily good approximation to the correct solution.

A problem with weak convergence. To see the difficulty with the approximation and convergence of $\{X^\delta(\cdot)\}$, consider a particular case of (2.1). Let $1/\delta$ be an integer and define $J^\delta(\cdot)$ by

$$J^\delta(t) = \begin{cases} 0, & 0 \le t < 1 \\ J^\delta(n\delta) + \sqrt{\delta}, & 1 \le n\delta \le t \le \min\{(n+1)\delta, 1+\sqrt{\delta}\} \\ 1, & t \ge 1 + \sqrt{\delta}. \end{cases}$$

Define $\Delta J_n^\delta = J^\delta(n\delta+\delta) - J^\delta(n\delta)$. Then neither $\{J^\delta(\cdot)\}$ nor $\{X^\delta(\cdot)\}$ are tight in the Skorohod topology. Also, note that a Stieltjes integral interpretation of (1.2) is not appropriate if the solution of (1.1) is to be an approximation to $X^\delta(\cdot)$ and $g(\cdot)$ depends on x.

In order to see what is needed to properly define (1.1), note that all of the "action" in the example takes place on the interval $[1, 1+\sqrt{\delta}]$, where $w(\cdot)$ plays no important role for small δ. The difficulty in taking limits can be dealt with via a time scale change of the type used in [5] and we next give the appropriate definitions.

Definition of (1.1), (1.2) and the random time scale change. Let $\hat{T}(\cdot)$ be a non-decreasing process with $\hat{T}(0) = 0$ and with Lipschitz continuous paths, and where $\hat{T}(t) \to \infty$ w.p.1 as $t \to \infty$, and $\hat{T}(t_0)$ is an $\mathcal{F}_t$-stopping time for each $t_0 > 0$. Let $\hat{J}(\cdot)$ be a right continuous process of bounded variation on each finite interval, whose paths are the sum of a pure jump process and a process with Lipschitz continuous paths, and where $\hat{J}(t)$ is $\mathcal{F}_{\hat{T}(t)}$-adapted. Define $\hat{w}(t) = w(\hat{T}(t))$. Such triples $(\hat{T}(\cdot), \hat{J}(\cdot), w(\cdot))$ are said to be *admissible*. If $w(\cdot)$ is fixed, we simply refer to the pair $(\hat{T}(\cdot), \hat{J}(\cdot))$ as admissible. We will not lose any generality by supposing that $\hat{T}(\cdot)$ is differentiable.

For admissible $(\hat{T}(\cdot), \hat{J}(\cdot), w(\cdot))$, there is a unique strong sense solution to

$$\hat{x}(t) = x + \int_0^t b(\hat{x}(u))d\hat{T}(u) + \int_0^t \sigma(\hat{x}(u))d\hat{w}(u) \tag{2.3}$$

$$+ \int_0^t g(\hat{x}(u^-))d\hat{J}(u) + \hat{L}(t) - \hat{U}(t),$$

where $\int_0^t g d\hat{J}$ is an ordinary Stieltjes integral, the jumps of $\hat{J}(\cdot)$ are bounded in magnitude by C and $\hat{L}(\cdot)$ and $\hat{U}(\cdot)$ are the reflection terms at the boundaries 0 and B, resp. Define the cost function

$$\begin{aligned} \hat{W}(x, \hat{J}, \hat{T}) = \; & E_x^{\hat{J},\hat{T}} \int_0^\infty e^{-\beta \hat{T}(s)} k(\hat{x}(s)) d\hat{T}(s) + \\ & E_x^{\hat{J},\hat{T}} \int_0^\infty e^{-\beta \hat{T}(s)} |d\hat{J}|(s). \end{aligned} \tag{2.4}$$

Define process $T(\cdot)$, by: $T(t) = \min\{s : \hat{T}(s) = t\}$. We say that $(x(\cdot), J(\cdot), w(\cdot))$ solves (1.1) if there are admissible $(\hat{T}(\cdot), \hat{J}(\cdot), w(\cdot))$ and an associated solution $\hat{x}(\cdot)$ to (2.3), such that $x(t) = \hat{x}(T(t))$ and $J(t) = \hat{J}(T(t))$. Note that $x(\cdot)$ and $J(\cdot)$ might be left continuous, but this is unimportant. If $W(x, J)$ is defined by (1.3), then

$$W(x, J) = \hat{W}(x, \hat{J}, \hat{T}).$$

Clearly, the time change $\hat{T}(\cdot)$ is not unique. Suppose that there is a $\tau(\cdot)$ which, together with its inverse, satisfies the conditions on $\hat{T}(\cdot)$ above. For admissible $(\hat{J}_1, (\cdot), \hat{T}_1(\cdot), w(\cdot))$, let $\hat{x}_1(\cdot)$ denote the solution to (2.3). Define $\hat{J}_2(\cdot) = \hat{J}_1(\tau(\cdot))$, etc. Then $(\hat{J}_2(\cdot), \hat{T}_2(\cdot), \hat{x}_2(\cdot), w(\cdot))$ also solve (2.3) and the two solutions are said to be equivalent. Also $\hat{W}(x, \hat{J}_1, \hat{T}_1) = \hat{W}(x, \hat{J}_2, \hat{T}_2)$.

Example of "simultaneous" jumps. Let $a < c_1 < c_2 < b$, and define $\hat{T}(\cdot)$ such that it is constant on $[a, b]$ and $\hat{T}'(t) = 1$ otherwise. Let $\hat{J}(\cdot)$ be piecewise constant with jumps only at c_1 and c_2. Then $J(\cdot)$

will have "simultaneous" jumps (which occur at time 1), but the jump corresponding to scaled time c_1 occurs "just before" that corresponding to scaled time c_2. Define

$$\tilde{x}(a) = x(a^-) + g(x(a^-))[\hat{J}(c_1) - \hat{J}(c_1^-)].$$

Then

$$x(a^+) = \tilde{x}(a) + g(\tilde{x}(a))[\hat{J}(c_2) - \hat{J}(c_2^-)].$$

3 The Numerical Method.

The approximating Markov chain. Following the idea in [3], [4], [5], for a scalar approximation parameter $h > 0$, we construct a controlled finite state Markov chain $\{\xi_n^h, n < \infty\}$ on a state space S_h. Let J^h denote the generic control policy for the chain. Then we construct a cost functional $W^h(x, J^h)$, such that the computational problem for the associated optimal control problem is feasible and the optimal cost functional $V^h(x) = \inf_{J^h} W^h(x, J^h)$ converges to $V(x)$ as $h \to 0$.

For expository simplicity, we use a simple "regularly spaced" state space. Let B be an integer multiple of h and define the sets $G_h = \{0, h, 2h, \cdots B\}$, $\partial G_h^+ = \{-h, B+h\}$ and $S_h = G_h \cup \partial G_h^+$. The set ∂G_h^+ is the "reflecting boundary". We are guided by the fact that the reflection terms $L(\cdot)$ and $U(\cdot)$ keep $x(\cdot)$ from leaving $[0, B]$. Once the chain leaves G_h, which is the discretization of the set $[0, B]$, it must be returned immediately. The idea is to choose the controlled transition probability for the chain such that the "local" behavior of the chain mimics that of $x(\cdot)$, in a sense to be described, for the various cases of reflection and control.

Let E_n^h denote the expectation given all the control actions taken and the states up to time n. Let $E_{n,x}^h$ denote the expectation given the above data and $\xi_n^h = x$. Consider the *unreflected and uncontrolled form of the process* (1.1), where $J(t) = U(t) = L(t) \equiv 0$. We say that the Markov chain with transition probability $p^h(x, y)$ is locally consistent with the *unreflected and uncontrolled form of the process* (1.1) if there is an "interpolation interval" $\Delta t^h(x) > 0$ with $\Delta t^h(x) \to 0$ as $h \to 0$ such that

$$E_{n,x}^h \Delta\xi_n^h = b(x)\Delta t^h(x) + o(\Delta t^h(x)) \tag{3.1}$$

$$E_{n,x}^h (\Delta M_n^h)^2 = \sigma^2(x)\Delta t^h(x) + o(\Delta t^h(x)),$$

$$|\xi_{n+1}^h - \xi_n^h| = O(h),$$

where

$$\Delta M_n^h = \Delta\xi_n^h - E_n^h \Delta\xi_n^h.$$

We now turn to a discussion of the properties which the Markov chain must have, if it is to provide a suitable approximation to (1.1). At each time step for the chain, we can distinguish three possibilities. (Case 1) Let ξ_n^h equal either $-h$ or $B+h$. Then it must be reflected back into G_h. If $\xi_n^h = -h$, set $\xi_{n+1}^h = 0$, and define $\Delta L_n^h = h$, otherwise set $\Delta L_n^h = 0$. If $\xi_n^h = B + h$, set $\xi_{n+1}^h = B$, and define $\Delta U_n^h = h$. Otherwise set $\Delta U_n^h = 0$. This is called a *reflection step.* If $\xi_n^h = x \in G_h$, there are two possibilities; either we (case 2) do not exert control, or (case 3) we do exert control. If we choose not to exert control, then the transition probability is to be "locally consistent" with the uncontrolled (1.1) in the sense of (3.1). This is called the *diffusion step.*

Let $\xi_n^h = x \in G_h$. Under the third possibility, the so-called "control step", let ΔJ_n^h denote the control effort and choose ΔJ_n^h subject to $x+g(x)\Delta J_n^h \in [0,B]$. If $x+g(x)\Delta J_n^h \in G_h$, then define $\xi_{n+1}^h = x+g(x)\Delta J_n^h$. If not, let ξ_{n+1}^h take random values among the two grid points which neighbor $x+g(x)\Delta J_n^h$ in such a way that $E_{n,x}^h \Delta\xi_n^h = g(x)\Delta J_n^h$. The randomization is used to achieve arbitrary average step sizes, but it is not really important in the one dimensional problem. Analogous randomizations play a crucial role in the multidimensional case [5]. For either a reflection or control step, let the interpolation interval $\Delta t^h(x)$ take the value zero. The sequence $\{\Delta J_n^h\}$ is said to be admissible if the Markov property is preserved under that control; in particular, if

$$P\{\xi_{n+1}^h = y|\xi_i^h, \Delta J_i^h, i \le n\} = P\{\xi_{n+1}^h = y|\xi_n^h, \Delta J_n^h\}.$$

Define $\Delta\tilde{J}_n^h = [\Delta\xi_n^h - g(\xi_n^h)\Delta J_n^h]/g(\xi_n^h)$ for $g(\xi_n^h) \neq 0$, and zero otherwise. Thus $\Delta\tilde{J}_n^h = 0$ if $x+g(x)\Delta J_n^h \in G_h$. If control is not used at time n, then set $\Delta J_n^h = \Delta\tilde{J}_n^h = 0$. Define $\Delta t_i^h = \Delta t^h(\xi_i^h)$ and the sums

$$U_n^h = \sum_{i=0}^{n-1} \Delta U_i^h, L_n^h = \sum_{i=0}^{n-1} \Delta L_i^h, J_n^h = \sum_{i=0}^{n-1} \Delta J_i^h, M_n^h = \sum_{i=0}^{n-1} \Delta M_i^h, B_n^h = \sum_{i=0}^{n-1} b(\xi_i^h)\Delta t_i^h$$

$$\tilde{G}_n^h = \sum_{i=0}^{n-1} g(\xi_i^h)\Delta\tilde{J}_i^h, G_n^h = \sum_{i=0}^{n-1} g(\xi_i^h)\Delta J_i^h.$$

Analysis of $\tilde{J}_n^h$. The "error" process $\tilde{J}_n^h$ disappears in the limit, and we next give some calculations which will be useful to show that fact later. Let $\xi_n^h = x$ and suppose that step n is a control step and that $x+g(x)\Delta J_n^h \notin G_h$. Let p denote the probability that ξ_{n+1}^h = next largest grid point to $x+g(x)\Delta J_n^h$. Then

$$E_n^h|g(x)\Delta\tilde{J}_n^h|^2 = h^2 p(1-p). \tag{3.2}$$

Also, there is a grid point $y \in G_h$ such that

$$y-(1-p)h \le g(x)\Delta J_n^h \le y+ph.$$

Thus,

$$E\sup_{i\le n} |\tilde{G}_i^h|^2 \le E|G_n^h|O(h). \tag{3.3}$$

Define the interpolated time $t_n^h = \sum_0^{n-1} \Delta t_i^h$. For admissible $J^h = \{\Delta J_n^h\}$, an appropriate approximation to the cost function (1.3) is

$$W^h(x, J^h) = E_x^{J^h} \sum_{n=0}^{\infty} e^{-\beta t_n^h}[k(\xi_n^h)\Delta t_n^h + |\Delta J_n^h|]. \tag{3.4}$$

Define $V^h(x) = \min_{J^h} W^h(x, J^h)$. A "reflection cost" can be added to (3.4) without altering the main results, but we wish to keep the development simple.

The dynamic programming equation. For $x \in G_h$, the dynamic programming equation for the Markov chain and cost (3.4) is

$$V^h(x) = \min\{\sum_y e^{-\beta\Delta t^h(x)} p^h(x,y)V^h(y) + k(x)\Delta t^h(x), \tag{3.5}$$

$$\min_{\substack{k,p \\ |kh+ph| \le C}} [(1-p)V^h(x+kh) + pV^h(x+kh+h) + (kh+ph)]\},$$

with the boundary condition

$$V^h(-h) = V^h(0), V^h(B+h) = V^h(B).$$

In (3.5), k is an integer (positive, negative or zero) such that $x + kh$ and $x + kh + h \in$ are in G_h, and $p \in [0, 1]$. We note that the minimizing p will always be either zero or one. Hence p can be eliminated from (3.5). I.e., in general, we need not randomize the control and can let $\delta \tilde{G}_n^h = 0$.

Since $\Delta t_n^h = 0$ if n is either a control or a reflection step, we need to be careful in defining an interpolation of $\{\xi_n^h\}$. Define the interpolated process $\xi^h(\cdot)$ by $\xi^h(t) = \xi_n^h$ for $t \in [t_n^h, t_{n+1}^h)$, and step n being a diffusion step. An equivalent definition is $\xi^h(t) = \sum_{n: t_{n+1}^h \leq t} \Delta\xi_n^h + \xi_0^h$. Thus, time does not "advance" at the control or reflection steps. $\xi^h(\cdot)$ will be shown to be an approximation to the solution to (1.1) for an appropriate control.

4 Time Scaling, Approximations and Interpolation.

Using the definitions in the last section, we can now write

$$\xi_n^h = x + B_n^h + M_n^h + L_n^h - U_n^h + G_n^h + \tilde{G}_n^h + \delta_n^h, \tag{4.1}$$

where for any $T < \infty$, $\sup_{n: t_n^h \leq T} |\delta_n^h| \xrightarrow{h} 0$. Define the interpolations $B^h(\cdot), M^h(\cdot), \cdots$ of the terms on the right side of (4.1) by $B^h(t) = B_n^h$ on $[t_n^h, t_{n+1}^h)$ for n a diffusion step. An equivalent definition is $B^h(t) = \sum_{t_{n+1}^h \leq t} b(\xi_n^h)\Delta t_n^h$, and similarly for the other terms. Define $\Delta w_n^h = \sigma^{-1}(\xi_n^h)\Delta M_n^h$ and $w_n^h = \sum_{i=0}^{n-1} \Delta w_n^h$. Define the interpolation $w^h(\cdot)$ as $B^h(\cdot)$ was defined. Note that $\Delta w_n^h = 0$ unless n is a diffusion step. It will turn out that $w^h(\cdot)$ converges weakly to a standard Wiener process (Theorem 4.1).

The major problem in getting the desired weak convergence of $\xi^h(\cdot)$ (or of a suitable subsequence) to a process $x(\cdot)$ satisfying (1.1) and the convergence $V^h(x) \to V(x)$ is the possibility of non-tightness (in the Skorohod topology) of $\{\xi^h(\cdot)\}$ or, equivalently, of $\{J^h(\cdot)\}$. The 'stretched out' time transformation scheme introduced in [5] can be used to circumvent this problem, and lead to (2.3) as well.

Define the sequence $\{\Delta\hat{t}_n^h\}$ by: $\Delta\hat{t}_n^h = \Delta t_n^h$ for n a diffusion step and set $\Delta\hat{t}_n^h = |\Delta J_n^h|$ otherwise. Thus $\Delta\hat{t}_n^h = 0$ if n is a reflection step. Define $\hat{t}_n^h = \sum_{i=0}^{n-1} \Delta\hat{t}_i^h$. Define $\hat{T}^h(\cdot)$ to be the continuous piecewise linear process which satisfies $\hat{T}^h(\hat{t}_n^h) = t_n^h$. Finally, define the "stretched out" process: $\hat{w}^h(t) = w^h(\hat{T}^h(t))$, and similarly define $\hat{\xi}^h(\cdot), \cdots$.etc.

Then we can write (4.1) as

$$\hat{\xi}^h(t) = x + \hat{B}^h(t) + \hat{M}^h(t) + \hat{L}^h(t) - \hat{U}^h(t) + \hat{G}^h(t) + \hat{\tilde{G}}(t) + \hat{\delta}_h(t). \tag{4.2}$$

Also,

$$W^h(x, J^h) = E_x^{J^h} \int_0^\infty e^{-\beta \hat{T}^h(s)}[k(\hat{\xi}^h(s))d\hat{T}^h(s) + |d\hat{J}^h|(s)] + \delta_h, \tag{4.3}$$

where $\hat{\delta}_h(t)$ and δ_h go to zero as $h \to 0$. The error term δ_h is due to the approximation of the sum in (3.4) by an integral. The time stretching was chosen to guarantee the tightness of the control processes. The idea of a time stretching is consistent with the definition of the solution to (1.1) which is given in Section 2, and with the relation between the "physical" processes (2.1) and (2.3).

The following theorem can be proved by a direct application of the methods Theorems 5.3 and 5.4 in [5].

Theorem 4.1. *Let $\{J^h(\cdot)\}$ be admissible controls. The set of processes $\{\hat{\xi}^h(\cdot), \hat{B}^h(\cdot), \hat{M}^h(\cdot), \hat{w}^h(\cdot), w^h(\cdot),$ is tight in the Skorohod topology. Let $(\hat{x}(\cdot), \hat{B}(\cdot), \hat{M}(\cdot), \hat{w}(\cdot), w(\cdot), \hat{L}(\cdot), \hat{U}(\cdot), \hat{J}(\cdot), \hat{T}(\cdot), \hat{\hat{G}}(\cdot))$ denote the limit of a weakly convergent subsequence. They have continuous paths w.p.1, $\hat{T}(\cdot)$ is non-decreasing and absolutely continuous, with $\hat{T}'(t) \leq 1$ and $\hat{T}(0) = 0$. $w(\cdot)$ is a standard Wiener process, $\hat{w}(t) = w(\hat{T}(t))$ and $\hat{\hat{G}}(t) = 0$. The 'hat' limit processes are non-anticipative with respect to $\hat{w}(\cdot)$ and satisfy*

$$\hat{x}(t) = x + \int_0^t b(\hat{x}(s))d\hat{T}(s) + \int_0^t \sigma(\hat{x}(s))d\hat{w}(s) + \int_0^t g(\hat{x}(s^-))d\hat{J}(s) \tag{4.4}$$

$$+\hat{L}(t) - \hat{U}(t).$$

The jumps of $\hat{J}(\cdot)$ are bounded in magnitude by C. Also, $\hat{x}(t) \in [0, B]$ and $\hat{L}(\cdot)$ ($\hat{U}(\cdot)$, resp.), increases only when $\hat{x}(t) = 0$ (resp., $\hat{x}(T) = B$). If h indexes a weakly convergent subsequence, then

$$\liminf_h W^h(x, J^h) \geq E_x^{\hat{J},\hat{T}} \int_0^\infty e^{-\beta\hat{T}(s)}[k(\hat{x}(s))d\hat{T}(s) + |d\hat{J}|(s)] = \hat{W}(x, \hat{J}, \hat{T}). \tag{4.5}$$

If $\sup_h W^h(x, J^h) < \infty$, then $\hat{T}(t) \to \infty$ w.p.1 as $t \to \infty$ and $(\hat{T}(\cdot), \hat{J}(\cdot), w(\cdot))$ is admissible.

Corollary. $\liminf_h V^h(x) \geq V(x)$.

Proof. The control $\hat{J}(t) \equiv 0$ yields $\sup_h W^h(x, J^h) < \infty$. Thus $\sup_h W^h(x, J^h) < \infty$ for an optimizing sequence $J^h(\cdot)$. The corollary then follows from (4.5) and the last sentence of the Theorem. Q.E.D.

5 Convergence of the costs $V^h(x) \to V(x)$.

In view of the last corollary, in order to prove $V^h(x) \to V(x)$, we need to show that

$$\overline{\lim}_h V^h(x) \leq V(x). \tag{5.1}$$

This inequality can be proved as follows, paralleling the line of argument in [5, Section 5]. An admissible pair $(\hat{J}(\cdot), \hat{T}(\cdot))$ is said to be δ-optimal if $\hat{W}(x, \hat{J}, \hat{T}) \leq V(x) + \delta$. Given arbitrary $\delta > 0$, one first shows that there is a "simple" δ-optimal "comparison" policy $(\hat{J}(\cdot), \hat{T}(\cdot))$ for (2.3), (2.4) (equivalently, for (1.1), (1.3)), which can be adapted for use on the approximating chain $\{\xi_n^h, n < \infty\}$. Then the minimality of $V^h(x)$ and a weak convergence argument yield the desired result.

The following sequence of lemmas establishes the existence of an appropriate comparison control. The proofs (after Lemma 5.1) all use weak convergence arguments to show that the sequence of processes and costs associated with the described approximations converge to that for the originally given control.

The next lemma is a consequence of the discounting.

Lemma 5.1. *For any $\delta > 0$, there is a $t_0 < \infty$ and a δ-optimal admissible $(\hat{J}(\cdot), \hat{T}(\cdot))$ for (2.3), (2.4) such that $\hat{J}(\cdot)$ is constant for $t \geq t_0$.*

Lemma 5.2. *Let $(\hat{J}(\cdot), \hat{T}(\cdot))$ be admissible for (2.3). Define $\hat{T}_\eta(\cdot)$ by*

$$\hat{T}_\eta(t) = \int_0^t I_{\{\hat{T}'(s) \geq \eta\}} d\hat{T}(s) + \eta \int_0^t I_{\{\hat{T}'(s) \leq \eta\}} ds.$$

Then $(\hat{J}(\cdot), \hat{T}_\eta(\cdot))$ is admissible. Given $\varepsilon > 0$, there is $\eta_0 > 0$ such that

$$\hat{W}(x, \hat{J}, \hat{T}_{\eta_0}) \leq \hat{W}(x, \hat{J}, \hat{T}) + \varepsilon.$$

Lemma 5.2 implies that the jumps of $\hat{J}(\cdot)$ can be considered to be "isolated" in the sense that time $\hat{T}_{n0}(\cdot)$ always increases.

Lemma 5.2 and a scale change imply:

Lemma 5.3. *There is admissible* $(\hat{J}_1(\cdot), \hat{T}_1(\cdot))$ *such that* $\hat{T}_1(t) = t$ *and*

$$\hat{W}(x, \hat{J}_1, \hat{T}_1) = \hat{W}(x, \hat{J}, \hat{T}_{n0}).$$

A simple approximation argument can be used to show that the control $\hat{J}_1(\cdot)$ in Lemma 5.3 can be assumed to be piecewise constant and with the jumps being bounded. In particular,

Lemma 5.4. *Fix* $\varepsilon > 0$. *There are* $t_0 < \infty, \delta > 0, K < \infty$ *and an* ε*-optimal admissible* $(J(\cdot), T(\cdot))$ *for* (1.1), (1.3), *such that:* (a) $J(\cdot)$ *is constant for* $t \geq t_0$; (b) *for* $t \leq t_0$, *it is piecewise constant with jumps in the set* $[-K, K]$; (c) *it can jump only at* $t = k\delta, k = 1, 2, \cdots$. *We can suppose that there is a continuous function* $h(\cdot)$ *such that* $d\hat{J}(k\delta) = h(\hat{x}(k\delta))$.

For the class of controls in Lemma 5.5, (1.2) can be interpreted as a Stieltjes integral. Finally, we can write:

Theorem 5.1. $V^h(x) \to V(x)$.

Proof. Only (5.1) needs to be proved. Let $J(\cdot)$ be an ε-optimal control of the type specified in Lemma 5.4, with $T(t) = t$. For k such that $k\delta < t_0$, define

$$n_k^h = \min\{j : t_j^h \geq k\delta\}.$$

For $j = n_k^h, k = 1, 2, \cdots$, use the control $\Delta J_j^h = h(\xi_j^h)$. If $\xi_j^h + g(\xi_j^h)h(\xi_j^h) \notin [0, B]$, then truncate, so that $\xi_{j+1}^h \in [0, B]$. For $j \neq n_k^h$, set $\Delta J_j^h = 0$. The sequence $\{\xi^h(\cdot), w^h(\cdot), J^h(\cdot)\}$ converges weakly to $(x(\cdot), w(\cdot), J(\cdot))$, where $x(\cdot)$ solves (1.1) under $(w(\cdot), J(\cdot))$, and (1.2) is a Stieltjes integral. Also $W^h(x, J^h) \to W(x, J)$. Since $V^h(x) \leq W^h(x, J^h)$, the ε-optimality of $J(\cdot)$ implies that $\lim_h V^h(x) \leq V(x)$. Q.E.D.

Comment on higher dimensions. A very similar proof can be used. The main problem is numerical; the analog of (3.5) is hard to solve, due to the "non local" search required.

References

[1] H.J. Kushner, Ramachandran, K.M., "Optimal and approximately optimal control policies for queues in heavy traffic", SIAM J. on Control and Optimization, **27**, 1293-1318, 1989.

[2] G. Ferreyra, "The optimal control problem for the Vidale-Wolfe advertising model revisited", Optimal control applications and methods, **11**, 1990.

[3] H.J. Kushner, "Probability Methods for Approximation in Stochastic Control and for Elliptic Equations", Academic Press, New York, 1977.

[4] H.J. Kushner, "Numerical methods for stochastic control in continuous time", SIAM J. Control and Optimiz., **28**, 999-1048, 1990.

[5] H.J. Kushner, L.F. Martins, "Numerical Methods for Stochastic Singularly Controlled Problems," to appear SIAM J. on Control and Optimization.

[6] H.J. Kushner, "Jump-diffusion approximations for ordinary differential equations with wideband random right hand sides", SIAM J. on Control and Optimization, **17**, 729-744, 1979.

[7] S.N. Ethier, T.G. Kurtz, "Markov Processes: Characterization and Convergence", Wiley, New York, 1986.

[8] P. Billingsley, "Convergence of Probability Measures", Wiley, New York, 1968.

[9] H.J. Kushner, Approximation and Weak Convergence Methods for Random Processes, with Applications to Stochastic Systems Theory, MIT Press, 1984.

AVERAGING FOR MARTINGALE PROBLEMS AND STOCHASTIC APPROXIMATION

Thomas G. Kurtz
Departments of Mathematics and Statistics
University of Wisconsin-Madison
Madison, WI 53706

0. Averaging Let $\{(X_n,Y_n)\}$ be a sequence of stochastic processes with values in a product space $E_1 \times E_2$, where E_1 and E_2 are complete separable metric spaces. We are interested in studying the limiting behavior of the sequence under the assumption that the "time scale" of the second component Y_n is much faster than that of the first component X_n in a sense that will be clear from our assumptions. The behavior of X_n will be related to that of Y_n through the assumption that there is an operator $A: \mathcal{D}(A) \subset C(E_1) \to C(E_1 \times E_2)$ such that for $f \in \mathcal{D}(A)$

$$f(X_n(t)) - \int_0^t Af(X_n(s),Y_n(s))\,ds \tag{0.1}$$

is a martingale, or more generally, that

$$f(X_n(t)) - \int_0^t Af(X_n(s),Y_n(s))\,ds + \epsilon_n^f(t) \tag{0.2}$$

is a martingale and $\epsilon_n^f \Rightarrow 0$. The fast process Y_n is generally specified in one of two ways. Y_n can simply be defined by setting $Y_n(t) = Y(\beta_n t)$ where Y is an ergodic, stationary process and $\beta_n \to \infty$, or we can assume that there is an operator $B: \mathcal{D}(B) \subset C(E_2) \to C(E_1 \times E_2)$ such that for $g \in \mathcal{D}(B)$

$$g(Y_n(t)) - \int_0^t \beta_n Bg(X_n(s),Y_n(s))\,ds + \delta_n^g(t) \tag{0.3}$$

is a martingale, $\beta_n \to \infty$, and $\beta_n^{-1}\delta_n^g \Rightarrow 0$. The goal in these problems is to show that $X_n \Rightarrow X$, where X can be characterized as a solution of a martingale problem for an operator $\bar{A}$ obtained by "averaging" A.

Problems of this type for deterministic systems ($Af(x,y) = b(x,y) \cdot \nabla f(x)$ in the present context) have a long history. (See, for example, Lochak and Meunier (1988).) Stochastic models considered by Khas'minskii (1966a,b) fit the above formulation as do random evolutions introduced by Griego and Hersh (1969) and studied by a variety of workers. See Hersh (1974), Pinsky (1974), Papanicolaou (1978), Kushner (1984), and Ethier and Kurtz (1986), for examples and further references.

In many of the examples, the hard work is in showing that the model satisfies the martingale conditions of (0.2). Our goal here, however, is to show how to carry the argument to completion, once the model is put into this form, so at this point we only give two simple examples. Khas'minskii (1966a) is concerned with models of the form

$$\dot{X}_n(t) = F(X_n(t),Y_n(t))$$

where $Y_n(t) = Y(nt)$. Setting $Af(x,y) = F(x,y) \cdot \nabla_x f(x,y)$, we see that

$$f(X_n(t)) - \int_0^t Af(X_n(s),Y_n(s))\,ds \equiv f(X_n(0))$$

which is trivially a martingale.

For a second example (a simple special case of Khas'minskii (1966b)), suppose that Y is an irreducible finite Markov chain with infinitesimal transition matrix $Q = ((q(y,y')))$ and stationary distribution π, $Y_n(t) = Y(nt)$, $\sum F(x,y)\,\pi(y) = 0$, and

$$\dot{X}_n(t) = \sqrt{n}\,F(X_n(t),Y_n(t)).$$

Employing the perturbation method of Kurtz (1973) (see Papanicolaou (1978) and Kushner (1979, 1984) for generalizations), let G satisfy $\sum q(y,y')\,G(x,y') = -F(x,y)$. (The solution will exist since π is the unique left eigenvector of the zero eigenvalue of Q, and $\sum F(x,y)\,\pi(y) = 0$.) Define $a_{ij}(x,y) = G_i(x,y)\,F_j(x,y)$, $b_i(x,y) = \sum_j F_j(x,y)\frac{\partial}{\partial x_j}G(x,y)$, and

$$Af(x,y) = \sum a_{ij}(x,y)\frac{\partial^2}{\partial x_i \partial x_j}f(x) + \sum b_i(x,y)\frac{\partial}{\partial x_i}f(x).$$

Then

$$f(X_n(t)) - \int_0^t Af(X_n(s),Y_n(s))\,ds + \frac{1}{\sqrt{n}}G(X_n(t),Y_n(t)) \cdot \nabla f(X_n(t))$$

is a martingale of the form (0.2).

Much of the material below exists at least implicitly in the earlier work, but some of the most useful ideas (e.g., Lemma 1.5) are, as far as we know, stated explicitly for the first time. Section 1 contains basic results on convergence of random measures. (See Kallenberg (1986) for more details.) Section 2 contains the basic limit theorem and a number of examples. We also introduce a notion of an *averaged martingale problem* corresponding to the operators A and B in (0.2) and (0.3). Section 3 applies the averaging arguments to stochastic approximation algorithms following results of Kushner and Schwartz (1984) and Dupuis and Kushner (1989).

Throughout, $\overline{C}(S)$ will denote the space of bounded, continuous real-valued functions on a metric space S, $C_E[0,\infty)$ will denote the space of continuous, E-valued functions on $[0,\infty)$, and $D_E[0,\infty)$ will denote the space of E-valued functions that are right continuous on $[0,\infty)$ and have left limits at each $t > 0$. The space E will always be a complete, separable metric space.

1. Convergence of random measures Let (S,d) be a complete separable metric space, and let $\mathcal{M}(S)$ be the space of finite measures on S with the weak topology. The Prohorov metric on $\mathcal{M}(S)$ is defined by

$$(1.1)\qquad \rho(\mu,\nu) = \inf\{\epsilon > 0: \mu(B) \le \nu(B^\epsilon)+\epsilon,\ \nu(B) \le \mu(B^\epsilon)+\epsilon,\ B \in \mathcal{B}(S)\}$$

where $B^\epsilon = \{x \in S: \inf_{y\in B} d(x,y) < \epsilon\}$. The following lemma is a simple consequence of Prohorov's theorem.

Lemma 1.1 Let $\{\Gamma_n\}$ be a sequence of $\mathcal{M}(S)$-valued random variables. Then $\{\Gamma_n\}$ is relatively compact if and only if $\{\Gamma_n(S)\}$ is relatively compact as a family of $\mathbb{R}$-valued random variables and for each $\epsilon > 0$ there exists a compact $K \subset S$ such that $\sup_n P\{\Gamma_n(K^c) > \epsilon\} < \epsilon$.

Proof The proof of necessity is left to the reader. To prove sufficiency, fix $\eta > 0$. Then there exist compact sets $K_k \subset S$ such that

$$\sup_n P\{\Gamma_n(K_k^c) > \frac{\eta}{2^{k+1}}\} < \frac{\eta}{2^{k+1}}$$

and a constant c such that

$$\sup_n P\{\Gamma_n(S) > c\} < \frac{\eta}{2}.$$

Define $\mathcal{K} = \{\mu \in \mathcal{M}: \mu(K_k^c) \le \eta/2^{k+1},\ k = 1,2,\ldots,\ \mu(S) \le c\}$, and observe that $P\{\Gamma_n \in \mathcal{K}\} \ge 1-\eta$. Prohorov's theorem implies that $\mathcal{K}$ is a compact subset of $\mathcal{M}(S)$, and consequently, again by Prohorov's theorem, $\{\Gamma_n\}$ is relatively compact. □

Corollary 1.2 Let $\{\Gamma_n\}$ be a sequence of $\mathcal{M}(S)$-valued random variables. Suppose that $\sup_n E[\Gamma_n(S)] < \infty$ and that for each $\epsilon > 0$ there exists a compact $K \subset S$ such that

$$\overline{\lim}_{n\to\infty} E[\Gamma_n(K^c)] \leq \epsilon .$$

Then $\{\Gamma_n\}$ is relatively compact.

Let $\mathcal{L}(S)$ be the space of measures on $[0,\infty)\times S$ such that $\mu([0,t]\times S) < \infty$ for each $t \geq 0$, and let $\mathcal{L}_m(S) \subset \mathcal{L}(S)$ be the subspace on which $\mu([0,t]\times S) = t$. For $\mu \in \mathcal{L}(S)$, let μ^t denote the restriction of μ to $[0,t]\times S$. Let ρ_t denote the Prohorov metric on $\mathcal{M}([0,t]\times S)$, and define $\hat{\rho}$ on $\mathcal{L}(S)$ by

$$\hat{\rho}(\mu,\nu) = \int_0^\infty e^{-t}\, 1 \wedge \rho_t(\mu^t,\nu^t)\,dt ,$$

that is, $\{\mu_n\}$ converges in $\hat{\rho}$ if and only if $\{\mu_n^t\}$ converges in ρ_t for almost every t. In particular, if $\hat{\rho}(\mu_n,\mu)\to 0$, then $\rho_t(\mu_n^t,\mu^t)\to 0$ if and only if $\mu_n([0,t]\times S)\to\mu([0,t]\times S)$. The following lemma is an immediate consequence of Lemma 1.1.

Lemma 1.3 A sequence of $(\mathcal{L}_m(S),\hat{\rho})$-valued random variables $\{\Gamma_n\}$ is relatively compact if and only if for each $\epsilon > 0$ and each $t > 0$ there exists a compact $K \subset S$ such that $\inf_n E[\Gamma_n([0,t]\times K)] \geq (1-\epsilon)t$.

Lemma 1.4 Let Γ be an $(\mathcal{L}(S),\hat{\rho})$-valued random variable adapted to a complete filtration $\{\mathcal{F}_t\}$ in the sense that for each t, $\Gamma([0,t]\times H))$ is $\mathcal{F}_t$-measurable for each $H \in \mathcal{B}(S)$. Let $\lambda(G) = \Gamma(G\times S)$. Then there exists an $\{\mathcal{F}_t\}$-optional, $\mathcal{P}(S)$-valued process γ such that

$$\int_{[0,t]\times S} h(s,y)\,\Gamma(ds\times dy) = \int_0^t\int_S h(s,y)\,\gamma_s(dy)\,\lambda(ds) \tag{1.2}$$

for all $h \in B([0,\infty)\times S)$ with probability one. If $\lambda([0,t])$ is continuous, then γ can be taken to be $\{\mathcal{F}_t\}$-predictable.

Proof Let $(\Omega,\mathcal{F},P)$ denote the probability space on which Γ is defined. Without loss of generality, we can assume that $E[\Gamma([0,\infty)\times S)] < \infty$. (Otherwise we replace Γ by the measure $\tilde{\Gamma}(C) = \int I_C(s,y)e^{-\lambda(s)}\Gamma(ds\times dy)$.) For $B \in \mathcal{F}$ and $C \in \mathcal{B}[0,\infty)\times\mathcal{B}(S)$, define $\nu(B\times C) = E[I_B\Gamma(C)]$. By Morando's theorem (see, for example, Appendix 8 of Ethier and Kurtz (1986)), ν extends to a measure on $\mathcal{F}\times\mathcal{B}[0,\infty)\times\mathcal{B}(S)$. Let $\mathcal{O}\subset\mathcal{F}\times\mathcal{B}[0,\infty)$ denote the optional σ-algebra, that is the smallest σ-algebra such that the mapping $(\omega,t)\to Z(\omega,t)$ is measurable for each right continuous, $\{\mathcal{F}_t\}$-adapted process Z. Let $\tilde{\nu}$ denote the restriction of ν to $\mathcal{O}\times\mathcal{B}(S)$ and η the restriction of $\nu(\cdot\times S)$ to $\mathcal{O}$. Then (see the appendix cited above) there exists $\mu:\Omega\times[0,\infty)\times\mathcal{B}(S)\to[0,1]$ such that for $(\omega,t)\in\Omega\times[0,\infty)$, $\mu(\omega,t,\cdot)\in\mathcal{P}(S)$ and for each $B\in\mathcal{B}(S)$, $\mu(\cdot,\cdot,B)$ is $\{\mathcal{F}_t\}$-optional, that is measurable with respect to $\mathcal{O}$, such that for each $C\in\mathcal{O}$

$$\tilde{\nu}(C\times B)=\int_C\mu(\omega,s,B)\,\eta(d\omega\times ds)$$

$$=E[\int_0^\infty I_C(s,\cdot)\,\mu(\cdot,s,B)\,\lambda(ds)].$$

Let τ be an $\{\mathcal{F}_t\}$-stopping time, $D\in\mathcal{F}_\tau$, and set $I_C(\omega,t)=I_D(\omega)I_{[\tau(\omega),\tau(\omega)+a)}(t)$. Note that I_C is right continuous and adapted, so $C\in\mathcal{O}$. Then, for $B\in\mathcal{B}(S)$,

$$E[I_D\Gamma([\tau,\tau+a)\times B)]=\tilde{\nu}(C\times B)$$

$$=E[I_D\int_{[\tau,\tau+a)}\mu(\cdot,s,B)\lambda(ds)],$$

and it follows that

$$E[\Gamma([\tau,\tau+a)\times B)\mid\mathcal{F}_\tau]=E[\int_{[\tau,\tau+a)}\mu(\cdot,s,B)\lambda(ds)\mid\mathcal{F}_\tau]. \tag{1.3}$$

By the right continuity and adaptedness of $\Gamma([0,t]\times B)$, for any stopping time τ, (1.3) implies

$$\Gamma(\{\tau\}\times B)=\lambda(\{\tau\})\,\mu(\cdot,\tau,B) \tag{1.4}$$

which in turn implies

$$M_B(t)\equiv\Gamma([0,t]\times B)-\int_{[0,t]}\mu(\cdot,s,B)\,\lambda(ds)$$

is continuous. But (1.3) also implies M_B is a martingale, and since it has sample paths of finite variation, M_B must be identically zero with probability one. The separability of $\mathcal{B}(S)$ then implies (1.2) holds for all $h\in B([0,\infty)\times S)$ with probability one.

If λ is continuous, replace the optional sets $\mathcal{O}$ by the predictable sets in the construction of μ. The continuity of λ implies M_B is continuous without the need to verify (1.4). The fact that M_B is a martingale follows by taking $I_C(\omega,t) = I_D(\omega)I_{(t_0,\infty)}(t)$ for $D \in \mathcal{F}_t$ (note that C will be predictable) and concluding, as in (1.3), that

$$E[\Gamma((t,\infty)\times B) \mid \mathcal{F}_t] = E[\int_{(t,\infty)} \mu(\cdot,s,B)\lambda(ds) \mid \mathcal{F}_t]$$

which implies that M_B is a martingale. □

Lemma 1.5 Let $\{(x_n,\mu_n)\} \subset D_E[0,\infty)\times\mathcal{L}(S)$, and $(x_n,\mu_n) \to (x,\mu)$. Let $h \in \bar{C}(E\times S)$. Define

$$(1.5)\quad u_n(t) = \int_{[0,t]\times S} h(x_n(s),y)\,\mu_n(ds\times dy),\quad u(t) = \int_{[0,t]\times S} h(x(s),y)\,\mu(ds\times dy)$$

$z_n(t) = \mu_n([0,t]\times S)$ and $z(t) = \mu([0,t]\times S)$.

a) If x is continuous on $[0,t]$ and $\lim_{n\to\infty} z_n(t) = z(t)$, then $\lim_{n\to\infty} u_n(t) = u(t)$.

b) If $(x_n,z_n,\mu_n) \to (x,z,\mu)$ in $D_{E\times\mathbb{R}}[0,\infty)\times\mathcal{L}(S)$, then $(x_n,z_n,u_n,\mu_n) \to (x,z,u,\mu)$ in $D_{E\times\mathbb{R}\times\mathbb{R}}[0,\infty)\times\mathcal{L}(S)$. In particular, $\lim_{n\to\infty} u_n(t) = u(t)$ at all points of continuity of z.

c) The continuity assumption on h can be replaced by the assumption that h is continuous a.e. ν_t for each t, where $\nu_t \in \mathcal{M}(E\times S)$ is the measure determined by $\nu_t(A\times B) = \mu\{(s,y):x(s) \in A, s \le t, y \in B\}$.

d) In both a) and b), the boundedness assumption on h can be replaced by the assumption that there exists a nonnegative convex function ψ on $[0,\infty)$ satisfying $\lim_{r\to\infty}\psi(r)/r = \infty$ such that

$$(1.6)\qquad \sup_n \int_{[0,t]\times S} \psi(|h(x_n(s),y)|)\,\mu_n(ds\times dy) < \infty$$

for each $t > 0$.

Proof Let $h \in \bar{C}(E\times S)$. For each $\epsilon > 0$ and $t > 0$, there exists a compact $K \subset S$ with $\sup_n \mu_n([0,t]\times K^c) \le \epsilon$. If x is continuous, then

$$\lim_{n\to\infty}\sup_{y\in K,\ s\le t}|h(x_n(s),y)-h(x(s),y)|=0,$$

and if $z_n(t)\to z(t)$, it follows that

$$\overline{\lim}_{n\to\infty}\left|\int_{[0,t]\times S}h(x_n(s),y)\,\mu_n(ds\times dy)-\int_{[0,t]\times S}h(x(s),y)\,\mu(ds\times dy)\right|\le 2\,\|h\|\,\epsilon$$

which verifies part a).

If $(x_n,z_n)\to(x,z)$ in the Skorohod topology, then there exist continuous, strictly increasing mappings η_n of $[0,\infty)$ onto $[0,\infty)$ such that $\eta_n(t)\to t$ for each t and $(x_n\circ\eta_n,z_n\circ\eta_n)\to(x,z)$ uniformly on bounded intervals. Define $\tilde{\mu}_n$ so that $\tilde{\mu}_n([0,t]\times H)=\mu_n([0,\eta_n(t)]\times H)$ and observe that $\tilde{\mu}_n\to\mu$ in $\mathcal{L}(S)$. But the uniformity of the convergence of $x_n\circ\eta_n$ to x and $z_n\circ\eta_n$ to z implies

$$(1.7)\quad \int_{[0,\eta_n(t)]\times S}h(x_n(s),y)\,\mu_n(ds\times dy)=\int_{[0,t]\times S}h(x_n\circ\eta_n(s),y)\,\tilde{\mu}_n(ds\times dy)$$

$$\to\int_{[0,t]\times S}h(x(s),y)\,\mu(ds\times dy)$$

for each fixed t. We want to show that the convergence is uniform on bounded intervals. Let $\tilde{u}_n(t)$ denote the integral on the left. It is sufficient to show that for any sequence satisfying $t_n\to t$, $\tilde{u}_n(t_n)-u(t_n)\to 0$. But this convergence holds if for any sequence satisfying $t_n\ge t$ and $t_n\to t$, we have $\tilde{u}_n(t_n)\to u(t)$, and for any sequence satisfying $t_n<t$ and $t_n\to t$, we have $\tilde{u}_n(t_n)\to u(t-)$. Since for all r, s, $|\tilde{u}_n(s)-\tilde{u}_n(r)|\le\|h\|\,|z_n\circ\eta_n(s)-z_n\circ\eta_n(r)|$, the pointwise convergence of $\tilde{u}_n$ and the uniformity of the convergence of $z_n\circ\eta_n$ imply, in the first case, that

$$\begin{aligned}
&\overline{\lim}_{n\to\infty}|\tilde{u}_n(t_n)-u(t)|\\
&\quad=\overline{\lim}_{n\to\infty}|\tilde{u}_n(t_n)-\tilde{u}_n(t)|\\
&\quad\le\overline{\lim}_{n\to\infty}\|h\|\,|z_n\circ\eta_n(t_n)-z_n\circ\eta_n(t)|\\
&\quad\le\overline{\lim}_{n\to\infty}\|h\|\,|z(t_n)-z(t)|\\
&\quad=0,
\end{aligned}$$

and in the second case, that

$$\overline{\lim}_{n\to\infty} |\tilde{u}_n(t_n) - u(t-)|$$

$$= \lim_{\epsilon\to 0} \overline{\lim}_{n\to\infty} |\tilde{u}_n(t_n) - \tilde{u}_n(t-\epsilon)|$$

$$\le \lim_{\epsilon\to 0} \overline{\lim}_{n\to\infty} \| h \| \; | z_n \circ \eta_n(t_n) - z_n \circ \eta_n(t-\epsilon) |$$

$$\le \lim_{\epsilon\to 0} \overline{\lim}_{n\to\infty} \| h \| \; | z(t_n) - z(t-\epsilon) |$$

$$= 0$$

which completes the proof of part b).

To obtain part c), define $\nu_t^n(A\times B) = \tilde{\mu}_n\{(s,y): x_n \circ \eta_n(s) \in A, s \le t, y \in B\}$. Then, for fixed t, the fact that (1.7) holds for each $h \in \overline{C}(E\times S)$ is just the assertion that $\nu_t^n \Rightarrow \nu_t$. We claim that this convergence is uniform on bounded time intervals. If not, then there exists a bounded sequence t_n and an $\epsilon > 0$ such that $\overline{\lim}_{n\to\infty} \rho(\nu_{t_n}^n, \nu_{t_n}) \ge \epsilon$ where ρ is the Prohorov metric on $\mathcal{M}(E\times S)$. As in the proof of part b), without loss of generality, we can assume that $t_n \to t$ and that either $t_n \ge t$ for all n or $t_n < t$ for all n. In the first case, $\nu_{t_n} \Rightarrow \nu_t$ and the uniformity of the convergence in (1.7) for $h \in \overline{C}(E\times S)$ implies that $\nu_{t_n}^n \Rightarrow \nu_t$, so $\rho(\nu_{t_n}^n, \nu_{t_n}) \to 0$. Similarly, in the second case, $\nu_{t_n} \Rightarrow \nu_{t-}$ and it follows that $\nu_{t_n}^n \Rightarrow \nu_{t-}$, so again $\rho(\nu_{t_n}^n, \nu_{t_n}) \to 0$. Note that if h is ν_t-almost surely continuous, then it is ν_{t-}-almost surely continuous. For $t_n \to t$ with $t_n \ge t$, the continuous mapping theorem gives $\tilde{u}_n(t_n) \to u(t)$; for $t_n \to t$ with $t_n < t$, the continuous mapping theorem gives $u_n(t_n) \to u(t-)$; part c) then follows as in the proof of part b).

Dropping the boundedness assumption on h and assuming (1.6), part d) follows by approximating h by $h_c = c \wedge ((-c) \vee h)$. □

2. Stochastic averaging

Theorem 2.1 Let E_1 and E_2 be complete, separable metric spaces, and set $E = E_1 \times E_2$. For each n, let $\{(X_n, Y_n)\}$ be a stochastic process with sample paths in $D_E[0,\infty)$ adapted to a filtration $\{\mathcal{F}_t^n\}$. Assume that $\{X_n\}$ satisfies the compact containment condition, that is, for each $\epsilon > 0$ and $T > 0$, there exists a compact

$K \subset E_1$ such that

(2.1) $$\inf_n P\{X_n(t) \in K, t \leq T\} \geq 1 - \epsilon,$$

and assume that $\{Y_n(t):t \geq 0, n = 1,2,\ldots\}$ is relatively compact (as a collection of E_2-valued random variables). Suppose there is an operator $A: \mathcal{D}(A) \subset \bar{C}(E_1) \to C(E_1 \times E_2)$ such that for $f \in \mathcal{D}(A)$ there is a process ϵ_n^f for which

(2.2) $$f(X_n(t)) - \int_0^t Af(X_n(s), Y_n(s))\, ds + \epsilon_n^f(t)$$

is an $\{\mathcal{F}_t^n\}$-martingale. Let $\mathcal{D}(A)$ be dense in $\bar{C}(E_1)$ in the topology of uniform convergence on compact sets. Suppose that for each $f \in \mathcal{D}(A)$ and each $T > 0$, there exists $p > 1$ such that

(2.3) $$\sup_n E\left[\int_0^T |Af(X_n(t), Y_n(t))|^p dt\right] < \infty$$

and

(2.4) $$\lim_{n \to \infty} E[\sup_{t \leq T} |\epsilon_n^f(t)|] = 0 .$$

Let Γ_n be the $\mathcal{L}_m(E_2)$-valued random variable given by

(2.5) $$\Gamma_n([0,t] \times B) = \int_0^t I_B(Y_n(s))\, ds .$$

Then $\{(X_n, \Gamma_n)\}$ is relatively compact in $D_{E_1}[0,\infty) \times \mathcal{L}_m(E_2)$, and for any limit point (X, Γ) there exists a filtration $\{\mathcal{G}_t\}$ such that

(2.6) $$f(X(t)) - \int_0^t \int_{E_2} Af(X(s), y)\, \Gamma(ds \times dy)$$

is a $\{\mathcal{G}_t\}$-martingale for each $f \in \mathcal{D}(A)$.

Proof The relative compactness of $\{X_n\}$ follows from (2.1), (2.3) and (2.4) by Theorems 3.9.1 and 3.9.4 of Ethier and Kurtz (1986). By the relative compactness of $\{Y_n(t):t \geq 0, n = 1,2,\ldots\}$, for each $\epsilon > 0$, there exists a compact $K \subset E_2$ such that $P\{Y_n(t) \in K\} \geq 1 - \epsilon$, and hence $E[\Gamma_n([0,t] \times K] \geq t(1-\epsilon)$. Consequently, the relative compactness of $\{\Gamma_n\}$ follows by Lemma 1.3. Let (X, Γ) be a limit point of $\{(X_n, \Gamma_n)\}$ and define $\mathcal{G}_t = \sigma\{X(s), \Gamma([0,s] \times H):s \leq t, H \in \mathcal{B}(E_2)\}$.

For $f \in \mathcal{D}(A)$,

$$M_n(t) = f(X_n(t)) - \int_0^t Af(X_n(s),Y_n(s))\,ds + \epsilon_n^f(t)$$
$$= f(X_n(t)) - \int_0^t \int_{E_2} Af(X_n(s),y)\,\Gamma_n(ds \times dy) + \epsilon_n^f(t)$$

is a martingale, and for each t

$$Z_n(t) = f(X_n(t)) - \int_0^t Af(X_n(s),Y_n(s))\,ds$$
$$= f(X_n(t)) - \int_0^t \int_{E_2} Af(X_n(s),y)\,\Gamma_n(ds \times dy)$$

is uniformly integrable and, applying Lemma 1.5, converges in distribution along an appropriate subsequence to (2.6). Since by (2.4), $\lim_{n\to\infty} E[|Z_n(t) - M_n(t)|] = 0$, it follows that (2.6) is a $\{\mathcal{G}_t\}$-martingale. □

Example 2.2 Suppose Y is stationary and ergodic, and Y_n in Theorem 2.1 is given by $Y_n(t) \equiv Y(nt)$. Then $\Gamma = m \times \pi$ where m denotes Lebesgue measure and π is the marginal distribution for Y. Consequently, under the other assumptions of Theorem 2.1, X is a solution of the martingale problem for C given by

$$Cf(x) = \int Af(x,y)\,\pi(dy), \quad f \in D^1.$$

Example 2.3 Suppose that there is an operator $B : \mathcal{D}(B) \subset \bar{C}(E_2) \to \bar{C}(E_1 \times E_2)$ such that for $g \in \mathcal{D}(B)$

$$g(Y_n(t)) - \int_0^t \beta_n Bg(X_n(s),Y_n(s))\,ds + \delta_n^g(t) \tag{2.7}$$

is an $\{\mathcal{F}_t^n\}$-martingale, $\beta_n \to \infty$, and for each $T > 0$

$$\lim_{n\to\infty} E[\sup_{t \le T} \beta_n^{-1} |\delta_n^g(t)|] = 0.$$

Then under the assumptions of Theorem 2.1, it follows that

$$\int_{[0,t] \times E_2} Bg(X(s),y)\,\Gamma(ds \times dy) \tag{2.8}$$

is a martingale. But (2.8) is continuous and of bounded variation and hence must be constant. Consequently, for each $g \in \mathcal{D}(B)$, with probability one

$$\int_{[0,t]\times E_2} Bg(X(s),y)\,\Gamma(ds\times dy) = 0 \tag{2.9}$$

for all $t > 0$.

Let γ be as in Lemma 1.4. Suppose that there exists a countable subset $\hat{D} \subset \mathcal{D}(B)$ such that the closure of $\{(g,Bg):g \in \hat{D}\}$ in $\bar{C}(E_2)\times\bar{C}(E_1\times E_2)$ is the same as the closure of $\{(g,Bg):g\in\mathcal{D}(B)\}$. For example, such a subset would exist if E_1 and E_2 were compact (by the separability of $\bar{C}(E_2)\times\bar{C}(E_1\times E_2)$), or if $\mathcal{D}(B) = C_c^\infty$ and B is a second order differential operator with bounded coefficients. By (2.9)

$$\int_0^t\int_{E_2} Bg(X(s),y)\,\gamma_s(dy)\,ds = 0$$

for all t a.s., and hence

$$\int_{E_2} Bg(X(s),y)\,\gamma_s(dy) = 0 \tag{2.10}$$

a.e. m a.s. Consequently, with probability one, there exists a single set $Q \subset [0,\infty)$ with $m(Q) = 0$ such that (2.10) holds for all $g \in \hat{D}$ and all $s \in [0,\infty) - Q$. But the choice of $\hat{D}$ ensures that (2.10) holds for all $g\in\mathcal{D}(B)$ and all $s \in [0,\infty) - Q$.

Define $B_x:\mathcal{D}(B) \to \bar{C}(E_2)$ by $B_xg(y) = Bg(x,y)$. Suppose that there is a unique measure π_x in $\mathcal{P}(E_2)$ satisfying

$$\int B_xg\,d\pi_x = 0$$

for all $g \in D^2$. (If B_x is the generator for an E_2-valued Markov process, this assumption is essentially the assertion that there is a unique stationary distribution corresponding to B_x.) Then we can take $\gamma_s = \pi_{X(s)}$, and defining C on D^1 by

$$Cf(x) = \int Af(x,y)\,\pi_x(dy),$$

it follows that X is a solution of the martingale problem for C. □

Example 2.4 Let $\sigma:\mathbb{R}^d\times\mathbb{R}^d \to \mathbb{M}^{d\times m}$ and $b:\mathbb{R}^d\times\mathbb{R}^d \to \mathbb{R}^d$ be continuous and suppose that $\sigma(x,y)$ and $b(x,y)$ are periodic with period 1 in the last d coordinates $(y_1,\dots,y_d)$. Suppose that X_n satisfies the Ito equation

$$dX_n = \sigma(X_n,nX_n)\,dW + b(X_n,nX_n)\,dt$$

and that $\{X_n(0)\}$ is relatively compact. Set $a = \sigma\sigma^T$, and assume that there exists a constant K such that

$$|a(x,y)| \leq K(1 + |x|^2), \quad x \cdot b(x,y) \leq K(1+|x|^2)$$

for all x, y. This assumption ensures that $\{X_n\}$ satisfies the compact containment condition.

Define $Y_n(t) \equiv nX_n(t) \bmod 1$. Let $\mathcal{D}(A)$ be the linear span of 1 and $C_c^2(\mathbb{R}^d)$, and for $f \in \mathcal{D}(A)$, define

$$Af(x,y) = \frac{1}{2}\sum_{i,j} a_{ij}(x,y)\frac{\partial}{\partial x_i \partial x_j}f(x) + \sum_i b_i(x,y)\frac{\partial}{\partial x_i}f(x).$$

Let $\mathcal{D}(B)$ be the collection of C^2 functions g on $[0,1]^d$ such that g and its first two derivatives satisfy periodic boundary conditions, that is, the periodic extension of g to all of $\mathbb{R}^d$ is C^2. For $g \in \mathcal{D}(B)$ define

$$B_n g(x,y) = \frac{1}{2}\sum_{i,j} a_{ij}(x,y)\frac{\partial}{\partial y_i \partial y_j}g(y) + \frac{1}{n}\sum_i b_i(x,y)\frac{\partial}{\partial y_i}g(y)$$

$$B_x g(y) = Bg(x,y) = \frac{1}{2}\sum_{i,j} a_{ij}(x,y)\frac{\partial}{\partial y_i \partial y_j}g(y)$$

and

$$Hg(x,y) = \sum_i b_i(x,y)\frac{\partial}{\partial y_i}g(y).$$

It follows from Ito's formula that

$$f(X_n(t)) - \int_0^t Af(X_n(s),Y_n(s))\,ds$$

and

$$g(Y_n(s)) - \int_0^t n^2 B_n g(X_n(s),Y_n(s))\,ds$$

$$= g(Y_n(s)) - \int_0^t n^2 Bg(X_n(s),Y_n(s))\,ds - \int_0^t nHg(X_n(s),Y_n(s))\,ds$$

are $\{\mathcal{F}_t^n\}$-martingales for $\mathcal{F}_t^n = \sigma(X_n(s):s \leq t)$.

Let E_1 be $\mathbb{R}^d$ and $E_2 = [0,1]^d$. Then the conditions of the previous example are satisfied. If

$$\int B_x g\, \pi_x(dy) = 0, \quad g \in \mathcal{D}(B),$$

then by Echeverria's theorem (Ethier and Kurtz (1986), Theorem 4.9.17), π_x is a stationary distribution for B_x. If B_x has a unique stationary distribution, for example, if a is positive definite, then any limit point X of $\{X_n\}$ is a solution of the martingale problem for C given by $Cf(x) = \int Af(x,y)\pi_x(dy)$.

For related work and further references on diffusions with rapidly varying periodic coefficients see Bensoussan, Lions, and Papanicolaou (1978) and Bhattacharya, Gupta, and Walker (1989). □

Let $A: \mathcal{D}(A) \subset \bar{C}(E_1) \to C(E_1 \times E_2)$ and $B: \mathcal{D}(B) \subset \bar{C}(E_2) \to C(E_1 \times E_2)$. A process (X,Γ) in $D_{E_1}[0,\infty) \times \mathcal{L}_m(E_2)$ will be called a solution of the *averaged martingale problem* for (A,B), if for each $f \in \mathcal{D}(A)$ and $g \in \mathcal{D}(B)$,

$$f(X(t)) - \int_{[0,t]\times E_2} Af(X(s),y))\Gamma(ds \times dy) \tag{2.11}$$

and

$$\int_{[0,t]\times E_2} Bg(X(s),y)\Gamma(ds \times dy) \tag{2.12}$$

are $\{\mathcal{G}_t\}$-martingales for a filtration $\{\mathcal{G}_t\}$ with respect to which X and Γ are adapted. Of course, as above, (2.12) being a martingale implies (2.12) is zero. Integrating $h(X(t))$ by (2.12), we get

$$\int_{[0,t]\times E_2} h(X(s))Bg(X(s),y)\Gamma(ds \times dy) \equiv 0. \tag{2.13}$$

We observe that a solution of the averaged martingale problem can be viewed as a solution of a controlled martingale problem (see Kurtz (1987)) in which E_2 is the control space and Γ is the relaxed control.

A solution of the averaged martingale problem is *stationary*, if X is stationary and Γ has stationary increments, i.e., $\Gamma([a+t,b+t] \times G)$ is stationary for all choices of $a < b$ and $G \in \mathcal{B}(E_2)$. If (X,Γ) is stationary, then the measure $\pi \in \mathcal{P}(E_1 \times E_2)$ determined by

$$\pi(G_1 \times G_2) = E[\int_0^1 I_{G_1}(X(s)) I_{G_2}(y) \Gamma(ds \times dy)] \tag{2.14}$$

satisfies

$$\int_{E_1 \times E_2} Af(x,y)\, \pi(dx \times dy) = 0, \quad \int_{E_1 \times E_2} h(x) Bg(x,y)\, \pi(dx \times dy) = 0 \tag{2.15}$$

for all $f \in \mathcal{D}(A)$, $g \in \mathcal{D}(B)$, and $h \in B(E_1)$. We now address the converse problem. When does a measure satisfying (2.15) correspond to a stationary solution of the averaged martingale problem? In particular, we extend Echeverria's theorem to this setting (see Ethier and Kurtz (1986), Theorem 4.9.17), or more precisely, we extend the analogue of Echeverria's theorem for controlled martingale problems given by Stockbridge (1990). For $f \in \mathcal{D}(A)$, define $A_y f(x) = Af(x,y)$, and for $g \in \mathcal{D}(B)$, define $B_x g(y) = Bg(x,y)$.

Theorem 2.5 Let E_1 and E_2 be locally compact and separable, and let $E_i^\Delta = E_i \cup \{\Delta_i\}$ denote their one-point compactifications. We assume the following conditions for A and B:

i. $\mathcal{D}(A)$ is an algebra and is dense in $\hat{C}(E_1)$.
ii. For each y, A_y satisfies the positive maximum principle.
iii. For each compact $K \subset E_2$, $\lim_{x \to \Delta_1} \sup_{y \in K} |Af(x,y)| = 0$.
iv. There exists $\varphi \in C(E_2)$, $\varphi \geq 0$ such that for each $f \in \mathcal{D}(A)$, there exist constants a_f and b_f satisfying $|Af(x,y)| \leq a_f + b_f \varphi(y)$.
v. There exists $\psi \in C(E_1)$, $\psi \geq 0$ such that for each $g \in \mathcal{D}(B)$, there exist constants a_g and b_g satisfying $|Bg(x,y)| \leq a_g + b_g \psi(x)$.

Suppose $\pi \in \mathcal{P}(E_1 \times E_2)$ satisfies (2.15) and

$$\int_{E_1 \times E_2} (\psi(x) + \varphi(y))\, \pi(dx \times dy) < \infty. \tag{2.16}$$

Then there exists a stationary solution of the averaged martingale problem satisfying (2.14).

Remark Note that Conditions iv and v and (2.16) ensure that Af and Bg are integrable with respect to π.

Proof Without loss of generality, we can assume that φ and ψ are strictly positive and that $\frac{1}{\psi} \in \hat{C}(E_1)$ and $\frac{1}{\varphi} \in \hat{C}(E_2)$. The proof of the theorem is much the same as the proof of Theorems 4.1 and 4.7 of Stockbridge (1990). In particular, by replacing A, B, and π by

$$(2.17)\qquad A_n f(x,y) = \frac{n}{n \vee \varphi(y)} Af(x,y),\quad B_n g(x,y) = \frac{n^2}{(n \vee \varphi(y))(n \vee \psi(x))} Bg(x,y)$$

and

$$(2.18)\qquad \pi_n(dx \times dy) = c_n(n \vee \varphi(y))\, \pi(dx \times dy),$$

where c_n is a constant that normalizes π_n to be a probability measure, we can first prove the theorem under the assumption that $\mathcal{R}(A)$, $\mathcal{R}(B) \subset \hat{C}(E_1 \times E_2)$ and then obtain the general case by taking the limit as $n \to \infty$. (In order to ensure that $A_n f$ and $B_n g$ are in $\hat{C}(E_1 \times E_2)$, one can find $\varphi^* \in C(E_2)$ and $\psi^* \in C(E_1)$ satisfying the integrability condition (2.16), but tending to ∞ faster than the original φ and ψ.) Under this assumption, all functions involved extend continuously to $E_1^\Delta \times E_2^\Delta$, so we may as well assume E_1 and E_2 are compact.

We need the following variation of the result in Lemma 4.2 of Stockbridge (1990). Assuming now that E_1 and E_2 are compact, let $f_1,\ldots,f_m \in \mathcal{D}(A)$ and $g_1,\ldots,g_m \in \mathcal{D}(B)$, and let H be a polynomial on $\mathbb{R}^m$ that is convex on $[-\alpha_1,\alpha_1] \times \cdots \times [-\alpha_m,\alpha_m]$ where $\alpha_i \geq \|f_i - \frac{1}{n}(Af_i + Bg_i)\| \vee \|f_i\|$. Then

$$(2.19)\quad \int_{E_1 \times E_2} H(f_1(x) - \tfrac{1}{n}(Af_1(x,y) + Bg_1(x,y)),$$
$$\ldots, f_m(x) - \tfrac{1}{n}(Af_m(x,y) + Bg_m(x,y)))\, \pi(dx \times dy)$$
$$\geq \int_{E_1 \times E_2} \Big(H(f_1(x),\ldots,f_m(x)) - \tfrac{1}{n}\nabla H(f_1(x),\ldots,f_m(x)) \cdot (Af_1(x,y),\ldots,Af_m(x,y))$$
$$- \tfrac{1}{n}\nabla H(f_1(x),\ldots,f_m(x)) \cdot (Bg_1(x,y),\ldots,Bg_m(x,y))\Big)\pi(dx \times dy)$$
$$\geq \int_{E_1 \times E_2} H(f_1(x),\ldots,f_m(x))\, \pi(dx \times dy)$$

where the first inequality follows from the convexity of H, and the second inequality follows from the fact that the dot product in the second term of the second integrand is negative by Lemma 3.3 of Stockbridge (1990) and the third term integrates to zero by (2.15). Note that the inequality between the first and third expressions can be extended to arbitrary convex functions.

Let $M \subset C(E_1 \times E_1 \times E_2)$ be the linear subspace of functions of the form

$$(2.20)\qquad F(x_1,x_2,y) = \sum_{i=1}^{m} h_i(x_1)(f_i(x_2) - \tfrac{1}{n}(Af_i(x_2,y) + Bg_i(x_2,y))) + h_0(x_2,y)$$

where $f_i \in \mathcal{D}(A)$, $g_i \in \mathcal{D}(B)$, $h_i \in C(E_1)$ for $i = 1,\ldots,m$ and $h_0 \in C(E_1 \times E_2)$. For $F \in M$, define the linear functional Ψ by

$$\Psi F = \int_{E_1 \times E_2} \Big(\sum_{i=1}^{m} h_i(x) f_i(x) + h_0(x,y)\Big) \pi(dx \times dy). \tag{2.21}$$

Define the convex function $H : \mathbb{R}^m \to \mathbb{R}$ by $H(r_1,\ldots,r_m) = \sup_{x \in E_1} \sum_{i=1}^{m} h_i(x)\, r_i$. Then

$$\begin{aligned}
\Psi F &= \int_{E_1 \times E_2} \Big(\sum_{i=1}^{m} h_i(x) f_i(x) + h_0(x,y)\Big) \pi(dx \times dy) \\
&\le \int_{E_1 \times E_2} \Big(H(f_1(x),\ldots,f_m(x)) + h_0(x,y)\Big) \pi(dx \times dy) \\
&\le \int_{E_1 \times E_2} \Big(H(f_1(x) - \tfrac{1}{n}(Af_1(x,y) + Bg_1(x,y)), \\
&\qquad \ldots, f_m(x) - \tfrac{1}{n}(Af_m(x,y) + Bg_m(x,y))) + h_0(x,y)\Big) \pi(dx \times dy) \\
&= \int_{E_1 \times E_2} \sup_{x_1} F(x_1,x_2,y)\, \pi(dx_2 \times dy) \\
&\le \|F\|.
\end{aligned} \tag{2.22}$$

If $F = 1$, then $\Psi F = 1$, so by the Riesz representation theorem, there exists a measure $\nu \in \mathcal{P}(E_1 \times E_1 \times E_2)$ such that

$$\Psi F = \int_{E_1 \times E_1 \times E_2} F(x_1,x_2,y)\, \nu(dx_1 \times dx_2 \times dy). \tag{2.23}$$

We can write $\nu(dx_1 \times dx_2 \times dy) = \pi_0(dx_1)\eta(x_1,dx_2 \times dy)$, where by taking $F(x_1,x_2,y) = h(x_1)$ we can see that $\pi_0(\Gamma) = \pi(\Gamma \times E_2)$. Consider η as a transition function on $E_1 \times E_2$. Then we have

$$\int_{E_1 \times E_2} \int_{E_1 \times E_2} h_0(x_2,y_2)\, \eta(x_1,dx_2 \times dy_2)\, \pi(dx_1 \times dy_1) = \int_{E_1 \times E_2} h_0(x_2,y_2)\, \pi(dx_2 \times dy_2), \tag{2.24}$$

and we see that π is a stationary measure for η. Finally, let $\{(U_k,V_k)\}$ be a Markov chain with transition function η and initial distribution π (hence it is stationary). Then, by definition of Ψ

$$E[h(U_k)(f(U_{k+1}) - \tfrac{1}{n}(Af(U_{k+1},V_{k+1}) + Bg(U_{k+1},V_{k+1}))] = E[h(U_k)f(U_k)], \tag{2.25}$$

and defining $X_n(t) = U_{[nt]}$ and $Y_n(t) = V_{[nt]}$, it follows from the Markov property that

$$f(X_n(t)) - \int_0^{\frac{[nt]}{n}} Af(X_n(s+\tfrac{1}{n}), Y_n(s+\tfrac{1}{n}))\,ds \tag{2.26}$$

and

$$\int_0^{\frac{[nt]}{n}} Bg(X_n(s+\tfrac{1}{n}), Y_n(s+\tfrac{1}{n}))\,ds \tag{2.27}$$

are martingales. Defining

$$\Gamma_n([0,t] \times G) = \int_0^{\frac{[nt]}{n}} I_G(Y_n(s+\tfrac{1}{n}))\,ds\,, \tag{2.28}$$

(X_n,Γ_n) converges to the desired process as in Theorem 2.1. □

3. Stochastic approximation Let S be a complete separable metric space. We consider a discrete-time process $\{(X_k,Y_k,U_k,a_k)\}$ in $\mathbb{R}^d \times S \times \mathbb{R}^d \times (0,\infty)$ adapted to a filtration $\{\mathcal{F}_k\}$. For each n and $t \geq 0$, let $\eta_n(t)$ satisfy

$$\sum_{k=n}^{\eta_n(t)-1} a_k \leq t < \sum_{k=n}^{\eta_n(t)} a_k \tag{3.1}$$

(we will assume $\sum_k a_k = \infty$ a.s.), and define Γ_n in $\mathcal{L}(S)$ by

$$\Gamma_n([0,t] \times H) = \sum_{k=n}^{\eta_n(t)} a_k I_H(Y_k)\,. \tag{3.2}$$

Following Kushner and Schwartz (1984), we assume the following:

A.1 $\{X_k\}$ satisfies the iterative scheme

$$X_{k+1} = X_k + a_k(f(X_k,Y_k) + U_{k+1})\,, \tag{3.3}$$

where f is a bounded, Borel measurable function on $\mathbb{R}^d \times S$.

A.2 There exists a one-step transition function $Q(x,y,H)$ such that

$$P\{Y_{k+1} \in H|\mathcal{F}_k\} = Q(X_k,Y_k,H)$$

and, defining Qh by $Qh(x,y) = \int h(x,y')\,Q(x,y,dy')$, $Q:\overline{C}(\mathbb{R}^d \times S) \to \overline{C}(\mathbb{R}^d \times S)$.

A.3 For each $x \in \mathbb{R}^d$, there exists a unique probability measure π_x satisfying

$$\int Q(x,y,H)\,\pi_x(dy) = \pi_x(H), \quad H \in \mathcal{B}(S).$$

A.4 For each $x \in \mathbb{R}^d$ and $C_x = \{y : f \text{ is continuous at } (x,y)\}$, $\pi_x(C_x) = 1$.

A.5 The following series converge almost surely:

$$\sum_{k=1}^{\infty} a_k^2 \tag{3.4}$$

$$\sum_{k=1}^{\infty} E[|a_{k+1} - a_k| \,|\, \mathcal{F}_k] \tag{3.5}$$

$$\sum_{k=1}^{\infty} a_k U_{k+1} \tag{3.6}$$

$$\sum_{k=1}^{\infty} a_k^2 E[|U_{k+1}| \,|\, \mathcal{F}_k] \tag{3.7}$$

A.6 For each u and $\gamma_u = \min\{m : \sum_{k=1}^{m} a_k^2 \geq u\}$ (so $\gamma_u = \infty$, if $\sum_{k=1}^{\infty} a_k^2 < u$)

$$E[\sum_{k=1}^{\gamma_u} a_k^2] < \infty. \tag{3.8}$$

A.7

$$\sup_k |X_k| < \infty \text{ a.s.} \tag{3.9}$$

A.8 The series $\sum_k a_k$ diverges almost surely, and for each $\epsilon > 0$, there exists a compact $K \subset \mathcal{L}(S)$ such that $P\{\Gamma_n \in K, n=1,2,\ldots\} > 1-\epsilon$.

3.1 Remark a) The iterative scheme in A.1 differs slightly from that studied by Kushner and Schwartz. They consider an iterative scheme of the form

$$X_{k+1} = X_k + a_k f(X_k, Y_{k+1}). \tag{3.10}$$

However, if we define $U_{k+1} = f(X_k, Y_{k+1}) - Qf(X_k, Y_k)$, then (3.10) can be written

(3.11) $$X_{k+1} = X_k + a_k(Qf(X_k,Y_k) + U_{k+1})$$

(which has the form in A.1), and since the U_k are bounded martingale differences, it is easy to check that the convergence of (3.4) and A.6 imply the convergence of (3.6) and (3.7). Kushner and Schwartz do not assume (3.4); however, most results in the literature giving almost sure convergence (with the notable exception of Dupuis and Kushner (1989)) do assume something similar to (3.4) (e.g., Metivier and Priouret (1984)).

b) In general, if $\{U_k\}$ is a sequence of martingale differences, bounded in L^2, then the convergence of (3.4) and A.6 imply the convergence of (3.6) and (3.7).

c) If the sequence $\{a_k\}$ is nonincreasing, then (3.5) will converge.

d) A.6 will hold if the a_k are bounded. □

For $h \in \bar{C}(\mathbb{R}^d \times S)$, define an $\{\mathcal{F}_k\}$-martingale by

(3.12) $$M_m = a_m h(X_m,Y_m) - \sum_{k=0}^{m-1} E[a_{k+1}h(X_{k+1},Y_{k+1}) - a_k h(X_k,Y_k)|\mathcal{F}_k]$$

$$= a_m h(X_m,Y_m) - \sum_{k=0}^{m-1} E[(a_{k+1}-a_k)h(X_{k+1},Y_{k+1})|\mathcal{F}_k]$$

$$-\sum_{k=0}^{m-1} E[a_k\big(h(X_{k+1},Y_{k+1}) - h(X_k,Y_{k+1})\big)|\mathcal{F}_k]$$

$$-\sum_{k=0}^{m-1} E[a_k\big(h(X_k,Y_{k+1}) - h(X_k,Y_k)\big)|\mathcal{F}_k]$$

By A.6, $E[\sum(M_{\gamma_n\wedge(k+1)} - M_{\gamma_n\wedge k})^2] < \infty$ for each n, and it follows that $M_{\gamma_n\wedge m}$ converges a.s. as $m \to \infty$. Since $\lim_{n\to\infty} P\{\gamma_n = \infty\} = 1$, it follows that M converges almost surely. Assume that h satisfies

(3.13) $$|h(x_1,y) - h(x_2,y)| \le L_h|x_1 - x_2|$$

for some constant L_h. Note that the first term on the right of (3.12) converges to zero as $m \to \infty$ and that the first two sums converge for all bounded h a.s. by (3.5) and (3.7). It follows that the third sum must also converge a.s. so we have

$$\sum_{k=0}^{\infty} a_k\big(Qh(X_k,Y_k) - h(X_k,Y_k)\big) \tag{3.14}$$

converges a.s. for all h satisfying (3.13).

Define $\check{X}_n(t) = X_{\eta_n(t)}$ and observe that

$$\sum_{k=n}^{\eta_n(t)} a_k h(X_k,Y_k) = \int_{[0,t]\times S} h(\check{X}_n(s),y)\,\Gamma_n(ds\times dy)\,.$$

The convergence of the series in (3.14) implies

$$\lim_{n\to\infty} \int_{[0,t]\times S}\big(Qh(\check{X}_n(s),y) - h(\check{X}_n(s),y)\big)\Gamma_n(ds\times dy) = 0, \quad \text{a.s.} \tag{3.15}$$

The convergence of the series in (3.6) and the boundedness of f imply

$$\overline{\lim}_{n\to\infty} |\check{X}_n(t+s) - \check{X}_n(t)| \le \|f\|\, s \tag{3.16}$$

for all s and $t > 0$, a.s., and (3.16) and assumption A.7 imply that for every $\epsilon > 0$ there exists a compact $K \subset D_{\mathbb{R}^d}[0,\infty)$ such that $P\{\check{X}_n \in K,\ n=1,2,\ldots\} \ge 1-\epsilon$. This conclusion along with assumption A.8 imply that for each $\epsilon > 0$ there exist compacts $K_1 \in D_E[0,\infty)$ and $K_2 \in \mathcal{L}(S)$ such that $P\{(\check{X}_n,\Gamma_n) \in K_1\times K_2, n=1,2,\ldots\} \ge 1-\epsilon$.

By (3.15), with probability one, any limit point (x,μ) of $\{(\check{X}_n,\Gamma_n)\}$ satisfies

$$\int_{[0,t]\times S}\big(Qh(x(s),y) - h(x(s),y)\big)\mu(ds\times dy) = 0 \tag{3.17}$$

for h satisfying the above assumptions. The collection of h for which (3.17) holds is closed under bounded pointwise convergence, and hence includes all of $B(\mathbb{R}^d\times S)$. By the separability of S there exists a countable subset $D \subset \overline{C}(\mathbb{R}^d\times S)$ such that the bounded, pointwise closure of D is $B(\mathbb{R}^d\times S)$. As in Example 2.3, there exists $J \subset [0,\infty)$ with $m(J) = 0$ and $\gamma_s \in \mathcal{P}(S)$ such that for $s \in [0,\infty) - J$

$$\int_S\big(Qh(x(s),y) - h(x(s),y)\big)\gamma_s(dy) = 0 \tag{3.18}$$

for all $h \in B(\mathbb{R}^d\times S)$. A.3 and (3.18) imply that $\gamma_s = \pi_{x(s)}$, and A.1 and Lemma 1.5 then give

(3.19) $$x(t) = x(0) + \int_0^t F(x(s))\,ds$$

where $F(x) = \int f(x,y)\,\pi_x(dy)$.

3.2 Remark Dupuis and Kushner (1989) obtain a.s. convergence of stochastic approximation algorithms using large deviation estimates. In the present context, results analogous to theirs can be obtained by observing that the limit in (3.15) holds with A.5 and A.6 replaced by

A.9 Let $\{c_k\}$ be a nonincreasing sequence of constants satisfying $c_k \geq \sup_{m \geq k} a_m$ and $\{\alpha_n\}$ a nonnegative sequence satisfying $\sum e^{-\alpha_n \epsilon} < \infty$ for all $\epsilon > 0$. For each $T > 0$, the following hold with probability one.

(3.4)′ $$\sup_n \alpha_n \sum_{k=n}^{\eta_n(T)} c_k^2 < \infty, \text{ for each } T > 0$$

(3.5)′ $$\lim_{n\to\infty} \sum_{k=n}^{\eta_n(T)} E[|a_{k+1} - a_k| | \mathcal{F}_k] = 0 \text{ a.s.}$$

(3.6)′ $$\lim_{n\to\infty} \sup_{t \leq T} \sum_{k=n}^{\eta_n(t)} a_k U_{k+1} = 0 \text{ a.s.}$$

(3.7)′ $$\lim_{n\to\infty} \sum_{k=n}^{\eta_n(T)} a_k^2 E[|U_{k+1}| | \mathcal{F}_k] = 0 \text{ a.s.}$$

The only complication in checking that A.9 can be used in place of A.5 and A.6 in the above argument is to verify that for each $T > 0$

(3.20) $$\lim_{n\to\infty} \sup_{n \leq m \leq \eta_n(T)} |M_m - M_n| = 0 \text{ a.s.}$$

Let $\zeta_n(R) = \min\{m : \alpha_n \sum_{k=n}^m c_k^2 \geq R\}$. Then, by (3.4)′, (3.20) will hold if

$$\lim_{n\to\infty} \sup_{n \leq m \leq \zeta_n(R)} |M_m - M_n| = 0 \text{ a.s.}$$

for each $R > 0$. Observing that $|M_{k+1} - M_k| \leq 4c_k \|h\|$, Lemma 4.1, below, can be applied using the assumption that $\sum e^{-\alpha_n \epsilon} < \infty$ to verify that for each $\epsilon > 0$ and $R > 0$

$$\sum_{n=1}^{\infty} P\{\sup_{n \leq m \leq \zeta_n(R)} |M_m - M_n| > \epsilon\} < \infty .$$

Theorem 3.3 Assume A.1-A.8 (or A.1-A4 and A.7-A.9). Let $F(x) = \int f(x,y)\pi_x(dy)$. Suppose uniqueness holds for the system $\dot{Y} = F(Y)$, and let $Y(t,x)$ denote the solution with $Y(0) = x$. Let O be the ω-limit set for this system, and let $G \subset O$ be closed. Suppose there exists an open set $U \supset G$ such that $\bar{U}$ is compact and that G is uniformly asymptotically stable in $\bar{U}$ in the sense that

$$\lim_{t\to\infty} \sup_{x \in \bar{U}} d(Y(t,x),G) = 0. \tag{3.21}$$

Then there exists an event Ω_0 with $P(\Omega_0) = 1$, such that on the event $\{X_k \in U,$ i.o.$\} \cap \Omega_0$

$$\lim_{k\to\infty} d(X_k,G) = 0 \tag{3.22}$$

Proof We follow an argument of Kushner and Clark (1978). Let $(\Omega,\mathcal{F},P)$ be the probability space on which the processes are defined, and let Ω_0 be an event with probability one on which all limit points of $\{\check{X}_n\}$ satisfy (3.19). Fix $\epsilon > 0$ such that $G^{2\epsilon} \equiv \{x : d(x,G) < 2\epsilon\} \subset U$. Note that by (3.21), there exists $T > 0$ such that for all $x \in \bar{U}$, $Y(T,x) \in G^\epsilon$, and hence if $\omega \in \Omega_0$ and along some subsequence, $\check{X}_n(t_n,\omega)$ converges to a point in $\bar{U}$, then $\check{X}_n(t_n+T,\omega)$ converges to a point in G^ϵ. Let $\tau_n = \inf\{t > 2T : d(\check{X}_n(t),G) \geq 2\epsilon\}$. Fix $\omega \in \{X_k \in U,$ i.o.$\} \cap \Omega_0$. Let $\Gamma = \{n : \tau_n(\omega) < \infty$ and $\check{X}_n(0,\omega) \in U\}$. If Γ is infinite, then $\overline{\lim}_{n \in \Gamma} d(\check{X}_n(T,\omega),G) \leq \epsilon$ and $\overline{\lim}_{n \in \Gamma} d(\check{X}_n(\tau_n(\omega)-T,\omega),G) \leq 2\epsilon$. But taking $t_n = \tau_n(\omega)-T$ in the above observation, this inequality implies $\overline{\lim}_{n \in \Gamma} d(\check{X}_n(\tau_n(\omega),\omega),G) \leq \epsilon$ contradicting the definition of τ_n. But if $\tau_n(\omega) = \infty$ for some n, then $d(X_k(\omega),G) < 2\epsilon$ for all $k \geq n$. Since ϵ can be taken to be arbitrarily small, (3.22) follows. □

4. Appendix

Lemma 4.1 Let $\{M_k\}$ be a martingale and $\{c_k\}$ be constants such that $|M_k - M_{k-1}| \leq c_k$. Then for each $\epsilon > 0$

$$P\{\sup_{k \leq m} |M_k - M_0| \geq \epsilon\} \leq 2\inf_{\lambda > 0} \exp\Big\{ \sum_{k=1}^{m} (e^{\lambda c_k} - 1 - \lambda c_k) - \lambda\epsilon\Big\}.$$

Proof The lemma follows as in the proof of Lemma 5.1 of Kurtz (1972). □

References

Bensoussan, A., Lions, J. L., and Papanicolaou, G. C. (1978). *Asymptotic Analysis for Periodic Structures*, North-Holland, Amsterdam.

Bhattacharya, R. N., Gupta, V. K., and Walker, H. F. (1989). Asymptotics of solute dispersion in periodic porous media. *SIAM J. Appl. Math.* 49, 86-98.

Dupuis, Paul and Kushner, Harold J. (1989). Stochastic approximation and large deviations: upper bounds and w.p. 1 convergence. *SIAM J. Control Optim.*, 27, 1108-1135.

Ethier, Stewart N. and Kurtz, Thomas G. (1986). *Markov Processes: Characterization and Convergence.* Wiley, New York.

Griego, Richard J. and Hersh, Reuben (1969). Random evolutions, Markov chains, and systems of partial differential equations. *Proc. Nat. Acad. Sci. USA* 62, 305-308.

Hersh, Reuben (1974). Random evolutions: a survey of results and problems. *Rocky Mt. J. Math.* 4, 443-477.

Kallenberg, Olav (1986). *Random Measures.* Academic Press, New York.

Khas'minskii, R. Z. (1966a). On stochastic processes defined by differential equations with a small parameter. *Theory Probab. Appl.* 11, 211-228.

Khas'minskii, R. Z. (1966b). A limit theorem for the solutions of differential equations with random right-hand sides. *Theory Probab. Appl.* 11, 390-406.

Kurtz, Thomas G. (1972). Inequalities for the law of large numbers. *Ann. Math. Statist.* 43, 1874-1883.

Kurtz, Thomas G. (1973). A limit theorem for perturbed operator semigroups with applications to random evolutions. *J. Funct. Anal.* 12, 55-67.

Kurtz, Thomas G. (1987). Martingale problems for controlled processes. *Stochastic Modeling and Filtering. Lecture Notes in Control and Information Sci.* 91, Springer-Verlag, Berlin.

Kushner, Harold J. (1979). Jump-diffusion approximation for ordinary differential equations with wide-band random right hand sides. *SIAM J. Control Optim.* 17, 729-744.

Kushner, Harold J. (1984). *Approximation and Weak Convergence Methods for Random Processes with Applications to Stochastic Systems Theory.* MIT Press, Cambridge, Mass.

Kushner, Harold J. and Clark, Dean S. (1978). *Stochastic Approximation Methods for Constrained and Unconstrained Systems.* Springer-Verlag, Berlin.

Kushner, Harold J. and Schwartz, Adam (1984). An invariant measure approach to the convergence of stochastic approximations with state dependent noise. *SIAM J. Control Optim.*, 22, 13-27.

Lochak, P. and Meunier, C. (1988). *Multiphase Averaging for Classical Systems.* Springer-Verlag, New York.

Metivier, Michel and Priouret, Pierre (1984). Applications of a Kushner and Clark lemma to general classes of stochastic algorithms. *IEEE Trans. Information Theory* 30, 140-151.

Papanicolaou, G. C. (1978). Asymptotic analysis of stochastic equations. *Studies in Probability Theory, MAA Studies 18*, M. Rosenblatt, ed. Mathematics Association of America, New York, 111-179.

Pinsky, Mark A. (1974). Multiplicative operator functionals and their asymptotic properties. *Advances in Probability* 3, Dekker, New York.

Stockbridge, Richard H. (1990). Time-average control of martingale problems: Existence of a stationary solution. *Ann. Probab.* 18, 190-205.

A NONLINEAR FILTER WITH TWO TIME SCALES

Jean Picard

Mathématiques Appliquées
Université Blaise Pascal (Clermont 2)
63177 Aubière Cedex, France

Abstract. We consider the nonlinear filtering problem defined as follows; the signal is the two-dimensional process (X_t, H_t) satisfying $dX_t = dW_t$, $dH_t = X_t dV_t$ for a standard Wiener process (V_t, W_t), and the observation is given by $dY_t = H_t dt + \varepsilon dB_t$ for an independent Wiener process B_t and a small positive parameter ε. We describe a suboptimal filter and estimate its difference with the optimal one as $\varepsilon \to 0$. It appears that the estimation of the two components of the signal involves two different time scales.

0. Introduction

In the last years, several papers have been devoted to the problem of finding suboptimal filters, which are demonstrably efficient when the observation noise is small. More precisely, one supposes that the signal process Z_t and the observation process Y_t are multidimensional processes satisfying

$$dZ_t = b(t, Z_t)dt + \sigma(t, Z_t)dW_t, \tag{0.1}$$

$$dY_t = h(t, Z_t)dt + \varepsilon dB_t \tag{0.2}$$

for a Wiener process (W_t, B_t) and a positive parameter ε; one lets $\mathcal{Y}_t$ be the filtration generated by Y_t, and tries to find an approximation of the conditional law of Z_t given $\mathcal{Y}_t$ (in particular of its conditional mean $\widehat{Z}_t$); the efficiency of the approximation is evaluated by estimating the difference with the optimal filter as $\varepsilon \to 0$. The first step generally consists of estimating the filtering error $Z_t - \widehat{Z}_t$; one looks for conditions ensuring that this error is small as $\varepsilon \to 0$ and in this case, one proves that it is of order ε^γ in the spaces L^q, for some $\gamma > 0$; this is done by finding a suboptimal filter M_t such that $Z_t - M_t$ is of order ε^γ. Then the second step consists of choosing M_t so that $\widehat{Z}_t - M_t$ is as small as possible; the aim is generally to find M_t so that the order of $\widehat{Z}_t - M_t$ in the spaces L^q is negligible with respect to the order of $\widehat{Z}_t - Z_t$. However, one has to notice that the behavior of the filter is not necessarily similar on all the components of Z_t.

When the function $z \mapsto h(t, z)$ is one-to-one, and under some other technical assumptions (including the ellipticity of $\sigma\sigma^*$), one can prove that the extended Kalman filter behaves nicely (see [8]); the conditional law of Z_t is approximately Gaussian with a variance of order ε (so that the filtering error $Z_t - \widehat{Z}_t$ is of order $\sqrt{\varepsilon}$) and the difference between the optimal and extended Kalman filters is of order ε. Moreover, one can prove that the memory of the filter is of order ε; this means that the influence of a measurement "dY_s" is forgotten when $t \gg s$ with an exponential rate of order $1/\varepsilon$. Let us also mention that in the homogeneous case, other Kalman-like filters are studied in [3], [6], [1], [7].

When $z \mapsto h(t,z)$ is not one-to-one, the behaviour is generally different, though the filtering error may still be small. For instance, when h is piecewise one-to-one, one can sometimes describe filters based on tests designed to choose the domain in which Z_t is (see [2]). Another basic family of problems is to take for Z_t a two-dimensional process (X_t, H_t) and to suppose that $h(t, Z_t) = H_t$. When ε is small, H_t is nearly observed, so the problem is to know whether X_t can be estimated. A first example is the case $\dot{H}_t = \phi(X_t, H_t)$ with $x \mapsto \phi(x,h)$ one-to-one; in this example, X_t can be estimated (see [9], [4]); intuitively, this is clear, because if one has a good estimation of the path of H_t, one can deduce an estimation of its derivative and thus of X_t; the filtering error is not the same on the two components ($\varepsilon^{3/4}$ on H_t and $\varepsilon^{1/4}$ on X_t) but the memory is of order $\sqrt{\varepsilon}$ on both components.

In this work, we want to consider a second example; we suppose that X_t is a Wiener process and that $dH_t = X_t dV_t$ where V_t is a Wiener process which is independent from X. This example is the prototype of a more general class of systems which will be studied elsewhere. Since one has a good estimation of the path of H_t, one can deduce an estimation of its quadratic variation

$$\langle H, H \rangle_t = \int_0^t X_s^2 ds \tag{0.3}$$

and thus an estimation of $|X_t|$; however, the direct computation of the approximate quadratic variation is not easily handled numerically, and probably does not give a good enough approximation of $|X_t|$; since $|X_t|$ can take infinitely many values, one cannot use tests as in [2]. The aim of this work is to explain how an estimation of $|X_t|$ can be obtained by solving a stochastic differential equation driven by Y_t, and to estimate its difference with the optimal filter. It will appear that our suboptimal filter involves two time scales; the component H_t is observed with an error of order $\sqrt{\varepsilon}$ and a memory of order ε, whereas $|X_t|$ is observed with an error of order $\varepsilon^{1/4}$ and a memory of order $\sqrt{\varepsilon}$. Notice that the observation does not give any information on the sign of X_t; since the behaviour of the filter seems complicated when X_t is close to 0, we will suppose that $X_0 > 0$ and will estimate X_t until it exits some subinterval of $(0, \infty)$; after that time, the filter should be reinitialized.

In §1 we set our framework, describe our suboptimal filter and list the results of this work. In §2, we give an intuitive justification for the choice of the filter. The results are proved in §3 and §4; we will use methods introduced in [7] and [8], as well as the analysis of multi-scale stochastic equations.

1. The results

We first make precise our assumptions and notation. The nonlinear filtering problem under consideration consists of a two-dimensional signal process (X_t, H_t) and a real observation Y_t which satisfy

$$\begin{cases} dX_t = dW_t, \qquad dH_t = X_t dV_t, \\ dY_t = H_t dt + \varepsilon dB_t. \end{cases} \tag{1.1}$$

We suppose that (V_t, W_t, B_t) is a three-dimensional standard Wiener process, that $Y_0 = 0$ and that $X_0 = x_0 > 0$ and $H_0 = h_0$ are deterministic. We want to find approximations of the conditional means $\widehat{X}_t$ and $\widehat{H}_t$ of X_t and H_t given $\mathcal{Y}_t$ until X_t comes near 0.

If ξ_t is a stochastic process depending on ε, we will say that ξ_t is of order ε^γ (γ real) if for any fixed $T > 0$ and $q \geq 1$, the L^q norm of ξ_t/ε^γ is bounded as $\varepsilon \to 0$, uniformly for $0 \leq t \leq T$; in this case we will write $\xi_t = O(\varepsilon^\gamma)$; if τ is a stopping time, we will say that ξ_t is of order ε^γ on $\{t \leq \tau\}$ if $\xi_t 1_{\{t \leq \tau\}}$ is of order ε^γ. Sometimes, the behavior will be different for small times; this means that the estimate will be uniform only for $t_0 \leq t \leq T$, for any fixed $t_0 > 0$.

The nonlinear filter which will be studied is given by the equations

$$d\overline{X}_t = \frac{\xi_t Q_t}{\varepsilon}(dY_t - \overline{H}_t dt), \qquad \overline{X}_0 = x_0, \tag{1.2}$$

$$d\overline{H}_t = \frac{\overline{X}_t}{\varepsilon}(dY_t - \overline{H}_t dt), \qquad \overline{H}_0 = h_0, \tag{1.3}$$

$$d\xi_t = -\frac{\overline{X}_t}{\varepsilon}\xi_t dt + \frac{1}{\varepsilon^{3/2}}(dY_t - \overline{H}_t dt), \qquad \xi_0 = 0, \tag{1.4}$$

$$\dot{Q}_t = \frac{1}{\sqrt{\varepsilon}}(-Q_t^2\xi_t^2 + 1), \qquad Q_0 = \sqrt{2x_0} \tag{1.5}$$

where x_0 and h_0 were given in the assumptions; the intuitive justification for this filter will be given in §2. As it was explained previously, we estimate X_t only until it exits some subinterval of $(0, \infty)$. Thus, let I be a compact subinterval of $(0, \infty)$ containing x_0 and put

$$\tau = \inf\{t \geq 0, \overline{X}_t \notin I\}. \tag{1.6}$$

The next result states that comparing t and τ is a good test for knowing whether X stayed in or near I up to time t. Theorems 1, 2 and 3 are the main results of this work.

Proposition 1. *Let I_* and I^* be two compact subintervals of $(0, \infty)$ such that I_* is included in the interior of I and I is included in the interior of I^*. Then for any q,*

$$\mathbb{P}\Big[\{t \leq \tau\} \cap \{\exists s < t,\ X_s \notin I^*\}\Big] = O(\varepsilon^q) \tag{1.7}$$

and

$$\mathbb{P}\Big[\{t > \tau\} \cap \{\forall s < t,\ X_s \in I_*\}\Big] = O(\varepsilon^q). \tag{1.8}$$

Theorem 1. *On $\{t \leq \tau\}$, $H_t - \overline{H}_t$ is of order $\sqrt{\varepsilon}$ and $X_t - \overline{X}_t$ is of order $\varepsilon^{1/4}$.*

Theorem 2. *For any fixed $t_0 > 0$, the processes $\widehat{H}_t - \overline{H}_t$ and $\widehat{X}_t - \overline{X}_t$, $t \geq t_0$ are respectively of order ε and $\sqrt{\varepsilon}$ on $\{t \leq \tau\}$.*

Theorem 3. *For any fixed $t_0 > 0$ and for $t \geq t_0$, the relations*

$$\mathbb{E}\big[(X_t - \widehat{X}_t)^2 \mid \mathcal{Y}_t\big] = \sqrt{\varepsilon} Q_t + O(\varepsilon^{3/4}), \tag{1.9}$$

$$\mathbb{E}\big[(X_t - \widehat{X}_t)(H_t - \widehat{H}_t) \mid \mathcal{Y}_t\big] = \varepsilon Q_t \xi_t + O(\varepsilon^{5/4}), \tag{1.10}$$

$$\mathbb{E}\big[(H_t - \widehat{H}_t)^2 \mid \mathcal{Y}_t\big] = \varepsilon \overline{X}_t + O(\varepsilon^{3/2}) \tag{1.11}$$

hold on $\{t \leq \tau\}$.

Higher-order conditional moments can be estimated with the same method, and in particular one can check that the conditional law of (X_t, H_t) is approximately Gaussian.

Remark. We have assumed that (X_0, H_0) is deterministic in order to simplify the proofs; however, the only important assumption for our approach is that the conditional law of H_0 given X_0 should be Gaussian. Let us explain briefly what should be done for non-deterministic initial conditions. First of all, in a time of order ε, one should get an approximation of order $\sqrt{\varepsilon}$ of H_t (this can be done easily with a Kalman-like filter); then one should use filter (1.2)-(1.5) with initial condition the obtained estimation for $\overline{H}$ and any $x_0 \in I$ for $\overline{X}$. Previous estimates can be proved for $t \gg \varepsilon^{1/4}$ up to time τ; notice, however, that if the sign of X_0 is not known, we only estimate $|X_t|$; notice also that if $|X_0|$ is not in I the time τ is very small. After time τ, one should reinitialize the filter and wait for $|X_t|$ to come again in the window I.

2. Intuitive derivation of the filter

In this section, we explain how the filter described in §1 is obtained. It is not difficult to verify that applying directly the extended Kalman filter to our system does not work. The idea is to introduce an auxiliary filtration $\mathcal{G}_t \supset \mathcal{Y}_t$ and to decompose the conditioning on $\mathcal{Y}_t$, first into a conditioning on $\mathcal{G}_t$, and then into a conditioning on $\mathcal{Y}_t$; the first conditioning will be trivial, and for the second one we shall use an extended Kalman filter. Thus, let $\mathcal{G}_t$ be the filtration generated by the process (X, Y). Then the conditional law of H_t given $\mathcal{G}_t$ is given by a Kalman filter (we are in the case of conditionally Gaussian processes); it is the Gaussian law with mean $\check{H}_t$ and variance εP_t given by

$$d\check{H}_t = \frac{P_t}{\varepsilon}(dY_t - \check{H}_t dt), \qquad \dot{P}_t = \frac{1}{\varepsilon}(-P_t^2 + X_t^2) \tag{2.1}$$

where $\check{H}_0 = h_0$ and $P_0 = 0$. Moreover, we know that the process

$$\check{B}_t = \frac{1}{\varepsilon}\Big(Y_t - \int_0^t \check{H}_s ds\Big) \tag{2.2}$$

is a standard $\mathcal{G}_t$ Wiener process (the innovation process). Thus our system can be written in the form

$$\begin{cases} dX_t = dW_t, \qquad d\check{H}_t = P_t d\check{B}_t, \qquad \dot{P}_t = \dfrac{1}{\varepsilon}(-P_t^2 + X_t^2), \\ dY_t = \check{H}_t dt + \varepsilon d\check{B}_t \end{cases} \tag{2.3}$$

where $(W_t, \check{B}_t)$ is a standard Wiener process, and the conditional mean of H_t given $\mathcal{Y}_t$ is equal to the conditional mean of $\check{H}_t$.

Let us now explain how X_t can be estimated from this system. Consider the family of processes $\mu_t(x)$, x real, given by

$$d\mu_t(x) = \frac{x}{\varepsilon}(dY_t - \mu_t(x)dt), \qquad \mu_0(x) = h_0. \tag{2.4}$$

For any t, $\mu_t(x)$ can be viewed as a smooth random function of x. If we suppose that X_t is still positive, one has $P_t \simeq X_t$, so it can be shown that $\check{H}_t$ is close to $\mu_t(X_t)$; suppose that we can replace $\check{H}_t$ by $\mu_t(X_t)$ in the filtering problem; then we obtain an approximate problem in which the observation at time t is

$$\int_0^t \mu_s(X_s)ds + \varepsilon\check{B}_t.$$

In this new problem, the derivative of the observation function $\mu_t(x)$ is given by

$$d\mu_t'(x) = -\frac{x}{\varepsilon}\mu_t'(x)dt + \frac{1}{\varepsilon}(dY_t - \mu_t(x)dt). \tag{2.5}$$

It is of order $\sqrt{\varepsilon}$, so the noise on X_t which is apparently of order ε is actually of order $\sqrt{\varepsilon}$; this explains why the memory in the filtering of X_t is of order $\sqrt{\varepsilon}$. The extended Kalman filter for the approximate problem can be written as

$$d\overline{X}_t = \frac{1}{\varepsilon^{3/2}}\mu_t'(\overline{X}_t)Q_t(dY_t - \mu_t(\overline{X}_t)dt), \tag{2.6}$$

$$\dot{Q}_t = \frac{1}{\sqrt{\varepsilon}}\left(-Q_t^2\frac{\mu_t'^2(\overline{X}_t)}{\varepsilon} + 1\right). \tag{2.7}$$

Then the estimation of $\widehat{H}_t$ should be $\overline{H}_t = \mu_t(\overline{X}_t)$. By putting $\xi_t = \mu_t'(\overline{X}_t)/\sqrt{\varepsilon}$, we recognize equations (1.2) and (1.5). However, the computation of $\mu_t(\overline{X}_t)$ and $\mu_t'(\overline{X}_t)$ is not easy; thus we replace x by $\overline{X}_t$ in the equations of $\mu_t(x)$ and $\mu_t'(x)/\sqrt{\varepsilon}$ and it appears that we obtain equations (1.3) and (1.4); this approximation works because $\overline{X}_t$ varies slowly (in a time scale of order $\sqrt{\varepsilon}$) with respect to $\overline{H}_t$ and ξ_t (which vary in a time scale of order ε). This gives an intuitive justification of the filter introduced in §1 (the reason for the choice of $\sqrt{2x_0}$ in (1.5) will appear in the proof).

3. Proof of Proposition 1 and Theorem 1

The basic lemma in our estimates is the following.

Lemma 1. *Suppose that μ_t is a family (indexed by ε) of real-valued semimartingales satisfying*

$$d\mu_t = -\frac{\rho_t\mu_t}{\varepsilon^\beta}dt + \frac{b_t}{\varepsilon^\beta}dt + \frac{g_t}{\varepsilon^{\beta/2}}dw_t \tag{3.1}$$

for some adapted processes ρ_t, b_t, g_t, some Wiener process w_t and some $\beta > 0$. Let τ be a stopping time, let γ be a real number and let $c > 0$; we suppose that on $\{t \le \tau\}$, b_t and g_t are of order ε^γ and $\rho_t \ge c$. If $\mu_0 = O(\varepsilon^\gamma)$, then μ_t is of order ε^γ on $\{t \le \tau\}$.

For a proof of this result without τ, look for instance at §1.3 of [8] (this is based on an application of Itô's formula to the estimation of the moments of μ_t); the localization by τ is not difficult. One can also verify that if conditions on μ_0, b_t and g_t are not valid for small times, then the conclusion holds for $t \ge t_0$, for any fixed $t_0 > 0$; more precisely, past values of μ_t are forgotten with an exponential rate of order $1/\varepsilon^\beta$. Here is another basic lemma.

Lemma 2. *Suppose that μ_t is a real-valued semimartingale satisfying*

$$d\mu_t = b_t dt + g_t dw_t \tag{3.2}$$

for adapted processes b_t, g_t and a Wiener process w_t. We suppose that for real β and γ and for a stopping time τ, b_t and g_t are of order ε^β and μ_t is of order ε^γ on $\{t \le \tau\}$. Then for any $\gamma' < \gamma$ and any bounded stopping time $\tau' \le \tau$, the variable $\mu_{\tau'}$ is of order $\varepsilon^{\gamma'}$.

Proof. Let $T > 0$ be such that $\tau' \le T$. If $\tau = \infty$, it is proved in Lemma 1.2.1 of [8] that $\sup_{t\le T} |\mu_t|$ is of order $\varepsilon^{\gamma'}$. By stopping μ_t at τ, we deduce that under our assumptions, $\sup_{t\le T} |\mu_{t\wedge\tau}|$ is of order $\varepsilon^{\gamma'}$. Since $\tau' \le T \wedge \tau$, we conclude. □

In particular, we immediately deduce from Lemma 1 that $H_t - \overline{H}_t$ is of order $\sqrt{\varepsilon}$ on $\{t \le \tau\}$, so in Theorem 1 we only have to estimate $X_t - \overline{X}_t$. To this end, fix some I^* satisfying the assumptions of Proposition 1, some $\alpha > 1$, and put

$$\tau_\alpha = \inf\{t \ge 0,\ X_t \notin I^* \quad \text{or} \quad \overline{X}_t \notin I \quad \text{or} \quad Q_t \notin [1/\alpha, \alpha]\}. \tag{3.3}$$

Our first aim will be to prove the estimate of Theorem 1 on $\{t \le \tau_\alpha\}$. Since the filter involves two time scales, the idea is to use a discretization of the time with a mesh size intermediate between the two time scales, more precisely $\varepsilon^{3/4}$. Thus we define $t_i = i\varepsilon^{3/4}$. The notation $O_i(\varepsilon^\gamma)$ will mean of order ε^γ with zero conditional mean given $\mathcal{G}_{t_i}$. We will use the processes $\breve{H}_t$ and $\breve{B}_t$ introduced in (2.1) and (2.2). Notice that $\breve{H}_t - \overline{H}_t = O(\sqrt{\varepsilon})$ and that $\xi_t = O(1)$ on $\{t \le \tau\}$ (apply Lemma 1).

Lemma 3. *For $1 \le i \le T/\varepsilon^{3/4}$, one has on $\{t_{i+1} \le \tau_\alpha\}$*

$$\frac{1}{\varepsilon(t_{i+1} - t_i)} \int_{t_i}^{t_{i+1}} (\breve{H}_s - \overline{H}_s)^2 ds = \frac{1}{2\overline{X}_{t_i}}(X_{t_i} - \overline{X}_{t_i})^2 + O_i(\varepsilon^{1/8}) + O(\varepsilon^{1/4}), \tag{3.4}$$

$$\frac{1}{\sqrt{\varepsilon}(t_{i+1} - t_i)} \int_{t_i}^{t_{i+1}} (\breve{H}_s - \overline{H}_s)\xi_s ds = \frac{X_{t_i}^2 - \overline{X}_{t_i}^2}{4\overline{X}_{t_i}^2} + O_i(\varepsilon^{1/8}) + O(\varepsilon^{1/4}), \tag{3.5}$$

$$\frac{1}{t_{i+1} - t_i} \int_{t_i}^{t_{i+1}} \xi_s^2 ds = \frac{X_{t_i}^2 + \overline{X}_{t_i}^2}{4\overline{X}_{t_i}^3} + O_i(\varepsilon^{1/8}) + O(\varepsilon^{1/4}). \tag{3.6}$$

Proof. Our estimates will be written on $\{t_{i+1} \leq \tau_\alpha\}$, though we will not repeat it. First notice that

$$\begin{cases} d(\check{H}_t - \overline{H}_t) = -\dfrac{\overline{X}_t}{\varepsilon}(\check{H}_t - \overline{H}_t)dt + (P_t - \overline{X}_t)d\check{B}_t, \\ d\xi_t = -\dfrac{\overline{X}_t}{\varepsilon}\xi_t dt + \dfrac{1}{\varepsilon^{3/2}}(\check{H}_t - \overline{H}_t)dt + \dfrac{1}{\sqrt{\varepsilon}}d\check{B}_t. \end{cases} \tag{3.7}$$

On the other hand one has from (1.2)

$$\overline{X}_t - \overline{X}_s = O(\sqrt{t-s} + (t-s)/\sqrt{\varepsilon}) \tag{3.8}$$

so if $t - s \leq \varepsilon^{3/4}$, $\overline{X}_t - \overline{X}_s$ is of order $\varepsilon^{1/4}$. From (2.3), one also has $P_t - X_t = O(\sqrt{\varepsilon})$ for $t \geq t_1$ (it may be larger for times of order ε). Thus we deduce that if we consider, for $i \geq 1$, $t_i \leq t < t_{i+1}$ the process $(\eta_t, \overline{\xi}_t)$ solution of

$$\begin{cases} d\eta_t = -\dfrac{\overline{X}_{t_i}}{\varepsilon}\eta_t dt + \dfrac{X_{t_i} - \overline{X}_{t_i}}{\sqrt{\varepsilon}}d\check{B}_t \\ d\overline{\xi}_t = -\dfrac{\overline{X}_{t_i}}{\varepsilon}\overline{\xi}_t dt + \dfrac{1}{\varepsilon}\eta_t dt + \dfrac{1}{\sqrt{\varepsilon}}d\check{B}_t \end{cases} \tag{3.9}$$

with $\eta_{t_i} = (\check{H}_{t_i} - \overline{H}_{t_i})/\sqrt{\varepsilon}$ and $\overline{\xi}_{t_i} = \xi_{t_i}$, one has from Lemma 1

$$\eta_t = \frac{\check{H}_t - \overline{H}_t}{\sqrt{\varepsilon}} + O(\varepsilon^{1/4}) \qquad \text{and} \qquad \overline{\xi}_t = \xi_t + O(\varepsilon^{1/4}). \tag{3.10}$$

Conditionally on $\mathcal{G}_{t_i}$, the process $(\eta_t, \overline{\xi}_t)$, $t_i \leq t < t_{i+1}$ is Gaussian; moreover for $t_i \leq s \leq t < t_{i+1}$, the conditional covariance of $(\eta_s, \overline{\xi}_s)$ and $(\eta_t, \overline{\xi}_t)$ is of order $\exp(-c(t-s)/\varepsilon)$; in particular, the conditional covariance of η_s^2 and η_t^2 satisfies an estimate of the same type, so the conditional variance of $\int_{t_i}^{t_{i+1}} \eta_s^2 ds$ is of order $\varepsilon^{7/4}$. Thus

$$\begin{aligned} &\frac{1}{\varepsilon(t_{i+1} - t_i)} \int_{t_i}^{t_{i+1}} (\check{H}_s - \overline{H}_s)^2 ds \\ &\quad = \frac{1}{t_{i+1} - t_i} \int_{t_i}^{t_{i+1}} \eta_s^2 ds + O(\varepsilon^{1/4}) \\ &\quad = \frac{1}{t_{i+1} - t_i} \int_{t_i}^{t_{i+1}} \mathbb{E}[\eta_s^2 \mid \mathcal{G}_{t_i}] ds + O_i(\varepsilon^{1/8}) + O(\varepsilon^{1/4}). \end{aligned} \tag{3.11}$$

By computing the conditional mean of η_s^2, we obtain from (3.9) that with an error of order $\exp(-c(t - t_i)/\varepsilon)$,

$$\mathbb{E}[\eta_s^2 \mid \mathcal{G}_{t_i}] \simeq \frac{1}{2\overline{X}_{t_i}}(X_{t_i} - \overline{X}_{t_i})^2 \tag{3.12}$$

and therefore deduce the estimate (3.4). The second and third estimates are proved with the same method and by applying the estimates

$$\mathbb{E}[\overline{\xi}_s \eta_s \mid \mathcal{G}_{t_i}] \simeq (X_{t_i}^2 - \overline{X}_{t_i}^2) \,/\, (4\overline{X}_{t_i}^2) \tag{3.13}$$

and

$$\mathbb{E}[\overline{\xi}_s^2 \mid \mathcal{G}_{t_i}] \simeq (X_{t_i}^2 + \overline{X}_{t_i}^2) \,/\, (4\overline{X}_{t_i}^3). \tag{3.14}$$

□

Lemma 4. *The process $X_t - \overline{X}_t$ is of order $\varepsilon^{1/4}$ on $\{t \leq \tau_\alpha\}$.*

Proof. For $t - s \leq \varepsilon^{3/4}$, $Q_t - Q_s$ is of order $\varepsilon^{1/4}$, so we deduce from (1.2) that

$$\overline{X}_{t_{i+1}} = \overline{X}_{t_i} + \frac{Q_{t_i}}{\varepsilon}\int_{t_i}^{t_{i+1}} \xi_s(dY_s - \overline{H}_s ds) + O(\sqrt{\varepsilon}). \tag{3.15}$$

By means of the decomposition of dY_s we obtain

$$\overline{X}_{t_{i+1}} = \overline{X}_{t_i} + \frac{Q_{t_i}}{\varepsilon}\int_{t_i}^{t_{i+1}} \xi_s(\check{H}_s - \overline{H}_s)ds + O_i(\varepsilon^{3/8}) + O(\sqrt{\varepsilon}). \tag{3.16}$$

Thus if we define

$$\psi_t = \frac{Q_t}{4}\frac{X_t + \overline{X}_t}{\overline{X}_t^2}, \tag{3.17}$$

we obtain from (3.5) in Lemma 3 that

$$\overline{X}_{t_{i+1}} = \overline{X}_{t_i} + \varepsilon^{1/4}\psi_{t_i}(X_{t_i} - \overline{X}_{t_i}) + O_i(\varepsilon^{3/8}) + O(\sqrt{\varepsilon}). \tag{3.18}$$

On the other hand

$$X_{t_{i+1}} = X_{t_i} + O_i(\varepsilon^{3/8}), \tag{3.19}$$

so we have

$$\overline{X}_{t_{i+1}} - X_{t_{i+1}} = (1 - \varepsilon^{1/4}\psi_{t_i})(\overline{X}_{t_i} - X_{t_i}) + m_i + v_i \tag{3.20}$$

where the $\mathcal{G}_{t_{i+1}}$-measurable variables m_i and v_i are respectively $O_i(\varepsilon^{3/8})$ and $O(\sqrt{\varepsilon})$. Then, in order to prove the lemma, it is sufficient to check that the process Z_t which is $\overline{X}_{t_i} - X_{t_i}$ on $[t_i, t_{i+1})$ is of order $\varepsilon^{1/4}$ on $\{t \leq \tau_\alpha\}$. This is a discrete-time analogue of Lemma 1; let us explain briefly how it can be proved from (3.20). We first modify Z_t after time τ_α so that (3.20) holds for any time. For any integer $p \geq 1$, by developing $Z_{t_{i+1}}^{2p}$ and by noticing that ψ_{t_i} is bounded below and above, we can check that there exists a $c > 0$ such that

$$Z_{t_{i+1}}^{2p} \leq Z_{t_i}^{2p}(1 - c\varepsilon^{1/4}) + 2pZ_{t_i}^{2p-1}m_i + O(\varepsilon^{(p/2)+(1/4)}). \tag{3.21}$$

We take the expectation in this expression and since Z_{t_i} is $\mathcal{G}_{t_i}$ measurable and $\mathbb{E}[m_i|\mathcal{G}_{t_i}]$ is zero, we deduce that $\mathbb{E}[Z_{t_i}^{2p}]$ is of order $\varepsilon^{p/2}$. □

Lemma 5. *Let*

$$\tau^* = \inf\{t \geq 0,\ X_t \notin I^* \quad \text{or} \quad \overline{X}_t \notin I\}. \tag{3.22}$$

Then $X_t - \overline{X}_t$ is of order $\varepsilon^{1/4}$ on $\{t \leq \tau^*\}$.

Proof. From Lemma 4, it is sufficient to prove that if α is chosen large enough, one has

$$\mathbb{P}[\tau_\alpha < t \leq \tau^*] = O(\varepsilon^q) \tag{3.23}$$

for any q. We deduce from (1.5) that on $\{t_{i+1} \leq \tau_\alpha\}$,

$$Q_{t_{i+1}} = Q_{t_i} + \varepsilon^{1/4} - \frac{Q_{t_i}^2}{\sqrt{\varepsilon}} \int_{t_i}^{t_{i+1}} \xi_s^2 ds + O(\sqrt{\varepsilon}) \tag{3.24}$$

so from (3.6) in Lemma 3

$$Q_{t_{i+1}} = Q_{t_i} + \varepsilon^{1/4}\Big(1 - Q_{t_i}^2 \frac{X_{t_i}^2 + \overline{X}_{t_i}^2}{4\overline{X}_{t_i}^3}\Big) + O_i(\varepsilon^{3/8}) + O(\sqrt{\varepsilon}). \tag{3.25}$$

By applying Lemma 4, we obtain

$$Q_{t_{i+1}} = Q_{t_i} + \varepsilon^{1/4} \frac{Q_{t_i} + (2\overline{X}_{t_i})^{1/2}}{2\overline{X}_{t_i}} \Big((2\overline{X}_{t_i})^{1/2} - Q_{t_i}\Big) + O_i(\varepsilon^{3/8}) + O(\sqrt{\varepsilon}) \tag{3.26}$$

on $\{t_{i+1} \leq \tau_\alpha\}$. Since $X_t - \overline{X}_t$ is of order $\varepsilon^{1/4}$, $\check{H}_t - \overline{H}_t$ is shown to be of order $\varepsilon^{3/4}$ (apply Lemma 1 to the first part of (3.7)), so by applying Itô's formula to (1.2), the increment of $(2\overline{X}_t)^{1/2}$ on $[t_i, t_{i+1}]$ is $O_i(\varepsilon^{3/8}) + O(\sqrt{\varepsilon})$. By proceeding as in the study of (3.20), one can deduce that $Q_t - (2\overline{X}_t)^{1/2}$ is of order $\varepsilon^{1/4}$ on $\{t \leq \tau_\alpha\}$. Thus from Lemma 2, for any bounded stopping time $\tau' \leq \tau_\alpha$, one has

$$Q_{\tau'} - (2\overline{X}_{\tau'})^{1/2} = O(\varepsilon^{1/5}). \tag{3.27}$$

On the other hand, on the event $\{\tau_\alpha < t \leq \tau^*\}$, one has $Q_{\tau_\alpha} \notin (1/\alpha, \alpha)$ and $2\overline{X}_{\tau_\alpha} \in [c_1, c_2]$ for some positive constant numbers c_1, c_2 which do not depend on α. Thus if α is greater than $\sqrt{c_2}$ and $1/\sqrt{c_1}$ and if one puts $\tau' = t \wedge \tau_\alpha$, one has

$$\{\tau_\alpha < t \leq \tau^*\} \subset \Big\{\big|Q_{\tau'} - (2\overline{X}_{\tau'})^{1/2}\big| \geq c\Big\} \tag{3.28}$$

for some $c > 0$. But from (3.27), the probability of this event is of order ε^q for any q. □

Proof of Proposition 1. We deduce from Lemmas 5 and 2 that for any bounded stopping time $\tau' \leq \tau^*$, the variable $X_{\tau'} - \overline{X}_{\tau'}$ is of order $\varepsilon^{1/5}$. On the other hand, on the event $\{\tau^* < t \leq \tau\}$, $X_{t\wedge\tau^*}$ is not in the interior of I^*, whereas $\overline{X}_{t\wedge\tau^*}$ is in I. Since their difference is of order $\varepsilon^{1/5}$, the probability of this event is of order ε^q for any q so (1.7) is proved. Similarly, by putting

$$\tau_* = \inf\{t \geq 0,\ X_t \notin I_*\}, \tag{3.29}$$

on the event $\{\tau < t \leq \tau_*\}$, $X_{t\wedge\tau}$ is in I_* whereas $\overline{X}_{t\wedge\tau}$ is not in the interior of I, so the probability of this event is also small. □

Proof of Theorem 1. The theorem is immediate from Lemma 5 and the fact that the probability of $\{\tau^* < t \leq \tau\}$ is of order ε^q for any q. □

4. Proof of Theorems 2 and 3

For $t \geq t_0 > 0$, we now want to estimate the conditional means and second-order moments of X_t and H_t on $\{t \leq \tau\}$. From the previous section, it is sufficient to estimate them on $\{t \leq \tau_0\}$ for some $\mathcal{Y}_t$-stopping time $\tau_0 \leq \tau$ such that Q_t is uniformly bounded below and above on $\{t \leq \tau_0\}$; one can indeed choose τ_0 so that the probability of $\{\tau_0 < t \leq \tau\}$ is small. There are three steps in the proof, a change of probability, a differential calculus and the integration by parts formula of Malliavin's calculus.

We first describe the change of probability. Let Λ_t be the process defined by

$$\begin{aligned}\Lambda_t = \exp\Big\{\frac{1}{\varepsilon^2}\int_0^t \check{H}_s dY_s - \frac{1}{2\varepsilon^2}\int_0^t \check{H}_s^2 ds \\ - \frac{1}{\sqrt{\varepsilon}}\int_0^t \frac{X_s - \overline{X}_s}{Q_s} dX_s + \frac{1}{2\varepsilon}\int_0^t \frac{1}{Q_s^2}(X_s - \overline{X}_s)^2 ds\Big\}.\end{aligned} \tag{4.1}$$

One can prove that the process Λ_t^{-1} stopped at τ_0 is a $\mathcal{G}_t$ martingale, so let $\widetilde{\mathbb{P}}$ be the probability which is absolutely continuous with respect to $\mathbb{P}$ with density $\Lambda_{t\wedge\tau_0}^{-1}$ on $\mathcal{G}_t$. Under $\widetilde{\mathbb{P}}$, the processes Y_t/ε and

$$\widetilde{W}_t = W_t - \frac{1}{\sqrt{\varepsilon}}\int_0^t \frac{1}{Q_s}(X_s - \overline{X}_s)ds \tag{4.2}$$

are independent standard Wiener processes. Note that

$$dX_t = \frac{1}{\sqrt{\varepsilon}Q_t}(X_t - \overline{X}_t)dt + d\widetilde{W}_t. \tag{4.3}$$

The variables X_t, $\check{H}_t$, ... can be viewed as functionals of $(\widetilde{W}, Y)$; we will now use a partial Malliavin calculus and differentiate them with respect to the Wiener process $\widetilde{W}$. For classical properties of the Malliavin calculus, we refer to [5]; we shall denote by ∇_s the differentiation operator; this means that $\int_0^\infty \nabla_s F u_s ds$ is the derivative of the variable $F(\widetilde{W}, Y)$ when $\widetilde{W}$ is perturbed in the direction of the adapted process $\int_0^\cdot u_s ds$; in particular, the derivatives of $\mathcal{Y}_t$-measurable variables will be zero. We will consider ∇_s for $0 \leq s \leq \sqrt{\varepsilon}$; in particular, since $t \geq t_0$, one has $s \leq t$ as soon as ε is small enough. Observe that the probability $\widetilde{\mathbb{P}}$ is used to compute the derivatives, but the processes are always estimated in the spaces L^q for the original probability $\mathbb{P}$; in particular, the notation $O(\varepsilon^\gamma)$ must be understood under $\mathbb{P}$. Our aim is to estimate the derivative of $\log \Lambda_t$.

Lemma 6. *On $\{t \leq \tau_0\}$, we have*

$$\frac{\nabla_s \log \Lambda_t}{\nabla_s X_t} = \zeta_t + O(1), \tag{4.4}$$

where ζ_t is the solution of

$$d\zeta_t = -\frac{1}{\sqrt{\varepsilon}Q_t}\zeta_t dt + \frac{1}{\sqrt{\varepsilon}}\xi_t d\check{B}_t - \frac{1}{\sqrt{\varepsilon}}\frac{dW_t}{Q_t}, \qquad \zeta_0 = 0. \tag{4.5}$$

Proof. By differentiating X_t in (4.3), we obtain

$$\frac{d}{dt}(\nabla_s X_t) = \frac{\nabla_s X_t}{\sqrt{\varepsilon} Q_t} \tag{4.6}$$

for $t \geq s$, with the initial condition $\nabla_s X_s = 1$. We also have from (2.1)

$$\frac{d}{dt}(\nabla_s P_t) = \frac{2}{\varepsilon}(-P_t \nabla_s P_t + X_t \nabla_s X_t), \tag{4.7}$$

so that

$$\frac{d}{dt}\Big(\frac{\nabla_s P_t}{\nabla_s X_t}\Big) = -\Big(\frac{2}{\varepsilon}P_t + \frac{1}{\sqrt{\varepsilon}Q_t}\Big)\Big(\frac{\nabla_s P_t}{\nabla_s X_t} - 1\Big) + \frac{2}{\varepsilon}(X_t - P_t) - \frac{1}{\sqrt{\varepsilon}Q_t}. \tag{4.8}$$

Since $X_t - P_t$ is of order $\sqrt{\varepsilon}$ and $P_t \geq c > 0$ with a large probability, we deduce

$$\frac{\nabla_s P_t}{\nabla_s X_t} = 1 + O(\sqrt{\varepsilon}) \tag{4.9}$$

on $\{t \leq \tau_0\}$. We also have

$$d(\nabla_s \check{H}_t) = -\frac{P_t}{\varepsilon}\nabla_s \check{H}_t dt + \frac{\nabla_s P_t}{\varepsilon}(dY_t - \check{H}_t dt) \tag{4.10}$$

so

$$d\Big(\frac{\nabla_s \check{H}_t}{\nabla_s X_t}\Big) = \frac{\nabla_s \check{H}_t}{\nabla_s X_t}\Big(-\frac{P_t}{\varepsilon} - \frac{1}{\sqrt{\varepsilon}Q_t}\Big)dt + \frac{\nabla_s P_t}{\nabla_s X_t} d\check{B}_t. \tag{4.11}$$

By comparing with (3.7), we get

$$\begin{aligned} d\Big(\frac{\nabla_s \check{H}_t}{\nabla_s X_t} - \sqrt{\varepsilon}\xi_t\Big) = &-\Big(\frac{P_t}{\varepsilon} + \frac{1}{\sqrt{\varepsilon}Q_t}\Big)\Big(\frac{\nabla_s \check{H}_t}{\nabla_s X_t} - \sqrt{\varepsilon}\xi_t\Big)dt + \frac{\overline{X}_t - P_t}{\sqrt{\varepsilon}}\xi_t dt \\ &- \frac{\xi_t}{Q_t}dt + \frac{\check{H}_t - \overline{H}_t}{\varepsilon}dt + \Big(\frac{\nabla_s P_t}{\nabla_s X_t} - 1\Big)d\check{B}_t. \end{aligned} \tag{4.12}$$

By means of the estimates $\check{H}_t - \overline{H}_t = O(\varepsilon^{3/4})$, $\overline{X}_t - P_t = O(\varepsilon^{1/4})$ and (4.9), one deduces from (4.12) that

$$\frac{\nabla_s \check{H}_t}{\nabla_s X_t} = \sqrt{\varepsilon}\xi_t + O(\varepsilon^{3/4}). \tag{4.13}$$

On the other hand, by differentiating (4.1), one obtains

$$\nabla_s \log \Lambda_t = \frac{1}{\varepsilon}\int_s^t \nabla_s \check{H}_u d\check{B}_u - \frac{1}{\sqrt{\varepsilon}}\int_s^t \frac{\nabla_s X_u}{Q_u} dW_u, \tag{4.14}$$

so that

$$d\Big(\frac{\nabla_s \log \Lambda_t}{\nabla_s X_t}\Big) = -\frac{1}{\sqrt{\varepsilon}Q_t}\frac{\nabla_s \log \Lambda_t}{\nabla_s X_t}dt + \frac{1}{\varepsilon}\frac{\nabla_s \check{H}_t}{\nabla_s X_t}d\check{B}_t - \frac{1}{\sqrt{\varepsilon}}\frac{dW_t}{Q_t}. \tag{4.15}$$

By comparing with (4.5) and applying (4.13), one proves (4.4). □

Lemma 7. *On $\{t \leq \tau_0\}$, one has*

$$\zeta_t - \frac{\overline{X}_t - X_t}{\sqrt{\varepsilon} Q_t} = O(1). \tag{4.16}$$

Proof. One can prove

$$\sqrt{\varepsilon}\xi_t(X_t - \overline{X}_t) = \check{H}_t - \overline{H}_t + O(\varepsilon) \tag{4.17}$$

by comparing the equations satisfied by both sides. Thus

$$\begin{aligned} \frac{d}{dt}\Big(\frac{\overline{X}_t - X_t}{\sqrt{\varepsilon}} - Q_t\zeta_t\Big) &= \frac{\xi_t Q_t}{\varepsilon^{3/2}}(\check{H}_t - \overline{H}_t) + \frac{Q_t^2\xi_t^2}{\sqrt{\varepsilon}}\zeta_t \\ &= -\frac{Q_t\xi_t^2}{\sqrt{\varepsilon}}\Big(\frac{\overline{X}_t - X_t}{\sqrt{\varepsilon}} - Q_t\zeta_t\Big) + O(\frac{1}{\sqrt{\varepsilon}}). \end{aligned} \tag{4.18}$$

Thus, if ϕ_t is the left hand side of (4.16) multiplied by Q_t, one has on $\{t_{i+1} \leq \tau_0\}$

$$\phi_{t_{i+1}} = \phi_{t_i} - \frac{Q_{t_i}\phi_{t_i}}{\sqrt{\varepsilon}}\int_{t_i}^{t_{i+1}} \xi_s^2 ds + O(\varepsilon^{1/4}) \tag{4.19}$$

so from (3.6) in Lemma 3,

$$\phi_{t_{i+1}} = \phi_{t_i} - \varepsilon^{1/4}\frac{Q_{t_i}\phi_{t_i}}{2\overline{X}_{t_i}} + O_i(\varepsilon^{1/8}) + O(\varepsilon^{1/4}). \tag{4.20}$$

By proceeding as in the study of (3.20), one proves that ϕ_t is of order 1 on $\{t \leq \tau_0\}$. □

Proof of Theorem 2. By using jointly the estimates of Lemmas 6 and 7, one deduces that

$$\frac{\nabla_s \log \Lambda_t}{\nabla_s X_t} = \frac{1}{\sqrt{\varepsilon}}\frac{\overline{X}_t - X_t}{Q_t} + O(1) \tag{4.21}$$

on $\{t \leq \tau_0\}$ and for $s \leq \sqrt{\varepsilon}$, so

$$\int_0^{\sqrt{\varepsilon}} \frac{\nabla_s \log \Lambda_t}{\nabla_s X_t} ds = \frac{\overline{X}_t - X_t}{Q_t} + O(\sqrt{\varepsilon}). \tag{4.22}$$

This estimate holds on $\{t \leq \tau_0\}$, and since the probability of $\{\tau_0 < t \leq \tau\}$ is small, it also holds on $\{t \leq \tau\}$. Define

$$\nabla_0 X_t = \exp\frac{1}{\sqrt{\varepsilon}}\int_0^t \frac{1}{Q_s} ds. \tag{4.23}$$

Notice that $\nabla_0 X_t$ is $\mathcal{Y}_t$ measurable and that we can write $\nabla_s X_t = \nabla_0 X_t / \nabla_0 X_s$. On the other hand, conditional expectations computed under $\mathbb{P}$ and $\widetilde{\mathbb{P}}$ are related by the Bayes formula

$$\widehat{Z}_t = \widetilde{\mathbb{E}}[Z_t \Lambda_t \mid \mathcal{Y}_t] \;/\; \widetilde{\mathbb{E}}[\Lambda_t \mid \mathcal{Y}_t]. \tag{4.24}$$

Therefore, we get

$$\begin{aligned}\mathbb{E}\Big[\int_0^{\sqrt{\varepsilon}} \frac{\nabla_s \log \Lambda_t}{\nabla_s X_t} ds \,\Big|\, \mathcal{Y}_t\Big] \\ = \widetilde{\mathbb{E}}\Big[\int_0^{\sqrt{\varepsilon}} \nabla_s \Lambda_t \nabla_0 X_s ds \,\Big|\, \mathcal{Y}_t\Big] \Big/ \Big(\widetilde{\mathbb{E}}[\Lambda_t \mid \mathcal{Y}_t] \nabla_0 X_t\Big).\end{aligned} \tag{4.25}$$

Under $\widetilde{\mathbb{P}}$, $\widetilde{W}$ and Y are independent, so the conditional expectation of the numerator can be viewed as an integration with respect to the Wiener process $\widetilde{W}$; we now apply the integration by parts formula of the Malliavin calculus to this integral (we cannot apply it directly since we do not know whether Λ_t is in the correct Sobolev space but we can use some localization technique; we omit the details). We obtain

$$\begin{aligned}\mathbb{E}\Big[\int_0^{\sqrt{\varepsilon}} \frac{\nabla_s \log \Lambda_t}{\nabla_s X_t} ds \,\Big|\, \mathcal{Y}_t\Big] \\ = \widetilde{\mathbb{E}}\Big[\Lambda_t \int_0^{\sqrt{\varepsilon}} \nabla_0 X_s d\widetilde{W}_s \,\Big|\, \mathcal{Y}_t\Big] \Big/ \Big(\widetilde{\mathbb{E}}[\Lambda_t \mid \mathcal{Y}_t] \nabla_0 X_t\Big) \\ = \mathbb{E}\Big[\int_0^{\sqrt{\varepsilon}} \nabla_0 X_s d\widetilde{W}_s \,\Big|\, \mathcal{Y}_t\Big] (\nabla_0 X_t)^{-1}.\end{aligned} \tag{4.26}$$

Then one applies (4.2) in order to decompose the integral with respect to $\widetilde{W}$, and deduces that the first term of the right-hand side is of order $\varepsilon^{1/4}$; moreover $(\nabla_0 X_t)^{-1}$ is of order $\exp -ct/\sqrt{\varepsilon}$, so the right hand side of (4.26) is very small. Thus, by taking the conditional expectation in (4.22), we obtain

$$\widehat{X}_t = \overline{X}_t + O(\sqrt{\varepsilon}). \tag{4.27}$$

The estimate on $\widehat{H}_t - \overline{H}_t$ then follows immediately by taking the conditional expectation in (4.17). □

Proof of Theorem 3. By multiplying (4.22) by $\overline{X}_t - X_t$, one gets

$$\frac{(\overline{X}_t - X_t)^2}{Q_t} = \frac{\overline{X}_t - X_t}{\Lambda_t} \int_0^{\sqrt{\varepsilon}} \frac{\nabla_s \Lambda_t}{\nabla_s X_t} ds + O(\varepsilon^{3/4}). \tag{4.28}$$

Since

$$(\overline{X}_t - X_t) \nabla_s \Lambda_t = \nabla_s (\Lambda_t (\overline{X}_t - X_t)) + \Lambda_t \nabla_s X_t \tag{4.29}$$

(∇_s is a differentiation operator), one has

$$\frac{(\overline{X}_t - X_t)^2}{Q_t} = \frac{1}{\Lambda_t \nabla_0 X_t} \int_0^{\sqrt{\varepsilon}} \nabla_s (\Lambda_t (\overline{X}_t - X_t)) \nabla_0 X_s ds + \sqrt{\varepsilon} + O(\varepsilon^{3/4}). \tag{4.30}$$

Then one takes the conditional expectation in this equation. By means of the integration by parts formula, one proves as in Theorem 2 that the conditional expectation of the first term of the right hand side is very small. Thus

$$\mathbb{E}[(\overline{X}_t - X_t)^2 \mid \mathcal{Y}_t] = \sqrt{\varepsilon} Q_t + O(\varepsilon^{3/4}). \tag{4.31}$$

On the other hand

$$\mathbb{E}[(\overline{X}_t - X_t)^2 \mid \mathcal{Y}_t] = \mathbb{E}[(X_t - \widehat{X}_t)^2 \mid \mathcal{Y}_t] + (\overline{X}_t - \widehat{X}_t)^2 \tag{4.32}$$

so (1.9) is proved. By conditioning first on $\mathcal{G}_t$ and secondly on $\mathcal{Y}_t$, one checks

$$\begin{aligned}\mathbb{E}[(X_t - \widehat{X}_t)(H_t - \widehat{H}_t) \mid \mathcal{Y}_t]&\\ &= \mathbb{E}[(X_t - \widehat{X}_t)(\check{H}_t - \widehat{H}_t) \mid \mathcal{Y}_t]\\ &= \mathbb{E}[(X_t - \overline{X}_t)(\check{H}_t - \overline{H}_t) \mid \mathcal{Y}_t] + (\overline{X}_t - \widehat{X}_t)(\overline{H}_t - \widehat{H}_t).\end{aligned} \tag{4.33}$$

The second term is of order $\varepsilon^{3/2}$ and for the first one, one deduces from (4.22) that

$$\begin{aligned}\frac{(\overline{X}_t - X_t)(\overline{H}_t - \check{H}_t)}{Q_t} =& \frac{1}{\Lambda_t \nabla_0 X_t} \int_0^{\sqrt{\varepsilon}} \nabla_s(\Lambda_t(\overline{H}_t - \check{H}_t)) \nabla_0 X_s ds\\ &+ \int_0^{\sqrt{\varepsilon}} \frac{\nabla_s \check{H}_t}{\nabla_s X_t} ds + O(\varepsilon^{5/4}).\end{aligned} \tag{4.34}$$

As before, the conditional expectation of the first term is very small and for the second one, one applies (4.13) and deduces (1.10). Finally, one notices

$$\mathbb{E}[(H_t - \widehat{H}_t)^2 \mid \mathcal{G}_t] = \varepsilon P_t + (\check{H}_t - \widehat{H}_t)^2 = \varepsilon X_t + O(\varepsilon^{3/2}), \tag{4.35}$$

so by taking the $\mathcal{Y}_t$-conditional expectation, one deduces (1.11). □

References

[1] A. Bensoussan, On some approximation techniques in non linear filtering, in: *Stochastic differential systems, stochastic control theory and applications*, IMA Vol. in Math. and Appl. **10**, Springer, 1988.

[2] W.H. Fleming and E. Pardoux, Piecewise monotone filtering with small observation noise, *SIAM J. Cont. Optim.* **27** (1989), 5, 1156–1181.

[3] R. Katzur, B.Z. Bobrovsky and Z. Schuss, Asymptotic analysis of the optimal filtering problem for one-dimensional diffusions measured in a low noise channel, *SIAM J. Appl. Math.* **44** (1984), Part I: 591–604, Part II: 1176–1191.

[4] P. Milheiro de Oliveira, *Etudes asymptotiques en filtrage non linéaire avec petit bruit d'observation*, Thèse de Doctorat, Université de Provence, 1990.

[5] D. Ocone, A guide to the stochastic calculus of variations, in: *Stochastic Analysis and Related Topics (Silivri 1986)*, Lect. N. in Math. **1316**, Springer, 1988.

[6] J. Picard, Nonlinear filtering of one-dimensional diffusions in the case of a high signal-to-noise ratio, *SIAM J. Appl. Math.* **46** (1986), 1098–1125.

[7] J. Picard, Nonlinear filtering and smoothing with high signal-to-noise ratio, in: *Stochastic Processes in Physics and Engineering (Bielefeld 1986)*, Reidel, 1988.

[8] J. Picard, Efficiency of the extended Kalman filter for nonlinear systems with small noise, *SIAM J. Appl. Math.* **51** (1991), 3, 843–885.

[9] I. Yaesh, B.Z. Bobrovsky and Z. Schuss, Asymptotic analysis of the optimal filtering problem for two dimensional diffusions measured in a low noise channel, *SIAM J. Appl. Math.* **50** (1990), 4, 1134–1155.

BOUNDS FOR THE PRICE OF OPTIONS

Monique JEANBLANC-PICQUE
CMLA. ENS DE CACHAN
F 94235 CACHAN CEDEX

Nicole EL KAROUI
Laboratoire de probabilités Tour 56 3ième étage
4,Place Jussieu
F 75230 PARIS CEDEX 05

Ravi VISWANATHAN
Mitsubishi Finance International
6,Broadgate
LONDON EC 2M 2AA

In this paper, we present an application of stochastic calculus to show " the robustness of the Black-Scholes formula". The Black-Scholes formula is extensively used in order to determine the price of financial products called options. This formula is valid only when the parameters (which can, in general, be stochastic and time-dependent) are constant or deterministic. When this is not the case, this formula is still used by means of an approximation at time t of these parameters, without theoretical justification. We prove in this paper that this methodology is correct in some sense.

1- The model

We recall the financial model and terminology regarding options.

As usual, one considers a probability space $(\Omega,\mathcal{F},P)$ on which is defined a Brownian motion $\hat{W}$.

We assume that the financial market consists of two assets :
(i) a riskless asset, called *the bond*, which produces interest at a random rate r (the return or the interest rate) : one dollar at time 0 gives $\$\exp[\int_0^t r(s)\, ds]$ at time t. Its price is given by

$$(1.1) \qquad \begin{cases} dS_0(t) = S_0(t)\, r(t)\, dt \\ S_0(0) = 1\ , \end{cases}$$

(ii) a risky asset, called *the stock,* whose price $S(t)$ at time t is a solution to the stochastic differential equation

$$(1.2) \qquad dS(t) = S(t)\,\{b(t)\, dt + \sigma(t)\, d\widehat{W}(t)\},$$

where $\widehat{W}$ is a Brownian motion.

The coefficient b is called the return of the stock, and σ is the *volatility* of the stock.
We shall denote $R_t = \exp[\, -\int_0^t r(s)\, ds]$.

A *European call* (resp. put) *option* on the stock is a contract which gives to its owner the right to purchase (resp. to sell) a share of the stock at the fixed price K (the exercise price) on the fixed expiration date T.
An *American call (or put) option* differs from a European option in that the holder can exercise the option and thus purchase (or sell) the stock at any time up to and including T. (The main books on this subject are Cox & Rubinstein [CR] and Jarrow & Rudd [JR])

A *portfolio* -or a *strategy*- is a pair $(\theta_0, \theta_1) = \{(\theta_0(t), \theta_1(t); t \geq 0\}$ where $\theta_0(t)$ and $\theta_1(t)$ are the number of shares of the bond and the stock held at time t. Borrowing and short-selling (i.e. $\theta_0 \leq 0$ or $\theta_1 \leq 0$) are possible.

The value of the strategy (θ_0, θ_1) at time t *(or the wealth)* is

$$(1.3) \qquad X_t^\theta = \theta_0(t)\, S_0(t) + \theta_1(t)\, S(t).$$

A *self-financing strategy* is a strategy which satisfies

$$(1.4) \qquad X_t^\theta = X_0^\theta + \int_0^t \theta_0(s)\, dS_0(s) + \int_0^t \theta_1(s)\, dS(s)\ .$$

We shall specify later the measurability assumptions on a strategy in order that the right-hand side of (1.4) be well-defined.

An *arbitrage opportunity* is a self-financing strategy θ such that $X_0^\theta \leq 0$, $E(X_T^\theta)>0$ and $X_T^\theta \geq 0$. We suppose that there are no arbitrage opportunities. Under mild conditions, this hypothesis implies that there exists a probability measure Q which is equivalent to P such that, under Q, the stock has the same return as the bond (and vice-versa). These facts are related with Girsanov's theorem. The precise study of this relation can be found in Stricker [ST].

Thus, in this paper, we shall assume that there exists a probability space $(\Omega,\mathcal{F},\mathcal{F}_t,Q)$ such that

$$dS_0(t) = S_0(t)\, r(t)\, dt$$
$$dS(t) = S(t)\,\{r(t)dt + \sigma(t)\, dW(t)\},$$

where $W(t)$ is an $(\mathcal{F}_t,Q)$ one-dimensional Brownian motion.
The coefficients r and σ are assumed to be $\mathcal{F}_t$-adapted and bounded.
The strategies are assumed to be $\mathcal{F}_t$-adapted and such that
$\int_0^T |\theta_0(t)|\, dt < \infty$ Q-a.s. and $Q(\int_0^T \{\theta_1(t)\}^2 dt < \infty) = 1$.

The problem is to price an option, i.e., to define the price (at time 0) of the option .

In a first part, we recall the well-known results on the Black-Scholes pricing formula. Then, we study a new problem: the volatility of the stock is not well known by the investor ; here we suppose that the volatility is bounded above and below by deterministic functions. We shall give bounds for the price of the option and give an hedging portfolio.

EUROPEAN OPTIONS

2. The Black-Scholes formula

2.1. Hedging strategies.

Let us recall some results about the Black-Scholes formula. We shall give the idea of the solution to the pricing problem with a "hedging strategy".
Suppose that h is a positive convex function such that $|h(x)| \leqslant C(1+|x|)$.

We want to give the price at time 0 of a contingent claim whose value is $h\{S(T)\}$ at time T.

Remark :
If $h(x) = (x-K)^+$, we obtain the price of a European call. If $h(x)=(K-x)^+$, it is a European put. The value $K\in\mathbb{R}^+$ is called the exercise price.

Suppose that θ is a self-financing strategy such that $X_T^\theta = h\{S(T)\}$. The price at time 0 of the contingent claim is defined

by the value X_0^θ.[It can be easily proved that this value does not depend on the strategy, by non arbitrage hypothesis].
Since θ is a self-financing strategy

$$X_t^\theta = \theta_0(t)S_0(t)+ \theta_1(t)S(t) = X_0^\theta + \int_0^t \theta_0(s)dS_0(s) + \int_0^t \theta_1(s)dS(s) \ .$$

This equality can be written in the form

$$(2.1) \qquad dX_t^\theta = \theta_0(t)\, dS_0(t) + \theta_1(t)\, dS(t).$$

Suppose that there exists a function $\mathcal{V} \in C^{1,2}(\mathbb{R}\times\mathbb{R}^+)$ such that

$$(2.2) \qquad \theta_0(t)\, S_0(t) + \theta_1(t)\, S(t) = \mathcal{V}(t,S(t)).$$

From Ito's formula, it follows that

$$(2.3)\, dX_t^\theta = \frac{\partial \mathcal{V}}{\partial t}(t,S(t))dt + \frac{\partial \mathcal{V}}{\partial x}(t,S(t))dS(t) + \frac{1}{2}\frac{\partial^2 \mathcal{V}}{\partial x^2}(t,S(t))[\sigma(t)S(t)]^2 dt$$

The function $\mathcal{V}$ is called the *value function*.

From now on, we shall denote by S^σ (resp.$\mathcal{V}^\sigma$) the price (resp the value function) corresponding to a volatility equal to σ.

Then, if we identify the coefficients of dW_t in the expressions (2.1) and (2.3) of dX_t, it follows that $\mathcal{V}^\sigma$ satisfies

$$\frac{\partial \mathcal{V}^\sigma}{\partial t}(t,S^\sigma(t)) + \frac{1}{2}\{\sigma(t)S^\sigma(t)\}^2 \frac{\partial^2 \mathcal{V}^\sigma}{\partial x^2}(t,S^\sigma(t)) + r(t)S^\sigma(t)\frac{\partial \mathcal{V}^\sigma}{\partial x}(t,S^\sigma(t))$$
$$= r(t)\, \mathcal{V}^\sigma(t,S^\sigma(t))$$
$$\mathcal{V}^\sigma(T,S^\sigma(T)) = h(S^\sigma(T)).$$

The hedging portfolio -i.e. a portfolio which has the same value as the option- is given by

$$(2.4) \quad \begin{cases} \theta_0(t) = \{\mathcal{V}^\sigma(t,S^\sigma(t)) - \dfrac{\partial \mathcal{V}^\sigma}{\partial x}(t,S^\sigma(t))\, S^\sigma(t)\}\, S_0^{-1}(t) \\[2ex] \theta_1(t) = \dfrac{\partial \mathcal{V}^\sigma}{\partial x}(t,S^\sigma(t)) \ . \end{cases}$$

2.2. Deterministic coefficients.

When the coefficients r and σ are deterministic, it follows that $\mathcal{V}$ satisfies (with $\Upsilon=\sigma$)

$$(2.5)\quad \begin{cases} \dfrac{\partial \mathcal{V}^\gamma}{\partial t} + \dfrac{1}{2}\gamma^2(t)x^2 \dfrac{\partial^2 \mathcal{V}^\gamma}{\partial x^2} + r(t)x \dfrac{\partial \mathcal{V}^\gamma}{\partial x} - r(t)\mathcal{V}^\gamma = 0, \quad t\in[0,T[,\ x>0 \\ \mathcal{V}^\gamma(T,x) = h(x), \end{cases}$$

and it is easy to check that $\mathcal{V}^\gamma(t,x) = E[R_T^t\ h\{S^{t,x}(T)\}]$ where $R_T^t = \exp[-\int_t^T r(u)\,du]$ and where $S^{t,x}$ is the solution to (1.2) such that $S^{t,x}(t)=x$. Furthermore, we can prove that there exists a solution to (2.5) and that the hedging portfolio (2.4) is a self-financing strategy.

F.Black and M.Scholes (BS) have solved this problem for a European option in the case where γ and r are constant. In this case, it is possible to give an explicit form for $\mathcal{V}^\gamma$. The famous Black-Scholes formula states that

(2.6) $\quad \mathcal{V}^\gamma(t,x) = x\,\Phi(d_1) - K\,e^{-r(T-t)}\,\Phi(d_2)$

where $\Phi(d) = \dfrac{1}{\sqrt{2\pi}} \displaystyle\int_{-\infty}^{d} \exp[-\frac{x^2}{2}]\,dx$

$$d_1 = \left(\log \frac{x}{K} + r(T-t) + \frac{1}{2}\gamma^2\ (T-t)\right) \{\gamma\sqrt{T-t}\}^{-1}$$

$$d_2 = d_1 - \gamma\sqrt{T-t}\ .$$

It is easy to prove that in this case, the function $\mathcal{V}^\gamma$ is a convex function with respect to x and that, if $\gamma_1 \leqslant \gamma_2$, then $\mathcal{V}^{\gamma_1} \leqslant \mathcal{V}^{\gamma_2}$.

If the coefficients are deterministic, there is also an explicit form. It suffices to change $r(T-t)$ (resp. $\gamma\sqrt{T-t}$) into $\int_t^T r(s)ds$ (resp. $\{\int_t^T \gamma^2(s)\,ds\}^{1/2}$).

2.3. General case.

If the volatility σ is not deterministic, (2.5) is no longer valid.

Suppose that there exists a self-financing pair (θ_0,θ_1) such that $X_T^\theta = h[S^\sigma(T)]$. From the definition of X_t^θ, it follows that

$$dX_t^\theta = X_t^\theta\ r(t)\ dt + \theta_1(t)\ S^\sigma(t)\ \sigma(t)\ dW_t\ .$$

From Ito's formula, we obtain $R_t X_t^\theta = X_t^\theta + \int_0^t R_s \theta_1(s)\ \sigma(s) S^\sigma(s) dW_s$ and thus $R_t X_t^\theta$ is an $\mathcal{F}_t$-local martingale and the price of $h[S^\sigma(T)]$

is defined by no-arbitrage arguments as equal to X_0^θ. If the local-martingale is a martingale, we have $X_0^\theta = E_Q[R_T h\{S^\sigma(T)\}]$. If $\mathcal{F}_t = \sigma\{W_s, s\le t\}$, we can construct such a self-financing strategy in the following way: since the process $E_Q[R_T h\{S^\sigma(T)\}|\mathcal{F}_t]$ is an $\mathcal{F}_t$-Q martingale, there exists (x, φ_1) such that

$$R_t X_t := E_Q[R_T h\{S^\sigma(T)\}|\mathcal{F}_t] = x + \int_0^t \varphi_1(s)\, dW(s) .$$

It is now easy to construct a self-financing strategy (θ_0,θ_1) such that

$$X_t = E_Q[R_T^t h\{S^\sigma(T)\}|\mathcal{F}_t] = x + \int_0^t \theta_0(s)dS_0(s) + \int_0^t \theta_1(s)dS^\sigma(s) .$$

The pair (θ_0,θ_1) is a hedging portfolio for the option against $h\{S^\sigma(T)\}$ and its t-time value is $E_Q[R_T^t h\{S^\sigma(T)\}|\mathcal{F}_t]$.

Theorem 2.1.
The price of a European option on $h\{S^\sigma(T)\}$ *is equal to* $E_Q[R_T h\{S^\sigma(T)\}]$. *The t-time value of the strategy is*
(2.7) $E_Q[R_T^t h\{S^\sigma(T)\}|\mathcal{F}_t]$.

Let us remark that we have no explicit form for the hedging portfolio (θ_0,θ_1).

2.4. Convexity

Proposition 2.2
Let us suppose that r *and* γ *are constant. Let* $\mathcal{V}^\gamma$ *be associated with a European call. The function* $\mathcal{V}^\gamma$ *is a convex function with respect to x and is increasing with respect to* γ.

These facts are well known and can be easily deduced from the explicit form (2.6). The financial meaning is explained in Cox-Rubinstein [CR] and Jarrow-Rudd [JR].

We are going to extend these results to a more general setting. Suppose that the interest rate and the volatility are deterministic functions. Let h be a convex function . We are studying the price of an option on $h(S_T^\gamma)$.

Proposition 2.3.
Let γ *be a deterministic function and* h *a convex function such that* $|h(x)| \le C(1+|x|)$.
Then, the value function $\mathcal{V}^\gamma(t,x)$ *is a convex function with respect to* x.

Proof: The convexity of $\mathcal{V}^\gamma$ follows easily from the convexity of the map $x \mapsto h(S^{\gamma,t,x}(T))$: it suffices to notice that the solution

$S^{\gamma,t,x}(.)$ depends on the initial condition in a linear way and that h is a convex function.

◁

The general problem of convexity is being studied by Bensoussan-Conze-Lasry. We study here the "increasing" property.

3. Bounds for European option

When the volatility is not deterministic, it is difficult to compute (2.7). Furthermore, at time t the investor does not know precisely the volatility $\sigma(s)$ for $s \geq t$. We suppose here that only bounds are known. We suppose that the interest rate is deterministic.

Lemma 3.1.

Let us suppose that r is a deterministic function.
Let $\sigma(t)$ be an $\mathcal{F}_t$-adapted process such that

(3.1) $\quad 0<\alpha(t) \leq |\sigma(t)| \leq \beta(t) \quad dQ\otimes dt$ a.s

where α and β are deterministic bounded coefficients.
Then $R_t\, \mathcal{V}^\beta(t,S^\sigma(t))$ is a positive supermartingale
$R_t\, \mathcal{V}^\alpha(t,S^\sigma(t))$ is a positive submartingale.

Proof:
Ito's formula applied to $R_t \mathcal{V}^\beta(t,S^\sigma(t))$ leads to

$$(3.2)\; R_t \mathcal{V}^\beta(t,S^\sigma(t)) - \int_0^t R_s \left[\frac{\partial \mathcal{V}^\beta}{\partial t}(s,S^\sigma(s)) + \frac{1}{2}\{\sigma(s)S^\sigma(s)\}^2 \frac{\partial^2 \mathcal{V}^\beta}{\partial x^2}(s,S^\sigma(s)) + r(s)\, S^\sigma(s)\, \frac{\partial \mathcal{V}^\beta}{\partial x}(s,S^\sigma(s)) - r(s)\, \mathcal{V}^\beta(s,S^\sigma(s))\right] ds$$

is equal to a local martingale.
Since $\mathcal{V}^\beta(t,.)$ is a convex function (proposition 2.3), it follows that the second derivative is non negative and

$$\sigma^2(s)\, \frac{\partial^2 \mathcal{V}^\beta}{\partial x^2}\, (s,S^\sigma(s)) \leq \beta^2(s)\, \frac{\partial^2 \mathcal{V}^\beta}{\partial x^2}\, (s,S^\sigma(s)).$$

Then, since $\mathcal{V}^\beta$ satisfies to (2.5), we obtain, from (3.2) that $R_t\, \mathcal{V}^\beta(t,S^\sigma(t))$ is equal to a local-martingale minus an increasing process and is positive, therefore it is a supermartingale.

If $|\sigma(t)| \geq \alpha(t)$, we have $R_t \mathcal{V}^\alpha(t,S^\sigma(t)) = M_t + A_t$ where M_t is a local martingale and A_t an increasing process. Since α is a deterministic function $\mathcal{V}^\alpha(t,x) = E_Q[R_T^t h(S^\alpha(T)|S^\alpha(t)=x]$ and it is easy to check that there exists a constant C_1 such that $\mathcal{V}^\alpha(t,x) \leq C_1(1+|x|)$. Therefore $|R_t \mathcal{V}^\alpha(t,S^\sigma(t))| \leq C_2(1+S^\sigma(t))$. The prices $S^\sigma(t)$ are bounded in $L^2(\Omega\times[0,T],dP\times dt)$ since the volatilities are bounded by β which belongs to L^2. Therefore, Fatou's lemma applies and $M_t + A_t$ is a submartingale.

◁

From this lemma, we now deduce bounds for the t-time value of the option $E_Q\left[R_T^t\ h\{S^\sigma(T)\}|\mathcal{F}_t\right]$.

<u>Theorem 3.2.</u>
Suppose that $0<\alpha(t) \leqslant |\sigma(t)| \leqslant \beta(t)$ where α and β are deterministic functions.
If the interest rate r is a deterministic function, we have

$$\mathcal{V}^\alpha(t,S^\sigma(t)) \leqslant E_Q(R_T^t h(S^\sigma(T))|\mathcal{F}_t) \leqslant \mathcal{V}^\beta(t,S_t^\sigma(t)).$$

Proof: Since $R_t \mathcal{V}^\beta(t,S^\sigma(t))$ is a supermartingale, we have

$$R_t\ \mathcal{V}^\beta(t,S^\sigma(t)) \geqslant E_Q(R_T\ \mathcal{V}^\beta(T,S^\sigma(T))|\mathcal{F}_t)$$

$$= E_Q(R_T h(S^\sigma(T))\ |\mathcal{F}_t)$$

and the right hand inequality follows.
The proof is the same for the left hand side.

◁

Our result is not only a comparison result on solutions to stochastic differential equations. The super-martingale property is an important one . Moreover, an interesting fact is that this method allows us to give a portfolio which controls the risk.

<u>Theorem 3.3</u>
The Black-Scholes portfolio constructed on the deterministic bounds hedges the price of the option.

The Black-Scholes portfolio constructed on $\mathcal{V}^\beta$ hedges the maximal price of the option ,i.e. (cf 2.4)
$\mathcal{V}^\beta(t,S^\sigma(t)) = \theta_0(t)S_0(t) + \theta_1(t)S^\sigma(t)$ where

$$\theta_0(t)= \{\mathcal{V}^\beta(t,S^\sigma(t)) - S^\sigma(t)\ \frac{\partial \mathcal{V}^\beta}{\partial x}(t,S^\sigma(t))\}\ R_t$$

$$\theta_1(t) = \frac{\partial \mathcal{V}^\beta}{\partial x}(t,S^\sigma(t)).$$

However, this portfolio is not self-financing, but the instantaneous risk is controlled: using Ito's formula, the hypothesis $\sigma \leqslant \beta$ and (2.5), it is easy to check that

$$\mathcal{V}^\beta(t,S^\sigma(t)) \leqslant x+\int_0^t \theta_0(s)\ dS_0(s) +\int_0^t \theta_1(s)\ dS^\sigma(s).$$

We emphasize that $\mathcal{V}^{\beta}$ and $\mathcal{V}^{\alpha}$ can be explicitly computed with the Black-Scholes formula.

Remark 3.2

Our result does not extend to the more general following case : Suppose that the volatilities are random processes such that $\sigma_1(t) \leqslant \sigma_2(t)$ Q-a.s.

It is not true that $E[f(S_T^{\sigma_1})] \leqslant E[f(S_T^{\sigma_2})]$ for each convex function f.

A counter example was given by Marc Yor (private communication).

To begin with, we establish the following lemma:

Lemma

Let N_t *a continuous local martingale such that*

$$N_1 \leqslant C \text{ (P-a.s) } \textit{and } M_t = \int_0^t H_s \, dN_s$$

where H *is a predictable process such that* $|H|\leqslant 1$.

If

$$E[f\{\exp (M_1 - \frac{1}{2} <M>_1)\}] \leqslant E[f\{\exp (N_1 - \frac{1}{2} <N>_1)\}]$$

for each convex function f, *then* $\text{ess sup}(M_1(\omega) - \frac{1}{2} <M>_1(\omega)) \leqslant C$

Proof:

It suffices to consider the inequality for $f(x)=x^p$. It follows that

$$\| \exp (M_1 - \frac{1}{2} <M>_1)\|_p \leqslant \| \exp (N_1 - \frac{1}{2} <N>_1)\|_p$$

and, letting $p \longrightarrow \infty$

$$\text{ess sup}\{\exp(M_1(\omega) - \frac{1}{2} <M>_1(\omega))\} \leqslant \text{ess sup}\{\exp (N_1(\omega) - \frac{1}{2} <N>_1(\omega))\}$$

and the result follows.

We can now give the counter example.

Let $T_a = \inf \{t \mid |W_t|\geqslant a\}$. Let us define the following processes:

$$\sigma_1(t) = \mathbb{1}_{W_t < 0} \, \mathbb{1}_{t < T_a} \; ; \; M_t = \int_0^t \sigma_1(s) \, dW_s = \int_0^{t \wedge T_a} 1_{W_s < 0} \, dW_s$$

$$\sigma_2(t) = \mathbb{1}_{t < T_a} \quad \text{and} \quad N_t = \int_0^t \sigma_2(s) \, dW_s = W_{t \wedge T_a} \; .$$

It is obvious that $\sigma_1(t)\leqslant\sigma_2(t)$.

Suppose that $E(f(S_T^{\sigma_1})) \leqslant E(f(S_T^{\sigma_2}))$ for each convex function f. From the previous lemma, it would follow that

$$\text{ess sup } M_1(\omega) \leqslant a + \frac{1}{2} \; .$$

Furthermore, from Tanaka's formula $M_t = \frac{1}{2} L_{t\wedge T_a} - W^-_{t\wedge T_a}$

where $(L_t\ ,t\geqslant 0)$ is the local time at 0 of the Brownian motion. It would follow that $1/2\ L_{1\wedge T_a} - W^-_{1\wedge T_a} \leqslant a+1/2$,thus

$$L_{1\wedge T_a} \leqslant 2\left\{(a + \frac{1}{2}) + W^-_{t\wedge T_a}\right\} \leqslant 4a + 1 \qquad \text{which is false.}$$

Remark 3.5

We have obtained some results in the case where r is not deterministic. Under a change of probability measure, this case reduces to the preceding one.

AMERICAN OPTIONS

An American option (or claim) on the stock with reward g is a financial security that pays $g\{S^\sigma(t)\}$ when exercised at time t (where t can be a stopping time). A reward function is a continuous, non-negative convex function.

Example :

An American *call* option is caracterized by $g(x)=(x-K)^+$, an American *put* option is $g(x)=(K-x)^+$ where $K\in\mathbb{R}^+$ is the exercise price.

4. American option pricing.

Let us recall the main results about American options.(See Bensoussan (B), Karatzas (K) or Myneni (M) for details)

We suppose that r is a deterministic function.

4.1.Generalities

Definition 4.1

Fix $x>0$. A pair (θ_0,θ_1) is called an *hedging strategy* against the American claim with initial value x if

(i) $\theta_0(0)\ S_0(0) + \theta_1(0)\ S^\sigma(0) = x$

(ii) the strategy (θ_0,θ_1) is self-financing

(iii) $\theta_0(t)\ S_0(t) + \theta_1(t)\ S^\sigma(t) \geqslant g(S^\sigma(t))$

(iv) $\theta_0(T)\ S_0(T) + \theta_1(T)\ S^\sigma(T) = g(S^\sigma(T))$.

We denote by $\mathcal{A}(x)$ the collection of such pairs.

We shall denote by X_t^θ the *wealth process* associated to the strategy (θ_0, θ_1), i.e.

$$X_t^\theta = \theta_0(t)S_0(t) + \theta_1(t)\, S^\sigma(t)\ .$$

Definition 4.2.

The number $\inf\{x>0\ ;\ \exists(\theta_0,\theta_1) \in \mathcal{A}(x)\}$ is called the fair price for the American claim.

The following existence result provides the basic theorem for pricing the American options.

Proposition 4.3.

The process

(4.1) $\quad X_t := \text{ess sup}\Big\{E_Q\big[R_U^t\ g\{S^\sigma(U)\}|\mathcal{F}_t\big]\ ;\ U\in\mathcal{U}_{t,T}\Big\}$

where $\mathcal{U}_{t,T}$ *is the set of stopping times* U *such that* $t\leq U\leq T$ *is a wealth process and the fair price of the American option is equal to* X_0.

Sketch of the proof: The smallest supermartingale which majorizes $R_t g(S^\sigma(t))$ is equal to (Snell envelope)

$$Y_t = \text{ess sup}\ (E_Q[(R_U g\{S^\sigma(U)\}|\mathcal{F}_t]\ ;\ U\in \mathcal{U}_{t,T}).$$

Using the Doob-Meyer decomposition and the martingale representation theorem, it is easy to prove that $X_t = R_t^{-1} Y_t$ is a wealth process, i.e. that there exists a self-financing strategy (θ_0, θ_1) such that

$$X_t = \theta_0(t)S_0(t) + \theta_1(t)S^\sigma(t).$$

Furthermore, θ is a hedging strategy against the American claim. It can be proved, using non arbitrage arguments that X_0 is the fair price.

We must notice that this process provides the optimal stopping time for the interval $[t,T]$ as the first instant Y drops to the level of the discounted reward

$$\tau_t^* = \inf\ \{s\in[t,T]\mid\ Y_s = R_s g(S^\sigma(s))\}.$$

Moreover, the stopped process $Y_{s\wedge\tau_t^*}$ is a martingale.

4.2. Deterministic coefficients

Let us now assume that the interest rate r and the volatility γ are deterministic functions.

Then, there exists $V^\gamma : [0,T]\times\mathbb{R}^+ \longrightarrow \mathbb{R}$ such that $X_t = V^\gamma(t, S^\gamma(t))$.

This function V^γ can be caracterized by means of variational inequalities.

Let L^γ be the operator

$$L^{\gamma}V(t,x) = \frac{\partial V}{\partial t}(t,x) + \frac{1}{2}[\gamma(t)]^2 x^2 \frac{\partial^2 V}{\partial x^2}(t,x) + r(t)\, x \frac{\partial V}{\partial x}(t,x).$$

Then, it can be proved ([B],[K]) that V^{γ} is "smooth" and is caracterized by

$$g(x) \leq V^{\gamma}(t,x)$$
$$L^{\gamma} V^{\gamma}(t,x) \leq 0$$
$$(L^{\gamma} V^{\gamma}(t,x))\,(g(x)-V^{\gamma}(t,x)) = 0.$$

5. American call option

It is well known that the price of an American call option with maturity T is the same as the price of a European call option with the same maturity (See [K]). We give here a proof of a more general result. Once more time, the important fact is that a process is a submartingale.

Theorem 5.1
Let g *be a convex function such that* $g(0) = 0$ *and* $|g(x)| \leq K(1+|x|)$.
Suppose that r *is a non negative bounded function.*
Then the price of an American claim against $g(S^0_T)$ *is equal to the price of a European option, i.e.*

$$\text{ess sup}(\ E_Q\{R(U)\, g(S^{\sigma}(U))|\mathcal{F}_t\};\ U\in \mathcal{U}_{t,T}) = E_Q\{R(T)\, g(S^{\sigma}(T))|\mathcal{F}_t\}.$$

Proof:
It suffices to remark that $R_t g(S^{\sigma}(t))$ is a submartingale. This fact follows directly from Ito's formula for convex function

$$R_t g(S^{\sigma}(t)) = g(S^{\sigma}_0) + \int_0^t R_s\ [-r(s)\, g(S^{\sigma}(s)) + r(s)S^{\sigma}(s)g'_-(S^{\sigma}(s))]ds$$
$$+ \frac{1}{2}\int_{\mathbb{R}} L^a_t(S)\, g''(da) + N_t$$

where N_t is a local-martingale and L^a_t is the local time of S . The hypotheses on g ensure that

$-\ g(x)+ x\, g'(x) \geq -g(0) = 0$ and that g'' is a positive measure. Therefore, as in lemma 3.1 Fatou's lemma ensures that the left member is a submartingale.◁
The theorem gives the result for American calls with $g(x)=(x-K)^+$.

6. American put option

We are now studying

(6.1) $P^{\sigma}(t) := \text{ess sup}\{E[R^t_U\ g\{S^{\sigma}(U)\}|\mathcal{F}_t]\ ;\ U\in\mathcal{U}_{t,T}\}$

where g is a convex function with $g(0)>0$ [and such that the discontinuities of g' are at points x_0 such that $g(x_0)=0$. This last hypothesis is made in order to apply the smooth-fit principle and to avoid first order discontinuities for the value function . It is then possible to use Ito's formula]

We suppose that $\alpha(t) \leqslant |\sigma(t)| \leqslant \beta(t)$.

Since $|\sigma(t)|\leqslant\beta(t)$, it is easy to prove that $R_t V^{\beta}(t,S^{\sigma}(t))$ is a supermartingale which majorizes $R_t g(S^{\sigma}(t))$.
Thus, we have established that $P^{\sigma}(t) \leqslant V^{\beta}(t,S^{\sigma}(t))$.

It is more difficult to prove the remaining inequality. From the convexity of V^{α} and the fact that $L^{\alpha}V^{\alpha}=0$ on $\{V^{\alpha}>g\}$, it follows that $L^{\sigma} V^{\alpha} \geqslant 0$ on $\{V^{\alpha} > g\}$.

Therefore if $D^{\alpha}_{\sigma}(t) = \inf\{u \, ; \, u \geqslant t | \, V^{\alpha}(u,S^{\sigma}(u)) = g(S^{\sigma}(u))\}$, it follows from Ito's formula and Fatou's lemma that
$E_Q[R\{D^{\alpha}_{\sigma}(t)\} \, V^{\alpha}\{D^{\alpha}_{\sigma}(t), \, S^{\sigma}(D^{\alpha}_{\sigma}(t))\} | \mathcal{F}_t] \geqslant R(t) \, V^{\alpha}\{t,S^{\sigma}(t)\}$.
Therefore, since $D^{\alpha}_{\sigma}(t)$ is a stopping time greater than t, we have proved that

$$V^{\alpha}(t,S^{\sigma}(t)) \leqslant P^{\sigma}(t).$$

Theorem 6.1

Let us suppose that $\alpha(t) \leqslant |\sigma(t)| \leqslant \beta(t)$ *where* α *and* β *are deterministic functions.*
Let us assume that r is a deterministic function.
Then, the t-time value $P^{\sigma}(t)$ *of an American put against* $S^{\sigma}(t)$ *satisfies* $V^{\alpha}(t,S^{\sigma}(t)) \leqslant P^{\sigma}(t) \leqslant V^{\beta}(t,S^{\sigma}(t))$.

Acknowledgement: We are indebted to the referee for interesting remarks

BIBLIOGRAPHY

[B] A. Bensoussan (1984)
On the theory of option pricing
Acta Applicandae Mathematicae 2 pp 139-158.

[BS] F. Black, M. Scholes (1973)
The pricing of options and corporate liabilities
Journal of Political Economy 81 pp 637-654.

[CR] J. Cox, M. Rubinstein (1985)
Options markets. Prentice Hall, New Jersey.

[E.J.V] N. El Karoui, M. Jeanblanc-Picqué, R. Viswanathan (1991)
On the robustness of Black-Scholes equation
Preprint.

[J.R] R.A.Jarrow, A.Rudd (1983)
Option pricing (Irwin) Chicago.

[K] I. Karatzas (1988)
On the pricing of American options
Appl. Math. Optim. 17 pp 37-60.

[M] R. Myneni (1990)
The pricing of the American option
Forthcoming in the Annals of Applied Probability.

[S] C. Stricker (1990)
Arbitrage et lois de martingale
Ann. Inst. Henri Poincaré vol 26 n°3 pp 451-460.

[S.A] C. Stricker , J.P. Ansel (1991)
Lois de martingale, densité et décomposition de Föllmer-Schweizer
Preprint

BROWNIAN AND DIFFUSION DECISION PROCESSES

J. P. QUADRAT
INRIA Domaine de Voluceau Rocquencourt 78153
LE CHESNAY Cedex (FRANCE)

Abstract

We show the analogy between probability calculus and dynamic programming. In the first field, iterated convolutions of probability laws play a central role, in the second field the role is played by inf-convolution of cost functions. The main analysis tools are: the Fourier transform for the first situation, the Fenchel transform for the second. To gaussian laws — stable by convolution — correspond quadratic forms, stable by inf-convolution. To the law of large number and the central limit theorem correspond asymptotic theorems for the value function of dynamic programming — convergence of the value function of an averaged state towards the characteristic function of the minimum of the instantaneous cost function, convergence of the normalized deviation from this minimum, towards a quadratic form. To Brownian motion trajectories correspond straight lines. To the operator $D_t + D_{x^2}$ corresponds the operator $D_t - (D_x)^2$ which must be seen as a min-plus linear operator. To the Green function $1/\sqrt{(2\pi t)}\exp(-x^2/2t)$ corresponds the min-plus Green function $x^2/2t$. To the diffusion process of generator $D_t + b(x)D_x + a(x)D_{x^2}$correspond a diffusion decision process of generator $D_t - b(x)D_x - a(x)(D_x)^2$.

1 Inf-Convolutions of Quadratic Forms

For $m \in \mathbb{R}$ et $\sigma \in \mathbb{R}^+$, let us denote $Q_{m,\sigma}(x)$ the quadratic form in x defined by:

$$Q_{m,\sigma}(x) = \frac{1}{2}\left(\frac{x-m}{\sigma}\right)^2 \quad \text{for } \sigma \neq 0,$$

$$Q_{m,0}(x) = \delta_m(x) = \begin{cases} +\infty & \text{for } x \neq m, \\ 0 & \text{for } x = m. \end{cases}$$

These quadratic forms take a null value at m.

Given two mappings f and g from $\overline{\mathbb{R}} = \mathbb{R} \cup \{\infty, -\infty\}$ into $\overline{\mathbb{R}}$ we call inf-convolution of f and g (Rockafellar [14]) the mapping from $\overline{\mathbb{R}}$ into $\overline{\mathbb{R}}$ (with the convention $\infty - \infty = \infty$) defined by:

$$z \to \inf_{x+y=z} [f(x) + g(y)]$$

that we denote $f * g$.

Proposition 1.1

$$Q_{m,\sigma} * Q_{m',\sigma'} = Q_{m+m',\sqrt{\sigma^2+\sigma'^2}}.$$

This result is the analogue of the convolution of Gaussian laws:

$$\mathcal{N}(m,\sigma) * \mathcal{N}(m',\sigma') = \mathcal{N}(m+m', \sqrt{\sigma^2+\sigma'^2})$$

where $\mathcal{N}(m,\sigma)$ denotes the Gaussian law of mean m and standard deviation σ. Therefore there exists a morphism between the set of quadratic forms endowed with the inf-convolution operator and the set of exponentials of quadratic forms endowed with the convolution operator.

Clearly this result can be generalized to the vector case.

2 Dynamic Programming

Given the simplest decision process:

$$x_{n+1} = x_n - u_n, \quad x_0 \text{ given},$$

for $x_n \in \mathbf{R}, u_n \in \mathbf{R}, n \in \mathbf{N}$, and the particular additive criterium:

$$\min_{u_0,u_1,\ldots,u_{N-1}} \sum_{i=0}^{N-1} c(u_i) + \phi(x_N),$$

with c and $\phi : \mathbf{R} \to \mathbf{R}$ convex, lower semi continuous (l.s.c), positive, null at their minimum. We denote m the abscissa where c takes its minimum.

$$\min_x c(x) = c(m) = 0.$$

The assumptions done here are not minimal but they simplify the discussion.

The value function defined by:

$$v_n(x) = \min_{u_n,\ldots,u_{N-1}} \left\{ \sum_{p=n}^{N-1} c(u_p) + \phi(x_N) | x_n = x \right\}$$

satisfies the dynamic programming equation:

$$v_n(x) = \min_u \{c(u) + v_{n+1}(x-u)\}, \quad v_N(x) = \phi(x).$$

It can be written using the inf convolution:

$$v_n = c * v_{n+1}, \quad v_N = \phi,$$

that is, (with the change of time index $p = N - n$, and the choice $\phi = \delta_0$):

$$v_p = c^{*p}.$$

This, in words, means that the solution of the dynamic programming equation in this particular case of "independent increment decision process" is obtained by reiterated convolutions of the instantaneous cost function.

In a more general case, the instantaneous cost $c(x_n, x_{n+1})$ depends on the initial and final state of a decision period (and not only on the state variation $u_n = x_{n+1} - x_n$), the dynamic is a general Markovian one $x_{n+1} \in \Gamma(x_n)$ (where Γ denotes a set valued function from $\mathbf{R}$ into the parts of $\mathbf{R}$). Then the dynamic programming equation becomes:

$$v_n(x) = \min_{y \in \Gamma(x)} \{c(x,y) + v_{n+1}(y)\}, \quad v_N(x) = \delta_0(x),$$

the solution of which can be written, with the same change of time, as:

$$v_n = c^n,$$

where the product of two kernels means:

$$[c_1 c_2](x,z) = \min_{y \in \Gamma(x)} \{c_1(x,y) + c_2(y,z)\}.$$

This more general case is the analogue of the general Markov chain case that we will study elsewhere.

Knowing that the analogues of the law of large numbers and the central limit theorem, that we recall here, have been given in Quadrat[13], *what is the analogue of the brownian motion and the diffusion processes ?*

Before answering the first question let us recall that the role of the Fourier transform in probability theory is played by the Fenchel transform in dynamic programming (Bellman-Karush [3]).

3 Fenchel and Cramér Transform

Let f be a mapping from $\overline{\mathbf{R}} \to \overline{\mathbf{R}}$ convex, l.s.c. and proper (i.e. never equal to $-\infty$). We define its Fenchel transform $\mathcal{F}(f)$ as the mapping $\hat{f} : \overline{\mathbf{R}} \to \overline{\mathbf{R}}$ such that:

$$\mathcal{F}(f)(p) = \hat{f}(p) = \sup_x [px - f(x)].$$

Then it can be shown that $\hat{f}$ is convex l.s.c. and proper.

Example 3.1 *The formula:*

$$\mathcal{F}(Q_{m,\sigma}) = \frac{1}{2} p^2 \sigma^2 + pm$$

is the analogue of the characteristic function of a Gaussian law.

$\mathcal{F}$ is an involution, that is, $\mathcal{F}(\mathcal{F}(f)) = f$ for all convex, proper, l.s.c. function f.

The main interest, for us, of the Fenchel transform is its ability to transform inf-convolutions into sums:

$$\mathcal{F}(f * g) = \mathcal{F}(f) + \mathcal{F}(g).$$

Applying the Fenchel transform to the dynamic programming equation in the case, c independent of x, we obtain:

$$v_N = \mathcal{F}(\hat{\phi} + N\hat{c}).$$

Using the fast Fenchel algorithm Brenier [4] this formula gives a fast algorithm to solve this particular case of the dynamic programming equation.

Moreover, let us recall that the Fenchel transform is continuous for the epigraph topology, that is, the epigraphs of the transformed functions converge if the epigraphs of the source functions converge for a well chosen topology. We can use, for example, the topology of Hausdorff on the epigraphs which are closed convex sets of $\mathbf{R}^2$, but this may be too strong (see Joly [8], Attouch-Wets [1] for discussions of these topological aspects). In this paper, we shall be formal on this point. We are, here, more concerned with the analogies between probability and deterministic control.

Example 3.2

$$\mathcal{F}(\varepsilon x)(p) = \delta_\varepsilon(p).$$

When $\varepsilon \to 0$, $\delta_\varepsilon \to \delta_0$ in the epigraph sense but does not converge pointwise even if $\varepsilon x \to 0$ pointwise.

Moreover, the pointwise convergence of numerical convex l.s.c. functions, towards a function in the same class, implies the convergence of their epigraphs.

The Cramér transform is defined as $\mathcal{F} \circ \log \circ \mathcal{L}$ where $\mathcal{L}$ denotes the Laplace transform. Therefore, it transforms the convolutions into inf-convolutions. Thus it is exactly the morphism in which we are interested. Unfortunately it is only a morphism for a set of functions endowed with one operation, the convolution. It is not a morphism for the sum (the pointwise sum of two functions is not transformed by the Cramér transform in the poinwise min of the transformed functions). Moreover the Cramér transform convexifies the functions but the inf-convolution is defined on a more general set of functions. Nevertheless the mapping $\lim_{\varepsilon\to 0} \log_\varepsilon$ defines an algebraic morphism of algebra between the asymptotic (around zero) of positive real functions of a real number and the real numbers endowed with the two operations min and plus (that is the classical asymptotic calculus), indeed:

$$\lim_{\varepsilon\to 0} \log_\varepsilon(\varepsilon^a + \varepsilon^b) = \min(a, b),$$

$$\log_\varepsilon(\varepsilon^a \varepsilon^b) = a + b.$$

We can now study the analogues of the limit theorems of the probability calculus.

4 Law of Large Numbers in Dynamic Programming

Suppose we are given two numerical mappings c and ϕ positive, convex, l.s.c., null in a unique munimum. To simplify the discussion let us suppose that $c \in C^2$ and $|1/c''|_\infty < \infty$. Let us denote by m the abscissa where c takes this null value. Let us denote $w_N(x)$ the mapping $x \to v_N(Nx)$. This change of scaling corresponds, on the value function, to the conventional averaging of the sampling.

Theorem 4.1 (Weak law of large numbers for dynamic programming) *Given the previous assumptions we have:*

$$\lim_{N\to\infty} v_N(Nx) = \delta_m(x),$$

the limit being in the sense of convergence of the epigraph.

Proof We have:

$$\hat{w}_N(p) = \hat{\phi}(p/N) + N\hat{c}(p/N),$$
$$\lim_{N\to\infty} \hat{\phi}(p/N) = \hat{\phi}(0) = 0,$$

since ϕ admits a null minimum by assumptions. Moreover, $\hat{c}(0) = 0$ for the same reasons. Then $\hat{c}(p)$ admits a Taylor expansion around 0 of the form $pm + O(p^2)$. Indeed:

$$\hat{c}'(p) = x_*(p) + x_*'(p)(p - c'(x_*(p))) = x_*(p) = m + O(p),$$

where $x_*(p)$ denotes the point realizing the maximum in the definition of the Fenchel transform of c. Therefore

$$\hat{w}_N(p) = pm + O(1/N).$$

Then using the continuity of the Fenchel transform we obtain:

$$\lim_{N\to\infty} \mathcal{F}(\hat{w}_N) = \mathcal{F}(pm) = \delta_m \quad \bullet$$

5 Central Limit Theorem in Dynamic Programming

We have the analogue of the central limit theorem of the probability calculus. The value function centered and normalized with the good scaling ($\sqrt{N}$) is asymptotically quadratic. More precisely, we have:

Theorem 5.1 (Central Limit) *Given the same assumptions as in Theorem 4.1, we have:*

$$\lim_{N\to\infty} v_N(\sqrt{N}(y + Nm)) = \frac{1}{2}c''(m)y^2.$$

The limit is in the sense of convergence of the epigraphs.

Proof

We make the expansion up to the second order of $p \to \hat{r}_N(p)$ where r_N is the mapping: $y \to v_N(\sqrt{N}(y + Nm))$.

But:

$$\hat{r}_N(p) = \hat{\phi}(\frac{p}{\sqrt{N}}) + N\hat{c}_m(\frac{p}{\sqrt{N}}),$$

where $c_m(y) = c(y + m)$. Then we have $\hat{\phi}(0) = 0$ and $\hat{c}_m(0) = 0$ because the minima of ϕ and c_m are zero.

Let us develop $\hat{c}_m$ up to the second order. We have seen that :

$$\hat{c}_m'(p) = x_*(p),$$

and therefore :

$$\hat{c}''_m(p) = x'_*(p).$$

Moreover, we know that $x_*(p)$ is defined by: $p - c'_m(x_*(p)) = 0$, and therefore,

$$1 - c''_m(x_*(p))x'_*(p) = 0,$$

that is,

$$x'_*(p) = \frac{1}{c''_m(x_*(p))}.$$

Therefore,

$$\hat{r}_N(p) = \frac{1}{2}\frac{p^2}{c''_m(0)} + o(1).$$

We obtain the result by passing to the limit using the continuity of the epigraph of the Fenchel transform •

These results can be extended to the vector case, when c depends on time (index n, in section 2) etc...

6 The Brownian Decision Process

Let us consider the discrete time decision process:

$$\min_u \sum_{i=0}^{(T/h)-1} \frac{(u_{ih})^2}{2h} + \Phi(x_T), \quad x_{t+h} = x_t - u_t.$$

It satisfies the dynamic programming equation:

$$v_t(x) = \min_u\{\frac{u^2}{2h} + v_{t+h}(x-u)\}, \quad v_T = \Phi.$$

The cost function, $Q_{0,\sqrt{h}}$, is therefore the analogue of the increment of Brownian motion on a time step of h. The analogue of the independence of the increments of the Brownian motion is the independence of the instantaneous cost function u^2/h from the state variable x.

Let us make the change of control $u = wh$ in the dynamic programming equation. We obtain:

$$v_t(x) = \min_w\{\frac{hw^2}{2} + v_{t+h}(x - wh)\}.$$

Passing to the limit, when h goes to 0, we obtain the Hamilton-Jacobi-Bellman equation:

$$D_t v + \min_w\{-wD_x v + \frac{w^2}{2}\} = 0, \quad v_T = \Phi.$$

That is:

$$D_t v - \frac{1}{2}(D_x v)^2 = 0, \quad v_T = \Phi,$$

which is the analogue of the heat equation:

$$D_t v + \frac{1}{2}D_{xx} v = 0, \quad v_T = \Phi.$$

Therefore, we can see the Brownian decision process as the Sobolev Space $H^1(0,T)$ endowed with the cost function $W(\omega) = \int_0^T (\omega')^2 dt$ for any function $\omega \in H^1(0,T)$. Then the decision problem can be written:

$$\mathbf{M}_W \Phi(x_T) \stackrel{\text{def}}{=} \min_{\omega \in H^1(0,T)} \{W(\omega) + \Phi(x_T(\omega))\}$$

by analogy with the probability theory. W is the analogue of the Brownian measure and can be interpreted as the cost of choosing ω. Then $\Phi(x_T(\omega))$ is the cost when we have choosen ω, of a decision function $\Phi(x_T(.))$.

But the solution of the Hamilton Jacobi equation:

$$D_t v - \frac{1}{2}(D_x v)^2 = 0, \quad v_T = \delta_y,$$

is unique P.L.Lions [9], and known explicitly. It is:

$$v_t(x) = \frac{(y-x)^2}{2(T-t)}, \quad t \leq T.$$

It can be seen as the min-plus Green kernel of the dynamic programming equation and the analogue of the Green kernel of the Kolmogorov equation for the Brownian equation:

$$\frac{1}{\sqrt{2\pi(T-t)}} e^{-\frac{(y-x)^2}{2(T-t)}}.$$

Therefore, by min-plus linearity we can deduce the solution of:

$$\min\{D_t v - \frac{1}{2}(D_x v)^2, c - v\} = 0, \quad v_T = \Phi,$$

which is the solution of the control problem:

$$v_t(y) = \mathbf{M}_W\{\min\{\min_{t \leq s \leq T} c(x_s(\omega)), \Phi(x_T(\omega))\} | x(t) = y\},$$

where s denotes a stopping time that we also want to optimize. This cost is clearly the min-plus analogue of:

$$v_t(y) = \mathbf{E}_W\{\int_t^T c(x_s(\omega))ds + \Phi(x_T(\omega)) | x(t) = y\}.$$

The solution of the decision problem is:

$$v_t(x) = \min\{\min_y\{\Phi(y) + \frac{(y-x)^2}{2(T-t)}\}, \min_{t \leq s \leq T} \min_y\{c(y) + \frac{(y-x)^2}{2(s-t)}\}\}.$$

This formula is the analogue of:

$$v_t(x) = \int \Phi(y) e^{-\frac{(y-x)^2}{2(T-t)}} dy + \int_t^T ds \int c(y) e^{-\frac{(x-y)^2}{2(s-t)}} dy.$$

Using the change of time s=T-t, we can summarize this part by the following theorem:

Theorem 6.1 *We have:*

$$\lim_{h \to 0} (Q_{0,\sqrt{h}})^{[s/h]} = Q_{0,\sqrt{s}},$$

where [x] denotes integer part of x. Moreover, $Q_{0,\sqrt{s}}$ *is the unique solution of:*

$$D_s Q + \frac{1}{2}(D_x Q)^2 = 0, \; s \geq 0, \; Q_{0,0} = \delta_0.$$

7 Diffusion Decision Process

In the previous section the dynamics of the systems are trivial and the instantaneous cost is independent of the state. Let us generalize this situation to a more general instantaneous cost, which will induce more complex optimal trajectories and which is the complete analogue of the diffusion process.

We consider the discrete decision process:

$$\min_u \sum_{i=0}^{(T/h)-1} \frac{(u_{ih} - b(ih)h)^2}{2h(\sigma(ih))^2} + \Phi(x_T), \quad x_{t+h} = x_t - u_t.$$

It satisfies the dynamic programming equation:

$$v_t(x) = \min_u \{\frac{(u - b(x)h)^2}{2h\sigma^2} + v_{t+h}(x-u)\}, v_T = \Phi.$$

By the change of control $u = wh$ in the dynamic programming equation and the passing to the limit when h goes to zero, we obtain the Hamilton Jacobi equation defined for the $t \leq T$:

$$D_t v - b(x)D_x v - \frac{\sigma(x)^2}{2}(D_x v)^2 = 0, \quad v_T = \Phi.$$

It is the Hamilton Jacobi Bellman (H.J.B.) equation corresponding to the variational problem:

$$v(t,x) = \min_{x \in H^1} \int_t^T \frac{1}{2}(\frac{x' - b}{\sigma})^2 dt + \Phi(x_T). \qquad (1)$$

This H.J.B. equation is the analogue of the Kolmogorov equation:

$$D_t v + b(x)D_x v + \frac{\sigma(x)^2}{2} D_{xx} v = 0, v_T = \Phi.$$

It is not necessary that the instantaneous cost be quadratic in such a way that the discrete decision process converges to the diffusion decision process.

Theorem 7.1 *The discrete decision process:*

$$\min_u \sum_{i=0}^{(T/h)-1} c_h(u_{ih}, x_{ih}) + \Phi(x_T), \quad x_{t+h} = x_t - u_t.$$

admits the discrete dynamic programming equation:

$$v_t(x) = \min_u \{c_h(u,x) + v_{t+h}(x-u)\}, v_T = \Phi,$$

which converges to the continuous dynamic programming equation:

$$D_t v - b(x)D_x v - \frac{\sigma(x)^2}{2}(D_x v)^2 = 0, v_T = \Phi,$$

as long as:

$$\hat{c}_h(x,p) = (b(x)p + \frac{\sigma^2(x)}{2}p^2)h + o(h),$$

where $\hat{c}_h$ denotes the Fenchel transform of the mapping $u \mapsto c_h(u,x)$.

The variational problem (1) has been met by large deviation researchers when they have studied differential equations perturbed by a small brownian noise. For example we have the following estimate:

$$\lim_{\Delta\to 0}\lim_{\varepsilon\to 0}\varepsilon\log(P_\varepsilon(X(T)\in(z-\Delta,z+\Delta)|X(0)=y)=\min_{\{x\in H^1(0,T)|x(0)=y,x(T)=z\}}\int_t^T\frac{1}{2}(\frac{x'-b}{\sigma})^2dt,$$

where P_ε denotes the probability law of a diffusion process with drift term b and diffusion term $\varepsilon\sigma$.

8 Conclusion

Let us conclude by summarizing the analogy between probability calculus and dynamic programming in the following table:

Probability	Dynamic Programming
$+$	min
$\times$	$+$
$\mathcal{N}_{m,\sigma}$	$\mathcal{Q}_{m,\sigma}$
$\int dF(x)=1$ $\mathbf{E}_F f=\int f(x)dF(x)$ Convolution	$\min_x c(x)=0$ $\mathbf{M}_c f=\inf\{f(x)+c(x)\}$ Inf-Convolution
Fourier or Laplace $\log(\hat{F})'(0)=i\int xdF(x)=im$ $-\log(\hat{F})''(0)=\int(x-m)^2dF(x)$	Fenchel $\hat{c}'(0)=m\ :\ c(m)=\min_x c(x)$ $\hat{c}''(0)=\frac{1}{c''(m)}$
Brownian Motion D_{x^2} $D_t+1/2D_{x^2}$ $1/\sqrt{(2\pi t)}e^{-x^2/2t}$	Brownian Decision Process $(D_x)^2$ $D_t-1/2(D_x)^2$ $x^2/2t$
Diffusion Process $D_t+b(x)D_x+a(x)D_{xx}$ Invariance Principle	Diffusion Decision Process $D_t-b(x)D_x-a(x)(D_x)^2$ Min-Plus Invariance Principle

The Cramér transform is an important tool for people interested in large deviations and researchers in this field know the morphism between the classical and the min-plus algebras. But in general they are more interested by the probability side than by the optimization one. Moreover, they are not much concerned with the algebraic point of view [2],[7],[15]. Bellman [3] was aware of the interest of the Fenchel transform (which he calls max transform). Maslov has also clearly understood the analogy between probability and dynamic programming and have developed a theory of idempotent integration [11]. The analogue of some jump processes can be found in [5].

The law of large numbers, the central limit theorem, the Brownian decision process, and the diffusion decision process and the min-plus invariance principle seem not to have been written down explicitly before.

Min-plus linearity of differential operators and its consequence for solving the Hamilton-Jacobi-Bellman equation is currently studied by M.Akian. M.Viot is working on the min-plus analogue of stochastic integrals. Min-plus linear, time invariant, finite dimensional, systems have been studied in M.Plus [10].

Thanks We thank all the Max Plus working group at INRIA, and in particular M. Viot and M. Akian who are also interested by this morphism between probability and dynamic programming. We refer to their future work, in preparation, on this subject. We want also thank P.L. Lions and R. Azencott for useful discussions with them which show me that they know also these results, and who explained to me the role of the Cramér transform as the morphism transforming the convolution into inf-convolution. Finally we thank the reviewers who do their best to improve the english of this paper.

References

[1] H. ATTOUCH, R.J.B. WETS : *Isometries of the Legendre Fenchel transform*, Trans. of The Am. Math. Soc. n°296, p.33-60, (1986).

[2] R.AZENCOTT, Y.GUIVARC'H, R.F.GUNDY *Ecole d'été de Saint Flour N 8*, Springer Verlag L.N.Math N.774 (1978).

[3] R. BELLMAN, W. KARUSH : *Mathematical programming and the maximum transform* SIAM Ap. Math. vol.10 n°3, (1962).

[4] Y. BRENIER : *Transformée de Fenchel rapide*. CRAS t.308,Série 1,p.587-589,(1989).

[5] P.DEL MORAL, T. THUILLET, G. RIGAL, G.SALUT : *Optimal versus random processes: the non-linear case*. Rapport LAAS Toulouse (1990).

[6] W. FENCHEL : *On the conjugate convex functions*, Canad. J. Math. n° 1 p.73-77, (1949).

[7] M.I. FREIDLIN, A.D. WENTZELL : *Random Perturbations of Dynamical Systems* Springer Verlag (1979).

[8] J.L. JOLY : *Une famille de topologies sur l'ensemble des fonctions convexes pour lesquelles la polarité est bicontinue*, J. Math. Pures et Ap. n° 52, p.421-441, (1973).

[9] P.L.LIONS :*Generalized Solutions of Hamilton-Jacobi Equation* Pitman R.N.Math. N. 69 (1982).

[10] M.PLUS : *A linear system theory for systems subject to synchronization and saturation constraints*, EEC (1991).

[11] V.MASLOV : *Méthodes opératorielles*, MIR, (1987).

[12] P.PLAZANET : *Thèse Toulouse*, (1990).

[13] J.P. QUADRAT : *Théorèmes asymptotiques en programmation dynamique*, CRAS,t.311,Série 1,p.745-748,(1990).

[14] R.T. ROCKAFELLAR : *Convex Analysis*, Princeton University Press, (1970).

[15] S.R.S. VARADHAN : *Large Deviations and Applications*, CBMS-NSF Conf. N.46,SIAM (1984).

KANTOROVICH'S FUNCTIONALS IN SPACE OF MEASURES

S.T. Rachev* and M. Taksar

Department of Statistics and Applied Probability
University of California, Santa Barbara
Santa Barbara, CA 93106

Department of Applied Mathematics and Statistics
State University of New York at Stony Brook
Stony Brook, New York 11794-3600

Abstract:

We investigate relationships between Kantorovich's functionals arising from mass transportation problems. Compactness, completeness and merging criteria w.r.t. Kantorovich's functionals are considered. Applications to the problems of best classification and best allocation policy are given.

AMS 1980 Subject Classifications: *Primary*: 60B05, 60B10 *Secondary:* 60E05, 28A35

Key Words and Phrases: translocation of masses, convergence of measures, compactness, completness, uniformity.

---

*Research supported by NATO Grant CRG900798.

I. Introduction

Let $S = (S,d)$ be a separable metric space (s.m.s.) with metric d. Let P_1 be a *mass distribution*, i.e. a finite nonnegative measure on the Borel σ-algebra $\mathscr{B} = \mathscr{B}(S)$ generated by the metric d. Let further P_2 be another mass distribution, such that $P_1(S) = P_2(S)$. In order to avoid unnecessary stipulations, we shall assume that P_i are probability measures,

(1.1) $\quad P_1(S) = P_2(S) = 1$.

By the *translocation of masses* we will understand the finite Borel measure P on $\mathscr{B}(SxS)$ such that

(1.2) $\quad P(AxS) = P_1(A)\,,\ P(SxA) = P_2(A)$

for any $A \in \mathscr{B}(S)$ (see Kantorovich (1942)). Denote the space of all translocations of masses by $\mathscr{P}(P_1, P_2)$. Under the *translocation of masses in transit*, we will understand the finite Borel measure Q on $\mathscr{B}(SxS)$ such that

(1.3) $\quad Q(A\times S) - Q(S\times A) = P_1(A) - P_2(A)$

for any $A \in \mathscr{B}(S)$ (see Kantorovich and Rubinstein (1958)). Denote the space of all Q satisfying (1.3) by $\mathcal{Q}(P_1, P_2)$. Let a continuous non-negative function $c(x,y)$ be given that represents the cost of transferring a unit mass from x to y. The total cost of transferring the given mass distributions P_1 and P_2 is given by

(1.4) $\quad \mu_c(P) := \int_{S\times S} c(x,y)\, P(dx, dy) \text{ if } P \in \mathscr{P}(P_1, P_2)$

or

(1.5) $\quad \mu_c(Q) := \int_{S\times S} c(x,y)\, Q(dx, dy) \text{ if } Q \in \mathcal{Q}(P_1, P_2)$.

Hence,

(1.6) $\quad \hat{\mu}_c(P_1,P_2) = \inf\{\mu_c(P) : P \in \mathscr{P}(P_1,P_2)\}$

may be viewed as the *minimal translocation cost* while

(1.7) $\quad \mathring{\mu}_c(P_1,P_2) = \inf\{\mu_c(Q) : Q \in \mathcal{Q}(P_1,P_2)\}$

may be viewed as the *minimal translocation cost in case of resolved transit conveyances.* (See Kantorovich and Akilov (1977), Kemperman (1983), Rachev (1984a,b)).

The problem of calculating the exact value of $\hat{\mu}_c$ is known as the *Kantorovich problem* and $\hat{\mu}_c$ is called *Kantorovich functional*. Similarly the problem of evaluating $\mathring{\mu}_c$ is known as the *Kantorovich–Rubinstein problem* and $\mathring{\mu}_c$ is said to be the *Kantorovich–Rubinstein functional* (see the survey Rachev (1984b)). Dual representations and explicit solutions of the Kantorovich and Kantorovich–Rubinstein problems attracted the attention of many specialists (see for example, Levin and Miljutin (1979), Kellerer (1984b), Rachev (1984b), Cambanis, Simons and Stout (1976), Dudley (1976), (1989), Olkin and Pukelsheim (1982), Ruschendorf (1985), Smith and Knott (1987), Rachev and Shortt (1989), Cuesta and Matran (1989)).

It is well–known that in general $\mathring{\mu}_c \neq \hat{\mu}_c$. For example if $S = \mathbb{R}$, $d(x,y) = |x-y|$,

(1.8) $\quad c(x,y) = d(x,y)\max(1, d^{p-1}(x,a), d^{p-1}(y,a)), \quad p \geq 1, a \in S$

then for any P_i (i=1,2) on $\mathscr{B}(\mathbb{R})$ with distribution function (d.f.) F_i, we have the following explicit expressions:

(1.9) $\quad \hat{\mu}_c(P_1, P_2) = \int_0^1 c(F_1^{-1}(t), F_2^{-1}(t))dt$

and

(1.10) $\quad \mathring{\mu}_c(P_1, P_2) = \int_{\infty}^{\infty} |F_1(x) - F_2(x)| \max(1, |x-a|^{p-1})dx,$

where F_i^{-1} is the function inverse to the d.f. F_i. (See Rachev (1984b), Rachev and Shortt (1989)).

However, in the space $\mathscr{M}_p = \mathscr{M}_p(S)$ (S = (S,d) is s.m.s.) of all Borel probability measures P with finite $\int d^p(x,a)P(dx)$, the functionals $\hat{\mu}_c$ and $\mathring{\mu}_c$ (where c is given by (1.8)) metrize one and the same topology. Namely, the following $\hat{\mu}_c$ – and $\mathring{\mu}_c$ – convergence criterion holds.

Theorem 1.1 *Let* (S,d) *be a s.m.s.* c *is given by* (1.8) *and* $P \in \mathscr{M}_p$, $P_n \in \mathscr{M}_p$ (n=1,2,...).

Then the following are equivalent

(I) $\hat{\mu}_c(P_n,P) \to 0$,

(II) $\mathring{\mu}_c(P_n,P) \to 0$,

(III) P_n *converges weakly to* P ($P_n \xrightarrow{w} P$) *and*

$$\lim_{N\to\infty} \sup_n \int_S d^p(x,a)\, I\{d(x,a) > N\}\, P_n(dx) = 0,$$

(IV) $P_n \xrightarrow{w} 0$ and $\int_S d^p(x,a)P_n(dx) \to \int_S d^p(x,a)P(dx)$.

Indication: See Rachev (1984a, b). (The equivalence of (I), (III) and (IV) was shown also by Bickel and Freedman (1981) in the case of separable Banach space $(S,\|.\|)$).

Theorem 1.1 is a qualitative $\hat{\mu}_c$ ($\mathring{\mu}_c$)–convergence criterion. One can rewrite (III) as

(III*) $\pi(P_n,P) \to 0$ and $\lim_{\epsilon\to 0} \sup_n \omega(\epsilon\,; P_n\,; \lambda) = 0$,

where π is the *Prokhorov metric*

(1.11) $\pi(P,Q) := \inf\{\epsilon > 0\colon P(A) \le Q(A^\epsilon) + \epsilon \text{ for all } A \in \mathscr{B}(S)\}$

$(A^\epsilon := \{x\colon d(x,A) < \epsilon\})$

(see Dudley (1989)), and $\omega(\epsilon;P;\lambda)$ is the following modulus of λ–integrability,

(1.12) $\omega(\epsilon;P;\lambda) := \int_S \lambda(x)\, I\{d(x,a) > 1/\epsilon\}\ P(dx)$,

where $\lambda(x) := \max(d(x,a),\ d^p(x,a))$. Analogously, IV is equivalent to

(IV*) $\pi(P_n, P) \to 0$ and $D(P_n, P;\lambda) \to 0$,

where

(1.13) $D(P,Q;\lambda) = |\int \lambda(x)\,(P-Q)\,(dx)|$.

In this paper we investigate quantitative relationships between $\hat{\mu}_c$, $\mathring{\mu}_c$, π, ω and D in terms of inequalities between these functionals (see Section 2). These relationships yield convergence and compactness criteria in the space of measures w.r.t. the Kantorovich type functionals $\hat{\mu}_c$ and $\mathring{\mu}_c$ (see Section 3) as well as the $\mathring{\mu}_c$–completeness of the space of measures. In Section 4 we prove that $\mathring{\mu}_c \le \hat{\mu}_c \le p\mathring{\mu}_c$ where c, $\hat{\mu}_c$ and $\mathring{\mu}_c$ are given by (1.8), (1.9) and (1.10) respectively. Sections 5 and 6 are devoted to illustrations of the Kantorovich and Kantorovich–Rubinstein problems to the problems of best classification and best allocation policy.

2. Inequalities between $\hat{\mu}_c$, $\overset{\circ}{\mu}_c$, π, ω, D.

In the sequel, we assume that the cost function c has the following form

(2.1) $$c(x,y) = d(x,y)k_0(d(x,a),d(y,a)) \quad x,y \in S,$$

where $k_0(t,s)$ is a symmetrical, continuous function, nondecreasing in both arguments $t \geq 0$, $s \geq 0$, and satisfying the following conditions:

C1) $$\alpha := \sup_{s \neq t} \frac{|K(t) - K(s)|}{|t-s| \; k_0(t,s)} < \infty,$$

where $K(t) := tk_0(t,t)$, $t \geq 0$;

C2) $\beta := k(0) > 0$,

where $k(t) = k_0(t,t)$, $t \geq 0$; and

C3) $$\gamma := \sup_{t \geq 0,\ s \geq 0} \frac{k_0(2t,2s)}{k_0(t,s)} < \infty.$$

If c is given by (1.8) then c admits the form (2.1) with $k_0(t,s) = \max(1, t^{p-1}, s^{p-1})$ and in this case $\alpha = p$, $\beta = 1$, $\gamma = 2^{p-1}$. The form (2.1) of the cost function c naturally arises in various problems of probability theory and its applications (see Kantorovich (1942), Fortet and Mourier (1953), Rachev (1984ab), Rachev and Shortt (1989), Kalashinkov and Rachev (1988), Zolotarev (1983)). Recall only that Fortet–Mourier (1953) metric is $\overset{\circ}{\mu}_c$ with c given by (1.8), see Rachev (1984b), Rachev and Shortt (1989). Further, let $\mathcal{M} = \mathcal{M}(S)$ be the space of all probability measures on the s.m.s. (S,d) and

(2.2) $$\mathcal{M}(S,\lambda) = \{P \in \mathcal{M}: \int_S \lambda(x)P(dx) < \infty\}$$

where $\lambda(x) = K(d(x,a))$ and a is a fixed point of S.

In the next Theorems 2.1 – 2.4 we set $P_1 \in \mathcal{M}(S,\lambda)$, $P_2 \in \mathcal{M}(S,\lambda)$, $\epsilon > 0$, $\hat{\mu}_c := \hat{\mu}_c(P_1,P_2)$ (see (1.6)), $\overset{\circ}{\mu}_c = \overset{\circ}{\mu}_c(P_1,P_2)$ (see (1.7)), $\pi := \pi(P_1,P_2)$ (see (1.11)), $\omega_i(\epsilon) := \omega_i(\epsilon; P_i; \lambda)$ (see (1.12)), $D := D(P_1,P_2; \lambda)$ (see (1.13)). We also assume that the function c satisfies C1 – C3 with fixed constants α, β and γ.

Theorem 2.1 For any $\epsilon > 0$, $\hat{\mu}_c \leq \pi[4K(1/\epsilon) + \omega_2(1) + 2k(1)] + 5\omega_1(\epsilon) + 5\omega_2(\epsilon)$.

Proof: Recall $\mathcal{P}(P_1,P_2)$ to be the space of all $P \in \mathcal{M}(SxS)$ with prescribed marginals P_1 and P_2 (see (1.2)). Let $\underline{KF}(P) := \inf\{\delta > 0: P(d(x,y) > \delta) < \delta\}$, $P \in \mathcal{M}(SxS)$ be the "*Ky Fan (probability) metric*", see Dudley (1989), Rachev and Shortt (1989).

Claim 1. For any $N > O$ and $P \in \mathcal{P}(P_1, P_2)$

$$\int_{SxS} c(x,y)P(dx,dy) \leq \gamma\{\underline{KF}(P)[4K(N) + \int_S k(d(x,a))(P_1 + P_2)(dx)]$$
$$+ 5\omega_1(1/N) + 5\omega_2(1/N)\}.$$

Proof of Claim. Let $\underline{KF}(P) < \delta \le 1$, $P \in \mathcal{P}(P_1,P_2)$. Then by (2.1),

$$\int_{S\times S} c(x,y)P(dx,dy) \le I_1 + I_2,$$

where $I_1 := \int_{S\times S} d(x,y)k(d(x,a))\, P(dx,dy)$ and $I_2 := \int_{S\times S} d(x,y)k(d(y,a))P(dx,dy)$.

For I_1 we readily obtain

$$I_1 \le \delta\omega_1(1) + \delta k(1) + 3\omega_1(1/N) + 2\omega_2(1/N) + 2K(N)\delta.$$

By the symmetry, we have $I_2 \le \delta\omega_2(1) + \delta k(1) + 3\omega_2(\frac{1}{N}) + 2\omega_1(\frac{1}{N}) + 2K(N)\delta$.

Letting $\delta \to \underline{KF}(P)$ proves the claim .

Claim 2 (*Strassen–Dudley theorem*): $\inf\{\underline{KF}(P): P \in \mathcal{P}(P_1,P_2)\} = \pi(P_1, P_2)$

Indication: Dudley (1989), Corollary 11.6.4.

Claims 1 and 2 establish the theorem. Q.E.D.

Theorem 2.2 $\delta\pi^2 \le \overset{\circ}{\mu}_c \le \hat{\mu}_c$.

Proof: Obviously, for any continuous nonnegative function c,

$$\overset{\circ}{\mu}_c \le \hat{\mu}_c \tag{2.3}$$

and

$$\overset{\circ}{\mu}_c \ge \zeta_c, \tag{2.4}$$

where

$$\zeta_c := \zeta_c(P_1, P_2) \tag{2.5}$$
$$:= \sup\{\textstyle\int_S fd(P_1-P_2): f:S \to \mathbb{R},\ f(x) - f(y) \le c(x,y) \text{ for all } x,y \in S\}.$$

Now, using C2 we have that $c(x,y) \ge \beta d(x,y)$ and hence, $\zeta_c \ge \beta\zeta_d$. Thus, by (2.4), $\overset{\circ}{\mu}_c \ge \beta\, \zeta_d$. Finally, $\zeta_d \ge \pi^2$, see for example Dudley (1976). Q.E.D.

Theorem 2.3 For any $\epsilon > 0$, $\omega_1(\frac{\epsilon}{2}) \le \alpha(2\gamma + 1)\overset{\circ}{\mu}_c + 2(\gamma + 1)\omega_2(\epsilon)$.

Proof: For any $N > O$, $\omega_1(1/2N) := \int_S \lambda(x)I\{d(x,a) > 2N\}\, P_1(dx) \le T_1 + T_2$,

where $T_1 := |\int_S \lambda(x)\, I\{d(x,a) > 2N\}\, (P_1 + P_2)\, (dx)|$ and

$$T_2 := \int_S \lambda(x)\, I\{d(x,a) > N\}\, P_2(dx) = \omega_2(1/N).$$

Claim 1 $T_1 \le \alpha\overset{\circ}{\mu}_c + K(2N)\int_S I\{d(x,a) > 2N\}\, (P_1 + P_2)(dx)$.

Proof of Claim 1: Let $f_N(x) := \frac{1}{\alpha}\max(\lambda(x),k(2N))$. Since $\lambda(x) = K(d(x,a))$ then, by C1, $|f_N(x) - f_N(y)| \le \frac{1}{\alpha}|\lambda(x)-\lambda(y)| \le |d(x,a) - d(y,a)|k_0(d(x,a),d(y,a)) \le c(x,y)$ for any $x,y \in S$, see also Proposition 7.2, Dudley (1976). Hence, by (2.4) and (2.5)

$|\int_S f_N(x)(P_1 - P_2)(dx)| \le \zeta_c \le \overset{\circ}{\mu}_c$. Since $\alpha f_N(x) = \max(K(d(x,a)), K(2N))$ then

$$T_1 \leq \alpha\mathring{\mu}_c + K(2N)\int_S I\{d(x,a) > 2N\}\,(P_1 + P_2)(dx)$$

which completes the claim.

Claim 2. $A(P_1) := K(2N)\int_S I\{d(x,a) > 2N\}\,P_1(dx) \leq 2\alpha\gamma\mathring{\mu}_c + 2\gamma\omega_2(\frac{1}{N})$.

Proof of Claim 2. Let $g_N(x) = \frac{1}{2\alpha\gamma}\min\{K(2N), K(2d(x,O(a,N)))\}$, where $O(a,N) :=$ $\{x : d(x,a) \leq N\}$. By C1 and C3,

$$|g_N(x) - g_N(y)| \leq d(x,y)k_0(d(x,O(a,N)), d(y,O(a,N))) \leq c(x,y).$$

Hence, $|\int g_N(P_1 - P_2)(dx)| \leq \zeta_c \leq \mathring{\mu}_c$. Using the implications

$$d(x,a) > 2N \Rightarrow d(x,O(a,N)) > N \Rightarrow K(2d(x,O(a,N))) \leq K(2N)$$

we obtain

$$A(P_1) \leq 2\alpha\gamma\mathring{\mu}_c + 2\gamma\int_S K(d(x,O(a,N)))\, I\{d(x,a) \geq N\}\,P_2(dx) \leq$$
$$2\alpha\gamma\mathring{\mu}_c + 2\gamma\omega_2(1/N),$$

which proves the claim.

For $A(P_2)$ we have the following bound

$$A(P_2) \leq \int_S K(d(x,a))\, I\{d(x,a) > 2N\}\,P_2(dx) \leq \omega_2(1/N).$$

Summarizing the results of Claims 1 and 2, we obtain

$$\omega_1(1/2N) \leq \alpha\mathring{\mu}_c + A(P_1) + A(P_2) + \omega_2(1/N) \leq (\alpha + 2\alpha\gamma)\mathring{\mu}_c + (2\gamma + 2)\,\omega_2(1/N)$$

for any $N > 0$ as desired. Q.E.D.

Theorem 2.4. $D \leq \mathring{\mu}_c$.

Proof. By C1, for any $Q \in \mathcal{Q}(P_1, P_2)$,

$$D := |\int_S K(d(x,a))\,(P_1 - P_2)\,(dx)|$$
$$\leq \int_{S\times S} \alpha|d(x,a) - d(y,a)|\,k_0(d(x,a), d(y,a))\,Q(dx,dy)$$
$$\leq \alpha\int_{S\times S} c(x,y)\,Q(dx,dy)$$

Q.E.D.

3. Convergence, compactness and completeness in $(\mathcal{M}, \hat{\mu}_c)$ and $(\mathcal{M}, \mathring{\mu}_c)$

In this section we always assume that the cost function c satisfies C1–C3 and $\lambda(x) =$ $K(d(x,a))$.

Theorem 3.1. *If* $P_n, P \in \mathcal{M}(S,\lambda)$ (see (2.2)) *then the following are equivalent as* $n \to \infty$:

A) $\hat{\mu}_c(P_n,P) \to 0$;

B) $\mathring{\mu}_c(P_n,P) \to 0$;

C) $P_n \xrightarrow{w} P$ *and* $\int\lambda d(P_n - P) \to 0$.

Proof. A) <=> C) See Rachev and Shortt (1989). A) => B) See (2.3).

Claim 1. C) implies

(3.1) $P_n \xrightarrow{w} P$ and $\lim_{\epsilon \to 0} \sup_n \omega_n(\epsilon) = 0$,

where $\omega_n(\epsilon) := \omega(\epsilon; P_n : \lambda)$, see (1.12).

The proof uses routine arguments, cf. Billingsley (1968), Theorem 5.4 for the special case $S = \mathbb{R}$, $\lambda(x) = |x|$.

Claim 2. Relation (3.1) imples A).

Proof of Claim 2. By Theorem 2.1,

$$\mu_c(P_n,P) \le \pi(P_n,P)\,[4K(\frac{1}{\epsilon_n}) + \omega_n(1) + \omega(1) + 2k(1)] + 4\omega_n(\epsilon_n) + 5\omega(\epsilon_n),$$

where ω_n and ω are defined as in Claim 1 ($\omega_n(\epsilon) := \omega(\epsilon; P_n; \lambda)$, $\omega(\epsilon) := \omega(\epsilon, P; \lambda)$)and moreover $\epsilon_n > 0$ is such that

$$4K(1/\epsilon_n) + \sup_{n \ge 1} \omega_n(1) + \omega(1) \le 1/\pi(P_n, P)^{1/2}$$

Hence, using the above in equalities we get

$$\mu_c(P_n,P) \le \sqrt{\pi(P_n,P)} + 5 \sup_{n \ge 1} \omega_n(\epsilon_n) + 5\omega(\epsilon_n).$$

Now, obviously, (3.1) implies A) as we claim.

Claim 1 and 2 yield the desired implication C) => A). Q.E.D.

The Kantorovich–Rubinstein functional $\mathring{\mu}_c$ is a metric in $\mathcal{M}(S,\lambda)$ while $\hat{\mu}_c$ is not a metric except for the case c=d (see the discussion in Neveu and Dudley (1980), DeAcosta (1982), Dudley (1989), Sec. 11.8, Rachev and Shortt (1989)). Further, a set $A \subset \mathcal{M}(S,\lambda)$ is said to be *$\mathring{\mu}$–relatively compact* if any sequence of measures in A has a $\mathring{\mu}_c$–convergent subsequence and the limit belong to $\mathcal{M}(S,\lambda)$.

Theorem 3.2. *The set* $A \subset \mathcal{M}(S,\lambda)$ *is* $\mathring{\mu}_c$*–relatively compact if and only if* A *is weakly compact and*

(3.2) $\lim_{\epsilon > 0} \sup_{P \in A} \omega(\epsilon;P;\lambda) = 0$.

Proof. *"if" part:* If A is weakly compact, (3.2) holds and $\{P_n\} \subset A$ then we can choose a subsequence $\{P'_n\} \subset \{P_n\}$ which converges weakly to a probability measure P. Then $P \in \mathcal{M}(S,\lambda)$ by routine arguments as in Theorem 5.1 (Billingsley (1968). Now it is enough to show that $\mathring{\mu}_c(P_{n'},P) \to 0$. Using Theorem 2.1 we have, for $\delta > 0$,

$$\overset{\circ}{\mu}_c(P_{n'},P) \le \hat{\mu}_c(P_{n'},P)$$
$$\le \alpha\,\gamma\{\pi(P_{n'},P)\,[4K(1/\epsilon) + \omega_1(1) + \omega_2(1) + 2k(1)] + \delta\}$$

if $\epsilon = \epsilon(\delta)$ is small enough.

"Only if" part: If A is $\overset{\circ}{\mu}_c$–relatively compact then, by Theorem 2.2, A is weakly compact. Further, (3.2) follows readily from Theorem 2.3. Q.E.D.

In a similar way we obtain the following analog of the Prohorov completeness theorem.

Theorem 3.3. *If* (S,d) *is a complete s.m.s. then* $(\mathcal{M}(S,\lambda), \overset{\circ}{\mu}_c)$ *is also complete.*

4. $\overset{\circ}{\mu}_c$ – and $\hat{\mu}_c$ –uniformity of Fortet–Mourier metric

In the previous section we saw that $\overset{\circ}{\mu}_c$ and $\hat{\mu}_c$ induce one and the same convergence in $\mathcal{M}(S,\lambda)$. Here, we would like to analyze the uniformity of $\overset{\circ}{\mu}_c$– and $\hat{\mu}_c$–convergence. (The $\hat{\mu}_d$–uniformity (merging) was considered by Dudley (1989), Sec. 11.7 and d'Aristotile, Diaconis and Freedman (1988).) We shall be interested in the conditions that imply, for any $P_n \in \mathcal{M}(S,\lambda)$ and $Q_n \in \mathcal{M}(S,\lambda)$,

(4.1) $\quad \overset{\circ}{\mu}_c(P_n, Q_n) \to 0$ iff $\hat{\mu}_c(P_n, Q_n) \to 0$ as $n \to \infty$.

"If"–part holds by (2.3). Clearly, if

(4.2) $\quad \hat{\mu}_c(P,Q) \le \phi(\overset{\circ}{\mu}_c(P,Q)), \quad P,Q \in \mathcal{M}(S,\lambda)$

for a continuous nondecreasing function, $\phi(0) = 0$ then (4.1) holds. We are going to prove (4.2) for the special but important case when $\overset{\circ}{\mu}_c$ is the Fortet–Mourier metric on $\mathcal{M}(\mathbb{R},\lambda)$ (see Fortet and Mourier (1953), Dudley (1976), Rachev and Shortt (1989)).

Further,

(4.3) $\quad c(x,y) = |x-y| \max(1,|x|^{p-1}, |y|^{p-1}), \ p \ge 1,$

(see 1.8), and

(4.4) $\quad \lambda(x) = 2\max(|x|^p, |x|^p),$

Theorem 4.1. *If c is given by* (4.3) *then*

(4.5) $\quad \hat{\mu}_c(P,Q) \le p\overset{\circ}{\mu}_c(P,Q) \quad$ for all $P,Q \in \mathcal{M}(\mathbb{R},\lambda)$.

Proof. Denote $h(t) = \max(1,|t|^{p-1}), \ t \in \mathbb{R}$ and

(4.6) $\quad H(x) = \int_0^x h(t)dt, \quad x \in \mathbb{R}.$

Let X and Y be real-valued random variables on a nonatomic probability space (Ω,A,Pr) with distributions P and Q respectively.

Claim 1.

(4.7) $\quad \overset{\circ}{\mu}_c(P,Q) = \int_{\infty}^{\infty} |F_{H(X)}(x) - F_{H(Y)}(x)|\,dx.$

Indication: See (1.10) and use similar arguments as in Problem 1, p.333, Dudley (1989) where the case $p = 1$ is considered.

Claim 2. $\mathring{\mu}_c(P,Q) = \inf\{E|H(\tilde{X}) - H(\tilde{Y})| : F_{\tilde{X}} = F_X, F_{\tilde{Y}} = F_Y\}$.

Indication: See Dudley (1976), Proposition 20.11.

Finally, one can easily check the following:

Claim 3. For any $x,y \in \mathbb{R}$, $c(x,y) \le p\ |H(x) - H(y)|$.

Now, (4.5) is a consequence of Claims 2, 3 and 4. Q.E.D.

5. Kantorovich functionals and the problem of classification

In multivariate statistical analysis the problem of classification is well-known (see, for example Anderson (1984), Chapter 6). Let us give one popular example of the problem of classification.

Army recruits are given a bettery of tests to determine their fitness for different jobs: The scores are a set of measurements $x \in S$, where (S,d) is a separable metric space, for example, $S = \mathbb{R}^k$, $d(x,y) = \|x - y\|$. The distribution of scores is given by the measure P_1

$$P_1(A) = \frac{\text{Number of the recruits with scores in } A}{\text{Total number of the recruits}}$$

On the other hand the needs of the army can be expressed by a probability measure P_2 on S which represents the desired distribution of scores for the jobs needed to be filled in. The problem is to choose an optimal classification (or assignment) of the recruits to the jobs. A classification can be specified by choosing a bounded measure Q on $\mathscr{B}(S\times S)$. If the classification satisfies the balancing conditions,

$$Q(A\times S) = P_1(A),\ Q(S\times B) = P_2(B) \tag{5.1}$$

then we view $Q(A\times B)$ as the number of recruits with scores $x \in A$ which are classified as satisfying (after retraining) the needs of the jobs which require scores $y \in \mathscr{B}$. If we think that the training procedure might be a multi-staged one, in which the same individual gradually changes his scores (and fitness for different jobs respectively) in a sequence of retraining stages that the measure Q satisfies the balancing conditions

$$Q(AxS) - Q(SxA) = P_1(A) - P_2(A). \tag{5.2}$$

The interpretation of $Q(AxB)$ is the combined number of GI's at all stages who had scores x in A and who were trained to fit the jobs which require scores y in B. Let $c_0(x,y)$ be the cost of training the person with score x to fit the job which requires score y. Consider the "joint cost"

$c(x,y) = c_0(x,y) + c_0(y,x)$. An obvious assumption on c is that

(5.3) $\qquad c(x,x) = 0$.

Moreover we can assume that

(5.4) $\qquad d(x',y') \le d(x'',y'') \Rightarrow c(x',y') \le c(x'',y'')$,

i.e. the cost $c(x,y)$ increses with $d(x,y)$. In particular (5.4) implies

(5.5) $\qquad d(x',a) < d(x'',a) \Rightarrow c(x',a) < c(x'',a)$

(5.6) $\qquad d(a,y) < d(a,y'') \Rightarrow c(a,y') < c(a,y'')$

for a fixed point $a \in S$ which one can consider as the "center" of recruit possibilities and army's needs. The assumption (5.3) and (5.4) – (5.6) imply that one reasonable form of c is given by

(5.7) $\qquad c(x,y) = d(x,y) \max(h(d(x,a))), h(d(y,a))$,

where h is a continuous non–decreasing function on $(0,\infty)$, $h(0) \ge 0$. Determining the cost function c we conclude that the total cost involved in using the classification Q is calculated by the integral $TC(Q) = \int_{S\times S} c(x,y)\, Q(dx,dy)$. Therefore the folowing problems arise:

Problem 5.1. Considering the classifications $Q \in \mathcal{P}(P_1,P_2)$ (i.e. (5.1) holds) we seek to characterize the optimal $P^* \in \mathcal{P}(P_1,P_2)$ which $TC(P^*) = \inf\{TC(Q) : Q \in \mathcal{P}(P_1,P_2)\}$ as well as to evaluate the bound $\hat{\mu}_c(P_1,P_2) = \inf\{TC(Q) : Q \in \mathcal{P}(P_1,P_2)\}$.

Problem 5.2. In the classifications $Q \in \mathcal{Q}(P_1,P_2)$ (i.e., (5.2) holds) find the optimal $Q^* \in \mathcal{Q}(P_1,P_2)$ (if Q^* exists) for which $TC(Q^*) = \inf\{TC(Q) : Q \in \mathcal{Q}(P_1,P_2)\}$. Find $\mathring{\mu}_c(P_1,P_2) = \inf\{TC(P) : P \in \mathcal{Q}(P_1,P_2)\}$.

Problem 5.3. What kind of quantitative relationships exist between $\hat{\mu}_c$ and $\mathring{\mu}_c$?

The next three theorems give us particular answers to the Problems 5.1–5.3 respectively.

Theorem 5.1. i) *If*

(5.8) $\qquad \int_S d(x,a)\, h(d(x,a))\, (P_1 + P_2)(dx) < \infty$

then $\hat{\mu}_c(P_1,P_2) = \sup\{\int_S f dP_1 + \int_S g dP_2;\ f: S \to \mathbb{R},\ g: S \to \mathbb{R},$

$\qquad \|f\|_\infty := \sup_{x\in S} |f(x)| < \infty,\ \|g\|_\infty < \infty,$

$\qquad |f(x) - f(y)| \le \alpha_f d(x,y),\ \alpha_f > 0,$

$\qquad f(x) + g(y) \le c(x,y)$ for all $x,y \in S\}$.

ii) *If* $S = \mathbb{R}$, $d(x,y) = |x-y|$, *then* $\hat{\mu}_c$ *has an explicit representation*

$$\hat{\mu}_c(P_1,P_2) = \int_0^1 c(F_1^{-1}(t), F_2^{-1}(t))dt,$$

where F_i is the d.f. of P_i and F_i^{-1} is the function inverse to F_i. The optimal $P^* \in \mathcal{P}(P_1,P_2)$ has the d.f. $F^*(x,y) = \min(F_1(x), F_2(y))$.

Indication: See the survey Rachev (1984b).

Theorem 5.2. i) *If* (5.8) *holds then* $\overset{\circ}{\mu}_c(P_1,P_2)$ *enjoys a dual representation*

$$\overset{\circ}{\mu}_c(P_1,P_2) = \sup\{|\int f(x)\,(P_1-P_2)\,(dx)| : f\colon S \to \mathbb{R},$$
$$|f(x)-f(y)| \le c(x,y) \text{ for all } x,y \in S\}.$$

(ii) *If* P_1 *and* P_2 *are tight measures then an optimal* $Q^* \in Q(P_1,P_2)$ *exists.*

(iii) *If* $U = \mathbb{R}$, $d(x,y) = |x-y|$ *then*

$$\overset{\circ}{\mu}_c(P_1,P_2) = \int_{-\infty}^{\infty} h(|x-a|)\,|F_1(x)-F_2(x)|\,dx,$$

where F_i *is the d.f. of* P_i.

Indication: Rachev (1984b), Rachev and Shortt (1989).

Let ρ be the following metric in S, $\rho(x,y) = d(x,y)\,/\,(1+d(x,y))$, $x,y \in S$. Then $\hat{\mu}_\rho = \overset{\circ}{\mu}_\rho$ (see Neveu and Dudley (1980)). Moreover, there exists an inequality between $\hat{\mu}_\rho$ and the Prokhorov metric π:

$$\frac{\pi}{1+\pi} \le \hat{\mu}_\rho \le 2\pi \tag{5.9}$$

(see Dudley (1976), Theorem 3.5 and Corollary 18.3). In the next theorem we use the same notations as in Theorems 2.1 – 2.2 and assume $K(t) = th(t)$,

$$\alpha = \sup_{t>s>0} (K(t)-K(s))\,/\,(t-s),\ \ \beta = h(0),\ \ \gamma = \sup_{t>0} h(2t)/h(t).$$

Theorem 5.3 *Let* $\alpha < \infty$, $\beta > 0$, $\gamma < \infty$, $\hat{\mu}_\rho(P_1,P_2)$, $\epsilon > 0$. *Then*

$$\hat{\mu}_c \le [4K(1/\epsilon) + \omega_1(1) + \omega_2(1) + 2k(1)]\,\hat{\mu}_\rho/(1+\hat{\mu}_\rho) + 5\omega_1(\epsilon) + 5\omega_2(\epsilon)\},$$
$$1/4\,\beta\,(\hat{\mu}_p)^2 \le \overset{\circ}{\mu}_c \le \hat{\mu}_c$$

ii) *If* $S = \mathbb{R}$, $d(x,y) = |x-y|$, $c(x,y) = |x-y| \max(1,|x|^{p-1},|y|^{p-1})$, $p \ge 1$,

then $\overset{\circ}{\mu}_c \le \hat{\mu}_c \le p\,\overset{\circ}{\mu}_c$.

Indication: i) See Theorems 2.1, 2.2 and (5.19).

ii) See Theorem 4.1. Q.E.D.

In the nonsymmetric case $c(x,y) = c_0(x,y)$ dual representations for $\hat{\mu}_c$ are given by Kellerer (1984a); explicit solutions in the case $U = \mathbb{R}$, $d(x,y) = |x-y|$ and c satisfying the "Monge condition".

$$c(x',y') - c(x',y) - c(x,y') + c(x,y) \underset{(\le)}{\ge} 0 \text{ for all } x' \ge x,\ y' \ge y,$$

are determined by Cambanis, Simons and Stout (1976). Dual representations for $\overset{\circ}{\mu}_c$ are studied by Levin and Miljutin (1979), Levin (1984a,b), Kemperman (1983). Explicit representations for $\hat{\mu}_c$ are not known except for the trival case $c(x,y) = |x-y|$. Problem 5.3 is not studied in this context.

6. Kantorovich functionals and the problem of the best allocation policy.

Karatzas (1984) (see also the general discussion in Whittle (1982) p. 210–211) considers d "medical treatments" (or "projects" of "investigations") with the state of the jth of them (at time $t \geq 0$), denoted by $x_j(t)$. At each instant of time t, it is allowed to use only one medical treatment denoted by i(t) which then evolves according to some Markovian rule; meanwhile, the states of all other projects remain frozen. If $i(t) = j$, one acquires an instant reward equal to $h(j, x_j(t))$ per unit time, discounted by the factor $e^{-\alpha t}$. The stochastic control problem is then to choose the "allocation policy" $\{i(t), t \geq 0\}$ in such a way as to maximize the expected discounted reward

$$E \int_0^\infty e^{-\alpha t} h(i(t), x_{i(t)}(t))dt.$$

Now, we will consider the situation when allowed to use a combination of different medical treatments (say, for brevity, *medicines*) denoted by $M_1,\ldots,M_d$. Let d = 2 and (S,d) be a s.m.s. The space S may be viewed as the space of patient's parameters. Assume that for i = 1,2 and for any Borel set $A \in \mathscr{B}(S)$ the exact quantity $P_i(A)$ of the medicine M_i (which should be prescribed to the patient with parameters A) is known. Normalizing the total quantity $P_i(S)$ which can be prescribed by 1 we can consider P_i as a probability measure on $\mathscr{B}(S)$. Our aim is to handle an optimal policy of treatments with medicines M_1,M_2 . Such a treatment should be a *combination* of the medicine M_1 and M_2 varying on different sets $A \subset S$.

A policy can be specified by choosing a bounded measure Q on $\mathscr{B}(S\mathrm{x}S)$ and the quantity of medicines M_i on the case of "patient parameter's interval" A_i, i = 1,2, by following the policy Q. The policy may satisfy the balancing condition (5.1) i..e., $Q \in \mathscr{P}(P_1,P_2)$ or (in case of a multistaged treatment) it may satisfy (5.2) i.e., $Q \in \mathcal{Q}(P_1,P_2)$. Let $c(x_1,x_2)$ be the cost of treating the patient with parameters x_i with medicines M_i, i = 1,2. The $\hat{\mu}$ and $\mathring{\mu}$ (see (1.6), (1.7)) represent the minimal total costs under the balancing conditions (5.1) and (5.2) respectively; so, we see that there is a close bond between the problems 5.1 – 5.3 and the problem of the optimal policy.

Acknowledgement

We thank R. Dudley, R. Shortt, D. Ocone and the reviewer for a number of helpful comments.

References:

Anderson, T. W. (1984) *An Introduction to Multivariate Statistical Analysis,* Wiley & Sons, New York

Bickel, P.J. and Freedman, D.A. (1981). Some asymptotic theory for the bootstrap. *Ann. Statist.*, 9, 1196–1217.

Billingsley, P. (1968) *Convergence of Probability Measures*, John Wiley, New York.

Cambanis, S., Simons, G. and Stout, W., (1976) Inequalities for **k(X,Y)** when the marginals are fixed. Z. Wahrsch. verw. Geb. 36, 285–294.

Cuesta J.A. and Matran C. (1989) Notes on the Wasserstein metric in Hilbert spaces. Ann. Probab. Vol. 17, pp. 1264–1278.

De Acosta, A. (1982) Invariance principals in probability for triangle arrays of B–valued random vectors and same Applications, *Ann. Probab.*, 10, 346–373.

D'Aristotile A., Diaconis P. and Freedman D. (1988). On merging of probabilities. *Technical Report* No. 301, *Department of Statistics, Stanford University*.

Dudley, R.M. (1976) *Probability and Metrics.* Lecture Notes Ser. No. 45, Aarhus Univesitet, Aarhus.

Dudley, R.M. (1989) *Real Analysis and Probability.* Wadsworth & Brooks/Cole, California.

Fortet, R., and Mourier, E. (1953) Convergence de la repartition empirique vers la repartition theorique. *Ann. Sci. Ecole Norm. Sup.* 70, 266–285.

Kantorovich. L.V. (1942) On the translocation of masses, *Comptes Rendus (Doklady) de L'Academie des Sciences de L'USSR,* Vol. XXXVII, No. 7–8, 199–201.

Kanotrovich, L.V. and Akilov, G.P. (1977) *Functional Analysis,* Nauka, Moscow (in Russian).

Kantorovich, L.V. and Rubinstein G. Sh. (1958) On the space of completely additive functions, *Vestnik LGU, Ser. Mat., Mekh. i Astron.*, 7/2, 52–59 (in Russian).

Kalashnikov V.V. and Rachev (1988). *Mathematical Methods for Construction for Queueing Models.* Nauka, Moscow, (in Russian). English transl., Wadsworth & Brooks/Cole Advanced Books, to appear, April 1990.

Karatzas, I., (1984) Gittins indices in the dynamic allocation problem for diffusion processes. *Ann. Probab.* 12, 173–192.

Kellerer, H.G. (1984a) Duality theorems and probability metrics, *Proc. 7th Brasov Conf.* 1982, Bukuresti, pp. 211–220.

Kellerer, H.G. (1984b) Duality theorems for marginal problems. Z. Wahrsch. Verw. Geb., 67, 399–432.

Kemperman, J.H.B. (1983) On the role of duality in the theory of moments, *Proc. Semi–infinite Programming and Applications 1981, Lecture Notes in Economics and Math. Systems,* 215, Springer–Verlag, New York 1983, 63–72.

Levin, V.L. and Miljutin, A.S. (1979) The problem of mass transfer with a discontinuous cost function and a mass statement of the duality problem for convex extremal problems. *Russian Math. Surveys*, 34, No. 3, 1–78.

Levin, V.L. (1984a) The problem of mass transfer in a topological space, and probability measures having given marginal measures on the product of two spaces, *Soviet Math. Dokl*, 29, 638–643.

Levin, V.L. (1984b) The mass transfer problem in topological space and probability measures on the product of two spaces with given marginal measures. *Dokl. Akad. Nauk USSR,* 276, 1059–1064.

Neveu, J. and Dudley R.M. (1980). On Kantorovich–Rubinstein theorems. *Preprint.*

Olkin, I. and Pukelsheim, F. (1982). The distance between two random vectors with given dispersion matrices. *Linear Algebra Appl.*, Vol. 43, pp. 257–263.

Rachev, S.T. (1982) Minimal metrics in the minimal variables space. *Pub. Inst. Statist. Univ. Paris*, XXVII, fasc. I. 22–47.

Rachev, S.T. (1984a) On a class of minimal functionals on a space of probability measures. *Theory Prob. Appl.*, 29, 41–49.

Rachev, S.T. (1984b) The Monge–Kantorovich mass transference problem and its stochastic applications. *Theory Prob. Appl.*, 29, 647–676.

Rachev, S.T. and Shortt, R.M. (1989) Classification problem for probability metrics. *Contemporary Mathematics*, 96, 221–262.

Ruschendorf, L. (1985) The Wasserstein distance and approximation theorems. *Z. Wahrsch. verw. Geb.*, 70, 117–129.

Smith, C.S. and Knott, M(1987) Note on the optimal transportation of distributions. *Journal Opt. Th. Appl.*, 52, 323–329.

Whittle, P. (1982) *Optimization Over Time: Dynamic Programming and Stochastic Control*, Wiley, New York.

Zolotarev, V.M. (1976) Metric distances in spaces of random variables and their distributions. *Math. USSR Sbornik*, 30, 3, 373–401.

Zolotarev. V.M. (1983) Probability metrics. *Theory Prob. Appl.*, 28, 278–302.

PARTIALLY PARALLEL SIMULATED ANNEALING: LOW AND HIGH TEMPERATURE APPROACH OF THE INVARIANTE MEASURE

A. Trouvé
LMENS-DIAM, Ecole Normale Supérieure
45,rue d'Ulm
75230 Paris Cedex 05

Abstract

In this paper, we consider parallelization of simulated annealing for a product configuration space L^S. At every step of the algorithm, each site is activated with probability τ and all the activated sites update synchronously their value. Concerning reversibility conditions and behaviour in the low temperature region, we show that the fully parallel algorithm ($\tau = 1$) is a quite singular case and we prove that for the 2-D Ising Model the invariant probability measure at constant temperature T converges to the uniform probability measure on the two global minima when $T \to 0$ iff $\tau < 1$.

The high temperature region is studied for $S = \mathbb{Z}$ and translation invariant potential. We construct a sequence of approximations converging towards the invariant probability measure which can be derived from an implementable algorithm and we show that this sequence may be useful to compute valuable approximations of such statistical quantities as the mean energy

1. Introduction

Simulated annealing is now currently used in many practical optimization problems (cf [12],[7]) and theoretical aspects have been extensively studied during the second part of the last decade (cf [2],[3], [7],[8],[9],[10],[11], [16]). We know that the relaxation time increases exponentially with the inverse of the temperature so that a huge amount of computer time is needed for large scale optimization problems. However, parallelization techniques may succeed in reducing drastically this computer time in many applications ([1]). Some parallelization schemes do not affect the sequential scheme so that the convergence is guaranteed. However, especially in image processing, it can be very attractive to use massive parallel schemes where each pixel is attached to an elementary processor locally connected with a set of neighbours. Due to the interactions between sites, the convergence towards global minima of such algorithms is not established and many new problems arise when considering the underlying Markov chain. The article is organized as following :

The mathematical description of the algorithm is given in the next section in order to define the invariant probability measure of the Markov chain at constant temperature.

Then, the third section deals with low temperature properties of the invariant probability measure. In a first part, we study the reversibility conditions given by Kozlov and Vasilyev [13] and we show that roughly speaking, the fully parallel algorithm ($\tau = 1$) is the only interesting case of reversibility. In a second part, we use the graphs of Wentzell and Freidlin for studying the configurations loaded at low temperature. In particular, we prove that for the 2-D Ising model, the parallel ground states are the two standard "black" and "white" configurations as soon as the parallelization rate τ is strictly less than 1.

Finally, in the last section, we derive in the restricted 1-D framework, an implementable algorithm to approximate the invariant measure for an infinite system with translation invariant potential in the so called high noise regime. Some numerical results for the one-dimensional Ising model are given. This work is part of a doctoral thesis preprared at DIAM-LMENS under the supervision of Robert Azencott.

2. The class of Markov chains underlying the parallel algorithm

Let S and L be two finite sets and define $Y = L^S$. The set Y will be the state space for the class of algorithms to be studied. Let H be a real valued function on Y. This function will be the energy to be minimized. For $x \in Y$, $i \in S$ and $l \in L$, we will note $x^{i,l}$ the configuration defined by $x_k^{i,l} = x_k$ if $k \neq i$ and $x_i^{i,l} = l$.

Definition 2.1 *Let T be a strictly positive real valued number and τ be in the interval $]0,1]$. We denote by $Q_{T,\tau}$ the Markov kernel on Y defined by:*

$$Q_{T,\tau}(x,y) = \prod_{i \in S} (\tau q_{i,T}(y_i|x) + (1-\tau) \mathbb{1}_{x_i = y_i}), \tag{1}$$

$$where \quad q_{i,T}(y_i|x) = \frac{exp(-H(x^{i,y_i})/T)}{\sum_{l \in L} exp(-H(x^{i,l})/T)}.$$

The kernel $Q_{T,\tau}$ will define one step of the parallel algorithm whereas the one-site kernels $q_{i,T}$ define the well-known sequential dynamic. The complete algorithm will be defined after choosing a sequence of temperatures as a Markov chain on Y.

Definition 2.2 *Let $\mathcal{T} = (T_n)_{n \in N}$ be a decreasing sequence of strictly positive numbers, $\tau \in]0,1]$, and let μ be a probability on Y called the initial probability. We denote by $P^{\mathcal{T},\tau,\mu}$, the unique probability measure on Y^N such that the coordinate process $(X_n)_{n \in N}$ is a Markov chain satisfying:*

- $X_0 \circ P^{\mathcal{T},\tau,\mu} = \mu$,
- $P^{\mathcal{T},\tau,\mu}(X_{n+1} = y \mid X_n = x) = Q_{T_n,\tau}(x,y)$.

The sequence $\mathcal{T}$ is usually called the cooling schedule. The measure $P^{\mathcal{T},\tau,\mu}$ is the natural object to be considered for the following parallelization of the sequential annealing : instead of activating only one site $i \in S$ at each step of the classical algorithm, each site is activated independently with probability τ and all the activated sites update their value synchronously according to the usual sequential dynamic. The parameter τ controls the parallelization rate so that the mean value of the number of activated site is $|S|\tau$. We can expect *a priori* a high speed-up as far as the parallel algorithm converges towards the global minima. However, under $P^{T,\tau,\mu}$, the Markov chain $(X_n)_{n\in\mathbb{N}}$ differs from the sequential annealing. We define now the invariant probability measure of the Markov chain at constant temperature ($(X_n)_{n\in\mathbb{N}}$ is then an homogeneous Markov chain).

Definition 2.3 *Let T be a strictly positive real valued number and let $\tau\in]0,1]$. We denote by $\mu_{T,\tau}$ the unique invariant probability measure of the irreducible and aperiodic kernel $Q_{T,\tau}$.*

Note that $Q_{T,\tau}$ tends to the singular Markov kernel $Q_{T,0}(x,y) = \mathbf{1}_{x=y}$ when τ tends to zero. Consider the renormalized kernel $\tilde{Q}_{T,\tau}$ defined by:

$$\tilde{Q}_{T,\tau}(x,y) = (Q_{T,\tau}(x,y) - (1-\tau)^{|S|}\mathbf{1}_{x=y})/(1-(1-\tau)^{|S|})\ ,\tau\in]0,1] \tag{2}$$

An easy computation shows that

- $\mu_{T,\tau}\tilde{Q}_{T,\tau} = \mu_{T,\tau}$
- $\lim_{\tau\to 0}\tilde{Q}_{T,\tau}(x,y) = \tilde{Q}_{T,0}(x,y)$ where $\tilde{Q}_{T,0}(x,y) = 1/|S|\sum_{i\in S} q_{i,T}(y_i|x\ )$

The kernel $\tilde{Q}_{T,0}$ is obviously the irreducible one step-transition matrix of the sequential Gibbs sampler at temperature T with random choice of the activated site. Then, defining by $\mu_{T,0}$ the Gibbs probability measure:

$$\mu_{T,0}(x) = \frac{exp(-H(x)/T)}{Z}, \tag{3}$$

we have $\mu_{T,0}\tilde{Q}_{T,0} = \mu_{T,0}$ and we recall here two basic facts proved in a more general setting in [18]:

1. Let $x\in Y$, the function $(T,\tau)\to\mu_{T,\tau}(x)$ is continuous on $\mathbb{R}_+^*\times[0,1]$.
2. For each $\tau\in[0,1]$, there exists a probability measure $\mu_{0,\tau}$ such that:

$$\lim_{\tau\to 0}||\mu_{T,\tau}-\mu_{0,\tau}||_{Var} = 0.$$

Assume now, that the cooling schedule $\mathcal{T} = (T_n)_{n\in\mathbb{N}}$ is decreasing to zero and that $T_n\geq K/\ln(n)$ for $n\geq n_0$. Then it can be proved (cf [3],[11] and [17]) that for a sufficiently large constant K, the distribution of X_n at time n tends to $\mu_{0,T}$. Hence, one of the main issues for us is to study the configurations $x\in Y$ for which $\mu_{T,\tau}(x)$ does not vanish when T tends to zero, that is the support of the limit measure $\mu_{0,\tau}$.

3. Study of the invariant measure at low temperature

3.1 Reversibility case

We say that $Q_{T,\tau}$ is reversible iff for all $x,y\in Y$ we have:

$$\mu_{T,\tau}(x)Q_{T,\tau}(x,y) = \mu_{T,\tau}(y)Q_{T,\tau}(y,x). \tag{4}$$

The reversibility property is well-known to be a very pleasant setting to study the $\mu_{T,\tau}$. For an arbitrary H, $Q_{T,\tau}$ is not reversible. Necessary and sufficient conditions for reversibility are given in [13] and will allow us to study the reversibility problem in the framework of our parallel algorithms. Let us first give the decomposition of H as a sum of "clique" potentials:

$$H(x) = \sum_{C \subset S} u_C(x_C), \tag{5}$$

where $u_C : L^C \to \mathbb{R}$ and $x_C = (x_i)_{i \in C}$. The family (u_C) is uniquely determined if we impose the so-called "vacuum condition" that is $u_C(x_C) = 0$ if $x_i = l_0$ for some $i \in C$ where l_0 is a fixed element of L. We define then the set of "cliques" by $\mathcal{C} = \{C \subset S \mid u_C \neq 0\}$ and for each $i \in S$ the set $N(i)$ of the neighbours of i by :

$$N(i) = \{ j \in S \mid \exists C \in \mathcal{C}, i \in C \text{ so that } j \in C \text{ and } i \neq j\}. \tag{6}$$

Theorem 3.4 *Let H be defined by (5)*

1. *Let $\tau = 1$,*
   (a) *then $Q_{T,1}$ is reversible iff H contains only pairwise interactions, i.e. each cliques C contains at most two elements.*
   (b) *if $Q_{T,1}$ is reversible then:*

$$\mu_{T,1}(x) = \left\{ \prod_{i \in S} \left[\exp(-u_{\{i\}}(x_i)/T) \sum_{l \in L} \exp(- \sum_{j \in N(i)} (u_{\{i\}}(l) + u_{\{i,j\}}(l, xj)/T)) \right] \right\} /Z. \tag{7}$$

2. *Let $\tau \in]0,1[$ and $|L| = 2$, then a necessary condition for $Q_{T,\tau}$ to be reversible is that each site $i \in S$ has at most one neighbour.*

Proof: Part 1. is a straightforward application of Theorem 1 in [13] . Since part 2. is a bit technical, the theorem will be proved in the appendix.□

The theorem shows that except in very degenerate cases, the kernel $Q_{T,\tau}$ is never reversible as soon as $\tau < 1$, so that $\tau = 1$ and pairwise interactions is roughly speaking the only interesting case of reversibility for our parallel algorithm. If we restrict to this case, we can try to use the explicit expression (7) to study the configurations visited at low temperature.

For a moment, we assume that $\tau = 1$ and that the cliques of H have at most two elements. Then we deduce from (7) that:

$$\lim_{T \to 0} T \ln \mu_{T,1}(x) = -(H_p(x) - \inf H_p), \tag{8}$$

with

$$H_p(x) = \sum_{i \in S} \inf_{l \in L} \left(u_{\{i\}}(l) + \sum_{j \in N(i)} u_{\{i,j\}}(l, x_j) \right) + u_{\{i\}}(x_i). \tag{9}$$

The function H_p can be interpreted as a parallel energy associated with the parallel algorithm with $\tau = 1$. Furthermore, for each local minimum x (i.e $H(x^{j,l}) \geq H(x), j \in S$ and $l \in L$) we have $H_p(x) = 2H(x)$. This gives us the corollary :

Corollary 3.5 *Assume that H defined by (5) has no clique with more than two elements and let H_p be the function defined by (9). Then, let x be a global minima of H. We have*

$$\lim_{T \to 0} \mu_{T,1}(x) = \mu_{0,1}(x) \neq 0 \text{ iff } \inf H_p = 2 \inf H.$$

This necessary and sufficient condition may be useful in practical cases to establish the convergence of the parallel algorithm with $\tau = 1$ towards global minima. A surprising fact is that the unoriented graph $G = \{\{i,j\} \mid i \in S, j \in N(i)\}$ plays an important role. In particular, if G does not contain any cycle with an odd number of edges (that is the chromatic number of G is 2) then $\inf H_p = 2 \inf H$. In the remaining case, we generally have $\inf H_p < 2 \inf H$ (see [6] for a study on spin glass like energies).
We come back now to the general case.

3.2 General case and Wentzell and Freidlin graphs

Since we have no explicit expression for $\mu_{T,\tau}$, the study of the invariant measure may be achieved with the ideas of Wentzell and Freidlin. For small values of the temperature, the behaviour of $\mu_{T,\tau}$ is controlled by:

$$V_\tau(x,y) = -\lim_{T\to 0} T \ln \tilde{Q}_{T,\tau}(x,y) \ , \ \tau \in [0,1], x \neq y. \tag{10}$$

We briefly report here the powerful graphs they introduce in [19]. A graph on Y will be a set of arrows $x \to y$ with vertices in Y. For $R \subset Y$, we denote by $G(R)$ the set of all the graphs g on Y satisfying:

- For each $y \in Y \setminus R$, there exits a unique arrow in g starting at x.
- For each $y \in R$, there is no arrow in g starting at x.

Now, for each g in G(R), we note $V_\tau(g) = \sum_{x\to y \in g} V_\tau(x,y)$ and we define:

$$W_\tau(x) = \inf_{g \in G(\{x\})} V_\tau(g) \text{ and } W_\tau^* = \inf_{x\in Y} W_\tau(x). \tag{11}$$

The family $(W_\tau(x))_{x\in Y}$ is related with the asymptotic behaviour of $\mu_{T,\tau}$ by the relation:

$$\lim_{T\to 0} T \ln \mu_{T,\tau}(x) = -(W_\tau(x) - W_\tau^*). \tag{12}$$

The numbers $V_\tau(x,y)$ can be easily computed from the energy H. Let us define for all $x \in Y$, $l \in L$ and $i \in S$:

$$\Delta_i^* H(x,l) = H(x^{i,l}) - \inf_{l' \in L} H(x^{i,l'}). \tag{13}$$

Proposition 3.6 *Let $x, y \in Y$, $x \neq y$. Then, we have:*

- $V_1(x,y) = \sum_{i\in S} \Delta_i^* H(x,y_i)$,
- $V_\tau(x,y) = \sum_{i\in S} \Delta_i^* H(x,y_i) I_{x_i \neq y_i}$; $0 < \tau < 1$,
- $V_0(x,y) = \Delta_i^* H(x,y_i)$ *if* $y = x^{i,y_i}$ *for some* $i \in S$ *and* $+\infty$ *in the remaining cases.*

Now, define $Y_\tau = \{x \in Y \mid \lim_{T\to 0} \mu_{T,\tau}(x) = \mu_{0,\tau}(x) \neq 0\}$. It is proved by Wentzell and Freidlin in [19] that $Y_\tau = \{x \in Y \mid W_\tau(x) = W_\tau^*\}$. However, Proposition 3.6 shows that $V_\tau(x,y)$ does not depend on τ for $\tau \in]0,1[$. This implies that if Y_{τ_0} contains some global minima for some value $\tau_0 \in]0,1[$, then this will be true for all the values of the parameter in the open interval. Conversely, if for some $\tau_0 \in]0,1[$, Y_{τ_0} does not contain any global minimum, then a change of the parameter τ within the open interval will not remove this drawback. Unfortunately, even if $V_\tau(x,y)$ has a quite simple expression, we cannot compute $W_\tau(x)$. A more tractable issue is to look for a transformation $\phi_{x,y}$ from $G(\{x\})$ to $G(\{y\})$ such that if $g \in G(\{x\})$ then $W_\tau(\phi_{x,y}(g)) \leq W_\tau(g)$. This is the idea of the following proposition. First, for all $x \in Y$, we define:

$$V_\tau(x) = \inf_{y \neq x} V_\tau(x,y). \tag{14}$$

Following [11], we define on Y the preorder $\preceq$ by $x \preceq y$ iff $x = y$ or there exist $n \in \mathbb{N}$ and a path $p : \{1, \cdots, n\} \to Y$ of length n such that $p(1) = y$, $p(n) = x$ and $V_\tau(p(i), p(i+1)) = V_\tau(p(i))$ for all $1 \leq i \leq n$.

Proposition 3.7 *If $x \preceq y$ and $V_\tau(x) > V_\tau(y)$ then we have:*

$$W_\tau(x) < W_\tau(y).$$

Proof: Let $x, y \in Y$ such that $x \preceq y$. Now, consider a path of length n such that $p(1) = y$, $p(n) = x$ and $V_\tau(p(i), p(i+1)) = V_\tau(p(i))$. Let g be an arbitrary graph in $G(\{y\})$. We define by induction the family $(g_k)_{1 \leq k \leq n}$ by:

- $g_1 = g$,
- $g_{k+1} = \{u \to v \mid u \to v \in g_k \text{ and } u \neq p(k+1)\ \} \cup \{\ p(k) \to p(k+1)\ \}$.

One easily see that $g_k \in G(\{p(k)\})$ for $1 \leq k \leq n$ and that:

$$V_\tau(g_{k+1}) - V_\tau(g_k) = V_\tau(p(k), p(k+1)) - V_\tau(p(k+1), a_{k+1}), \tag{15}$$

where a_{k+1} is the ending point of the arrow starting at $p(k+1)$ in g_k. From (15) we deduce that:

$$\begin{aligned} V_\tau(g_n) - V_\tau(g) &= \sum_{k=1}^{n-1} V_\tau(p(k), p(k+1)) - V_\tau(p(k+1), a_{k+1}) & (16)\\ &= V_\tau(p(1), p(2)) + \sum_{k=2,n-1} V_\tau(p(k), p(k+1)) & (17)\\ &\quad -V_\tau(p(k), a_k) - V_\tau(p(n), a_n). & (18) \end{aligned}$$

The value of central summation is negative, so that $V_\tau(g_n) - V_\tau(g) \leq V_\tau(y) - V_\tau(x)$. Under the hypothesis of the proposition, we deduce that $V_\tau(g_n) - V_\tau(g) < 0$. Since $g_n \in G(\{x\})$ the proposition is proved. □

We can immediately deduce an important corollary. We say that a local minimum x is strict if for all $y = x^{i,l}, y \neq x$ we have $H(y) > H(x)$.

Theorem 3.8 *Assume that each local minimum is strict, then for $\tau < 1$ we have :*

$$\textit{if } \lim_{T \to 0} \mu_{T,\tau}(x) = \mu_{0,\tau}(x) \neq 0\ , \textit{ then } x \textit{ is a local minimum}.$$

Proof : Assume that y is not a local minimum; then there exists a path $p : \{1, \cdots, n\} \to Y$ such that $p(1) = y$, $p(n) = x$ is a local minimum, and for all $1 \le i < n$, the two configurations $p(i)$ and $p(i+1)$ differ only on one site and verify $H(p(i)) \ge H(p(i+1))$. One easily deduces that $x \preceq y$ and $V_\tau(y) = 0$. Furthermore, since all the local minima are strict, we have $V_\tau(x) > 0$ so that Proposition 3.7 gives that $W_\tau(x) < W_\tau(y)$.□

The theorem cannot be extended to the case $\tau = 1$. Furthermore, one can construct examples where the global minima are not in the support of $\mu_{0,\tau}$. However, for the simple 2-D Ising Model on a square lattice, the proposition 3.7 gives us a complete answer. We say that H is a 2-D Ising energy on a square lattice if $L = \{-1, 1\}$, $S = \{1, \cdots, n\} \times \{1, \cdots, n\}$ and H is defined by:

$$H(x) = -J \sum_{i,j \in S, |i-j|=1} x_i x_j \; ; \; J > 0 \tag{19}$$

where $|i - j| = |i_1 - j_1| + |i_2 - j_2|$ for $i = (i_1, i_2)$ and $j = (j_1, j_2)$.

Theorem 3.9 *Assume that H is an Ising energy defined by (19) and that $\tau < 1$. Then we have,*

$$\mu_{0,\tau} = 1/2(\delta_{x_+} + \delta_{x_-}),$$

where x_+ (resp. x_-) is the configuration where $x_i = 1$ (resp. $x_i = -1$) for all $i \in S$.

Proof: a completely rigorous proof would be very long even if the ideas are very simple, so that we will report here only the main steps.
Step1 : It is well known that the strict local minima of H are the alternate "black" and "white" strips of width more than 2. One computes that if x is a strict local but not global minimum, then $V_\tau(x) = 1$ (we can exit from x by flipping a site in a corner of a strip). Moreover, we have $V_\tau(x_+) = V_\tau(x_-) = 2$
Step 2 : One can show that if x is a strict local but not global minimum, then we have $x_+ \preceq x$ or $x_- \preceq x$ (we go from x to one of the global minima by a sequential path where the strips are removed columns by columns). Hence we deduce from Proposition 3.7 that $W_\tau(x) > W_\tau(x_+) = W_\tau(x_-)$.
Step 3 : Finally, if x is a configuration which is not a local minimum, then we have $V_\tau(x) = 0$ and there exists a strict local minimum y such that $y \preceq x$. This gives with Proposition 3.7 that $W_\tau(x) > W_\tau(y)$.
Step 4 : From the previous steps, we deduce that if x is not a global minimum, then $W_\tau(x) > W_\tau(x_+)$ so that the theorem is proved.□
This result has been proved independently by A. Ferrari, A. Frigessi and H. Schonmann in [5] using different arguments from renewal process theory. They give also a very beautiful proof of the convergence of $\mu_{T,\tau}$ when T vanishes towards the uniform measure on the global minima for spin glass like energies in finite one dimensional volume.
The case $\tau = 1$ is well known for the Ising energy : The limit measure is concentrated on four configurations which are the two global minima and the two checkerboard configurations.

4. High temperature study in the one-dimensional case

In the preceding section, we have put the emphasis on the dichotomy at low temperature of the two domains $\tau < 1$ and $\tau = 1$. However, for non-zero temperature, the invariant probability measure evolves continuously in the parameter τ but this continuity cannot be handled by the asymptotical computations given above. The high temperature behaviour can be studied by numerical simulations as in [6], which require however large amount of CPU time on a sequential computer, especially for low values of τ. Therefore, we would like to develop another approach, still numerical, but which should be more efficient than the traditional Gibbs sampler in the restricted context of one-dimensional translation-invariant energies and which is suited for dealing with the infinite-volume limit.

4.1 Extension to the infinite-volume limit

In order to deal with the infinite-volume limit, we shall consider only translation invariant energies on the new configuration space $Y = L^{\mathbb{Z}}$.

Definition 4.10 *We say that $U = (U_A)_{A\in\Delta}$ is a translation-invariant potential on Y with finite radius r, if there exists a finite family $\mathcal{A} = (A_i)_{1\leq i\leq k}$ of finite subsets of $\mathbb{Z}$ such that $\Delta = \{ A_i + t | 1 \leq i \leq k, t \in \mathbb{Z} \}$ and $U_A : L^A \to \mathbb{R}$ verifies $U_A(x_A) = U_{A+t}(x_{A+t})$. Furthermore, we impose that $diam(A_i) \leq r$ for $1 \leq i \leq k$.*

Let $U = (U_A)_{A\in\Delta}$ be a translation-invariant potential on Y with finite radius of interaction r_0. The definition of the parallel simulated annealing on a infinite number of sites as a Markov chain on Y presents no special difficulty. For each $T > 0$, we define the single site dynamic by the following family of kernels:

$$q_{t,T}(l|x) = \exp\left(-\sum_{A:t\in A} U_A(x_A^{t,l})/T\right)/Z_t(x) \; ; \; t \in \mathbb{Z}, \tag{20}$$

and as usual, the complete dynamic is given by the kernel $Q_{T,\tau}$ on Y defined on the cylinder sets by:

$$Q_{T,\tau}(x, C) = \prod_{t\in V}(\tau q_{t,T}(C_t|x) + (1-\tau)\mathbf{1}_{x_t\in C_t}), \tag{21}$$

where V is a finite set and $C = \{ x \in Y \mid x_t \in C_t \text{ for } t \in V \}$. The kernel $Q_{T,\tau}$ with an infinite set of sites is no longer ergodic for all the value of (T, τ) so that we will now introduce the Dobrushin coefficient :

$$\delta_{T,\tau} = \sum_{t\in\mathbb{Z}} \sup_{x\in Y, l\in l} 1/2 \sum_{l'\in L} |q_{0,T}(l'|x) - q_{0,T}(l'|x^{t,l})|. \tag{22}$$

Following [14], we say that $Q_{T,\tau}$ satisfies the high noise regime (H.N.R) if :

$$\delta_{T,\tau} < 1 \quad \text{H.N.R} . \tag{23}$$

Under condition (23), $Q_{T,\tau}$ is exponentially ergodic (cf [15]) so that we note $\mu_{T,\tau}$ the unique invariant probability measure.

4.2 The approximation scheme

We introduce now an increasing sequence of subspaces of the translation-invariant probability measures on Y which will be useful for defining our approximation of $\mu_{T,\tau}$.

Notation 4.11 *Let $N \in \mathbb{N}^*$. We denote by $\mathcal{G}_s^N(Y)$ the set of all the Gibbs states on Y associated with a translation invariant potential with radius of interaction less than $N+1$.*

In a general framework, the invariant probability measure $\mu_{T,\tau}$ does not belong to $\mathcal{G}_s^N(Y)$ for any $N \in \mathbb{N}^*$. However, we can try to look for an approximating probability measure of $\mu_{T,\tau}$ in $\mathcal{G}_s^N(Y)$, and this will be the underlying idea of the approximation scheme. The first step is given by the following proposition, which shows that $\mathcal{G}_s^N(Y)$ can be identified with the set of Markov chains of order N.

Proposition 4.12 *Let $\mathcal{M}_s(Y)$ be the set of translation-invariant probability measures on Y and let $N \in \mathbb{N}^*$. We denote by $\mathcal{M}_s^N(Y)$ the set of probability measures $\nu \in \mathcal{M}_s(Y)$ which are strictly positive on the cylinder sets and which satisfy :*

$$\nu(x_N \mid (x_t); t < N) = \nu(x_N \mid (x_t); 0 \leq t < N) \;\; \nu \text{ a.s.} \tag{24}$$

Then we have,

$$\mathcal{G}_s^N(Y) = \mathcal{M}_s^N(Y) \tag{25}$$

Proof: The proof of the proposition is straightforward □

Let $\nu \in \mathcal{G}_s^N(Y)$. Since we are working on $\mathbb{Z}$, the unique Gibbs state associated with a given translation-invariant potential is a translation-invariant probability measure so that $\nu \in \mathcal{M}_s(Y)$. Furthermore, since ν is a Gibbs state, ν is strictly positive on the cylinder sets. In addition, using again the one-dimensional framework and choosing a fixed boundary condition $y \in Y$, we have :

$$\nu(\, x_N \mid (x_t);\; -p \leq t < N\,) = \lim_{k \to \infty} \frac{\nu(\,(x_t);\; -p \leq t \leq N \mid (y_s);\; |s| \geq k\,)}{\nu(\,(x_t);\; -p \leq t < N \mid (y_s);\; |s| \geq k\,)} \tag{26}$$

However, for fixed $k \geq p$, the right hand member of the previous equality can be explicitly obtained using the potential associated with ν. A straightforward computation shows that the result depends only on (x_t); $0 \leq t < N$ so that we have proved :

$$\nu(\, x_N \mid (x_t);\; -p \leq t < N\,) = \nu(\, x_N \mid (x_t);\; 0 \leq t < N\,)$$

A standard martingale argument gives that :

$$\nu(\, x_N \mid (x_t);\; t < N\,) = \nu(x_N \mid (x_t)\;\; 0 \leq t < N\,)$$

Hence $\nu \in \mathcal{M}_s^N(Y)$.
Conversely, assume that $\nu \in \mathcal{M}_s^N(Y)$, then for $p > N$ we have :

$$\begin{aligned}
\nu((x_t);\; |t| \leq p\,) &= \nu((x_t);\; -p \leq t < -p+N\,) \prod_{k=-p+N}^{p} \nu(x_k \mid (x_s);\; k-N \leq s < k\,) \\
&= f_p((x_t); -p \leq t \leq p \text{ and } t \neq 0\,) g_p((x_t); -N \leq t \leq N\,)
\end{aligned}$$

where the last equality has been obtained after regrouping the terms involving x_0. Hence we deduce that :

$$\nu(\, x_0 \mid (x_t); |t| \leq p \text{ and } t \neq 0\,) = \nu(\, x_0 \mid (x_t);\; |t| \leq N \text{ and } t \neq 0\,)$$

Since ν is strictly positive on the cylinder sets, the previous conditional probability is strictly positive and this is sufficient to prove that $\nu \in \mathcal{G}_s^N(Y)$ so that the proposition is proved.□

We can deduce from the previous proposition that if $\nu \in \mathcal{G}_s^N(Y)$ then the coordinate process on $Y = L^{\mathbb{Z}}$ is, under ν, a doubly infinite stationary Markov chain of order N. Hence, an element of $\mathcal{G}_s^N(Y)$ is completely defined by its marginal on N+1 consecutive sites so that we can define a mapping $\phi_N : \mathcal{M}_s(Y) \to \mathcal{G}_s^N(Y)$ by :

$$\phi_N(\mu) = \nu \text{ if } \nu_{\Lambda_N} = (\mu Q_{T,\tau})_{\Lambda_N} \tag{27}$$

where $\Lambda_N = \{0, 1, \cdots, N\}$. We have denoted by ν_{Λ_N} and $(\mu Q_{T,\tau})_{\Lambda_N}$ the corresponding marginal distributions on Λ_N.
From the relation (27) we deduce that if $\mu_{T,\tau} \in \mathcal{G}_s^N(Y)$ for some $N \in \mathbb{N}^*$, then $\mu_{T,\tau}$ is a fixed point of ϕ_N. In the general case, this is no longer true but the existence of at least one fixed point is given by the Schauder's fixed point theorem. Such a fixed point will be called an approximation at the order N of $\mu_{T,\tau}$. We can now state our approximation theorem.
For each function $f : Y \to \mathbb{R}$ we note $\Delta(f) = \sum_{t \in \mathbb{Z}} \Delta_t(f)$ where:

$$\Delta_t(f) = \sup_{x \in Y, l \in L} |f(x) - f(x^{t,l})|.$$

Theorem 4.13 *For each $N \in N^*$ let $\nu_N \in \mathcal{G}_s^N(Y)$ be a fixed point (a priori not unique) of ϕ_N. Assume that $\delta_{T,\tau} < 1$ and let $f : Y \to \mathbb{R}$ be a function depending only on p consecutive coordinates. Then we have,*

$$|\mu_{T,\tau} f - \nu_N f| \leq (\delta_{T,\tau})^k \Delta(f) \; ; \; N \geq p, \tag{28}$$

where k is the integer part of $1 + \frac{N-p}{2(r_0-1)}$.

Before giving the proof we should notice that theorem 4.13 proves that the sequence of fixed points $(\nu_N)_{N \in \mathbb{N}}$ converges to $\mu_{T,\tau}$ exponentially fast on the finite marginal distributions.
Proof of Theorem 4.13 : Since r_0 is the radius of interaction , for any finite set $V \subset \mathbb{Z}$ and any $\nu \in \mathcal{M}_s(Y)$, the marginal distribution $(\nu Q_{T,\tau})_V$ depends only on $\nu_{\tilde{V}}$ where :

$$\tilde{V} = \{\, t \in \mathbb{Z} \mid |t - i| \leq r_0 - 1 \text{ for some } i \in V \,\}. \tag{29}$$

Now, let $N \in \mathbb{N}^*$, let k_0 be the integer part of $N/(2(r_0 - 1))$ and let $\lambda \in \mathcal{M}_s(Y)$ such that :

$$(\lambda Q_{T,\tau})_{\Lambda_N} = \lambda_{\Lambda_N}. \tag{30}$$

We define for $0 \leq m \leq k_0$ the set $\Lambda_N^m = \{m(r_0 - 1), \cdots, N - m(r_0 - 1)\}$ and one can easily prove by induction that :

$$(\lambda Q_{T,\tau}^{m+1})_{\Lambda_N^m} = \lambda_{\Lambda_N^m}. \tag{31}$$

Let $f : Y \to \mathbb{R}$ be a function depending only on p consecutive coordinates and let k be the integer part of $1 + (N - p)/(2(r_0 - 1))$. As far as translation-invariant measures are concerned, we can assume that $f(x)$ depends only on $(x_t)_{t \in \Lambda_N^{k-1}}$ ($|\Lambda_N^{k-1}| \geq p$). Then from (31) we deduce that:

$$\lambda f = \lambda Q_{T,\tau}^k f. \tag{32}$$

Since $\phi_N(\nu_N) = \nu_N$ we can apply (32) for $\lambda = \nu_N$ so that :

$$(\nu_N - \mu_{T,\tau})f = (\nu_N - \mu_{T,\tau})Q^k_{T,\tau}f. \tag{33}$$

Furthermore, it is known (cf lemma 3 in [14]) that :

$$\Delta(Q_{T,\tau}f) \leq \delta_{T,\tau}\Delta(f). \tag{34}$$

Since $|(\nu_n - \mu_{T,\tau})Q^k_{T,\tau}f| \leq \Delta(Q^k_{T,\tau}f)$, we deduce the inequality :

$$|(\nu_N - \mu_{T,\tau})f| \leq \Delta(Q^k_{T,\tau}f) \leq (\delta_{T,\tau})^k\Delta(f). \tag{35}$$

Hence the theorem is proved.□

4.3 Numerical derivation of the approximating sequence

In the preceding subsection, we have built a sequence of Gibbs states of increasing order, which converges to $\mu_{T,\tau}$. We will show now that this approximating sequence of fixed points $(\nu_N)_{N\in\mathbb{N}}$ can be effectively computed by an implementable algorithm. The main point here is that the elements of $\mathcal{G}^N_s(Y)$ are completely defined by their marginal distributions on N+1 consecutive sites (see proposition 4.12), and that the potential U has a finite radius of interaction r_0.

Notation 4.14 *Let V be a finite subset of $\mathbb{Z}$.*

1. *We denote by $\tilde{V}$ the set:*

$$\tilde{V} = \{\ t \in \mathbb{Z}\ |\ |t-i| \leq r_0 - 1 \textit{ for some } i \in V\ \}.$$

2. *We denote by M_V (resp. $M_{\tilde{V}}$) the set of all the probability measures on L^V (resp. $L^{\tilde{V}}$).*

3. *For all $T \in \mathbb{R}^*_+$ and $\tau \in]0,1]$, we denote by $Q^V_{T,\tau}$ the Markov kernel on $L^{\tilde{V}}$ defined by :*

$$Q^V_{T,\tau}(x,y) = \prod_{t\in V} q_{t,T}(y|x_{N(t)})\ ;\ x,y \in L^{\tilde{V}}.$$

Now, for all $N \in \mathbb{N}^*$, let $\Lambda_N = \{0,\cdots,N\}$. We can define a mapping $\psi_N : M_{\tilde{\Lambda}_N} \to M_{\tilde{\Lambda}_N}$ by $\psi_N(\mu) = \nu$ iff:

$$\nu((x_j); j \in \tilde{\Lambda}_N) = (\prod_{k=-(r_0-1)}^{r_0-1} \lambda((x_{j+k}); j \in \Lambda_N))/(\prod_{k=-(r_0-1)}^{r_0-2} \lambda((x_{j+k}); j \in \Lambda^*_N)), \tag{36}$$

where $\Lambda^*_N = \Lambda_N \setminus \{0\}$ and $\lambda = \mu Q^{\tilde{\Lambda}_N}_{T,\tau}$.
Given $\mu \in M_{\tilde{\Lambda}_N}$, one can compute, using (36), the value of $\psi_N(\mu)$. Both mappings ϕ_N and ψ_N are related by the following relation:

$$\psi_N(\mu_{\tilde{\Lambda}_N}) = \phi_N(\mu)_{\tilde{\Lambda}_N}; \mu \in \mathcal{M}_s(Y). \tag{37}$$

We deduce from (37) that if ν_N is a fixed point of ϕ_N then $\lambda_N = (\nu_N)_{\tilde{\Lambda}_N}$ is a fixed point of ψ_N. Hence it is sufficient to find the fixed points of ψ_N which is a finite dimensional problem. This can be achieved by an iterative algorithm starting from an arbitrary element of $M_{\tilde{\Lambda}_N}$. However, the convergence is not a priori guaranteed but the following theorem shows that ψ_N is Lipschitz with a Lipschitz coefficient depending on the temperature.

Theorem 4.15 *Let $N \in I\!N^*$ and let ψ_N be defined by (36). Then, we have:*

$$||\psi_N(\mu) - \psi_N(\nu)||_{\tilde{\Lambda}_N} \leq \delta_{T,\tau}(N+1)(2r_0 - 1)||\mu - \nu||_{\tilde{\Lambda}_N} \ ; \ \mu, \nu \in M_{\tilde{\Lambda}_N}, \tag{38}$$

where $|| \ ||$ is the total variation norm.

The proof of this theorem is given in the appendix. Since $\lim_{T \to \infty} \delta_{\tau,T} = 0$, the Lipschitz coefficient is strictly less than one for a sufficiently high temperature so that the fixed points can be computed by a standard iterative method. Hence we deduce from theorem 4.13 a new way to study the influence of the parameter τ on some statistical quantities such as the mean value of energy per spin.

4.4 Plotted results for 1D Ising model

We have used the previous approach for 1D Ising model given by the translation invariant potential defined on $\{-1,1\}^{\mathbb{Z}}$ by (see definition 4.10 for the notations):

$$\mathcal{A} = (\{0,1\}) \text{ and } U_{0,1}(x_0, x_1) = -x_0 x_1.$$

For this model, one can easily compute that:

$$\delta_{T,\tau} = (1-\tau) + \tau \mathrm{th}(2/T),$$

so that we are in the high noise regime for any temperature and any parallelization rate τ. We have computed the approximation of $\mu_{T,\tau}$ at the order 2 and 3 for different values of T and τ. The convergence is not guaranteed for low temperature, but the iterative method starting from the uniform probability measure seems to converge everywhere in the tested domain $[0.4, +\infty[\times [0.1, 1]$ for T and τ.

We have computed the mean energy per spin with the approximation at the order 2 and with a standard Metropolis method. The simulation has been performed for an Ising model with 100 sites and both results are reported in figure 1. One can notice on figure 1 that our approximation is quite accurate even at a fairly low temperature. The results obtained by the approximation at the order 3 are not reported here because they cannot be really distinguished from the order 2 in our graphical representation. The figures 2 gives the mean energy by spin obtained by the approximation at the order 2 for ten values $(\tau_k)_{1 \leq k \leq 10}$ of τ ($\tau_k = k/10$). We can easily see that for a fixed temperature, the mean energy per spin evolves continuously and increasingly with τ. Furthermore, concerning the low temperature domain, we have here clearly expressed the singular behaviour of the fully parallel algorithm ($\tau = 1$).

5 Conclusion

Concerning the parallelization of the simulated annealing, we have given some first insights into the theoretical issues we have to deal with. The most powerful tool seems to be the Wentzell-Freidlin graphs, because they are well suited for studying the invariant measure in a very general setting. However, those graphs cannot be handled directly, and a deeper understanding of the effects of the parallelization is needed in order to make further progress.

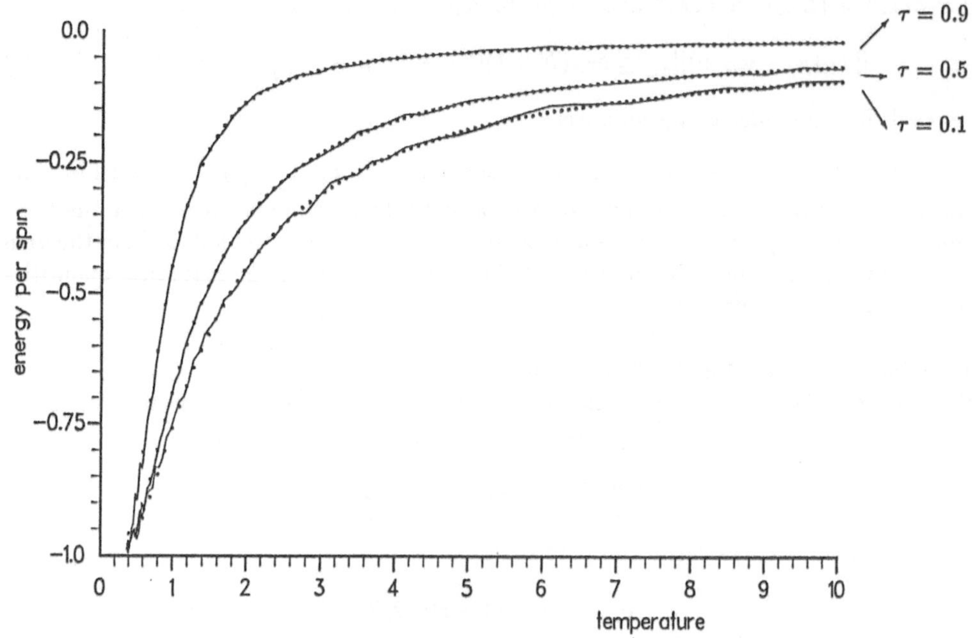

Figure 1: Comparative results for the mean energy per spin between the classical simulations (plain lines) and the approximation at the order 2 (dotted lines).

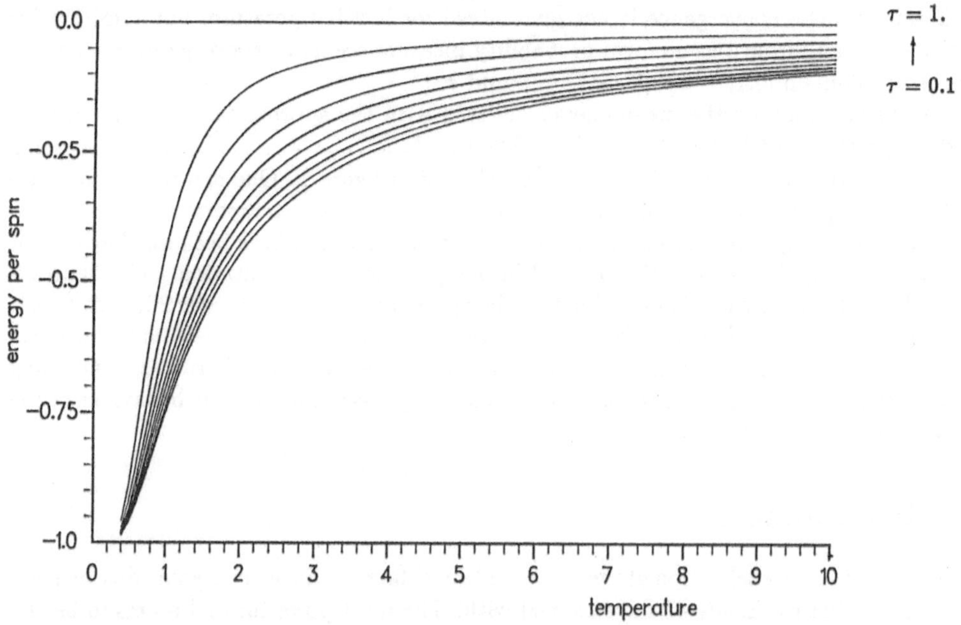

Figure 2: Mean energy per spin obtained by the approximation of $\mu_{T,\tau}$ at the order 2 for $\tau = k/10$, $1 \leq k \leq 10$.

5. Appendix

Proof of theorem 3.4

Theorem 1 in [13] says that $Q_{T,\tau}$ is reversible iff there exist two families $(v_i)_{i\in S}$ and $(v_{ij})_{i,j\in S}$ where $v_i : L \to I\!R$ and $v_{ij} : L \times L \to I\!R$, such that $v_{ij}(l,l') = v_{ji}(l',l)$ for $l,l' \in L$ and

$$Q_{T,\tau} = \prod_{i\in S}\{\frac{\exp(-v_i(y_i) - \sum_{j\in S} v_{ij}(y_i,x_j))}{\sum_{l\in L}\exp(-v_i(l) - \sum_{j\in S} v_{ij}(l,x_j))}\}. \tag{39}$$

Since for $\tau = 1$ we have $Q_{T,\tau}(x,y) = \prod_{i\in S} q_{i,T}(y_i|x)$ (see (1)), the proof of 1.a is obvious. To prove 1.b it is sufficient to verify that the probability measure on Y defined by (7) satisfies the reversibility equality (4).

Assume now that $L = \{-1,1\}$. Then, the reversibility condition given by (39) is equivalent to the existence of two families of real numbers $(h_i)_{i\in S}$ and $(J_{ij})_{i,j\in S}$ such that $J_{ij} = J_{ji}$ and

$$Q_{T,\tau} = \prod_{i\in S}\{\frac{\exp(-y_i(h_i + \sum_{j\in S} J_{ij}x_j))}{2\mathrm{ch}(h_i + \sum_{j\in S} J_{ij}x_j)}\}. \tag{40}$$

Assume that $Q_{T,\tau}$ is reversible and define:

$$f_i(x) = 1/2\ln(\frac{\tau q_{i,T}(1|x) + (1-\tau)\mathbf{1}_{x_i=1}}{\tau q_{i,T}(-1|x) + (1-\tau)\mathbf{1}_{x_i=-1}})\ ;\ x \in Y, i \in S. \tag{41}$$

We have $f_i(x) = -(h_i + \sum_{j\in S} J_{ij}x_j)$ so that:

$$1/2(f_i(x^{j,1}) - f_i(x^{j,-1})) = -J_{ij}\ ;\ x \in Y, j \in S. \tag{42}$$

We will use equation (42) for different choices of j to prove the second statement of the theorem. First, let $(J_C^H)_{C\subset S}$ be the unique family of real numbers such that $H(x) = \sum_{C\subset S}(J_C^H \prod_{i\in C} x_i)$ and denote $V_i(x) = \sum_{C:i\in C}(J_C^H \prod_{j\in C\setminus\{i\}} x_j)$. We have:

$$q_{i,T}(y_i|x) = \frac{\exp(-y_iV_i(x)/T)}{2\mathrm{ch}(V_i(x)/T)}. \tag{43}$$

Using (42) for $j = i$, a straightforward computation gives for all $x \in Y$:

$$1/4\ln(\frac{1 + (1-\tau)^2 + 2(1-\tau)\mathrm{ch}(2V_i(x)/T)}{\tau^2}) = -J_{ii},$$

so that $|V_i(x)|$ keeps a constant value noted a_i. Now, we will prove that V_i satisfies one of the two following conditions:

1. $V_i = 0$,

2. $V_i \neq 0$ and for all $j \in S$, if there exists $z \in Y$ such that $V_i(z^{j,1}) = -V_i(z^{j,-1})$ then for all $x \in Y$ we have $V_i(x) = x_jV_i(x^{j,1})$.

Assume that $V_i \neq 0$. Let $j \in S$, $z \in Y$ such that $V_i(z^{j,1}) = -V_i(z^{j,-1})$. Noting that $\mathrm{ch}(V_i(x)) = \mathrm{ch}(a_i)$ for $x \in Y$, the equation (42) gives for $x = z$:

$$1/4 \ln\left(\frac{\exp(-2\epsilon a_i) + (1-\tau)}{\exp(2\epsilon a_i) + (1-\tau)}\right) = -J_{ij} \text{ where } \epsilon a_i = V_i(z^{j,1}). \tag{44}$$

Now, let $x \in Y$. If $V_i(x^{j,1}) = V_i(x^{j,-1})$, then (42) shows that $J_{ij} = 0$ so that we deduce from (44) that $a_i = 0$. This is a contradiction with $V_i \neq 0$ so that $V_i(x^{j,1}) = -V_i(x^{j,-1})$. We can now deduce part 2. of the theorem. For $i \in S$, if $|N(i)| \geq 2$ then there exist $j \in N(i)$, x and $z \in Y$ such that $V_i(x^{j,1}) = -V_i(z^{j,1})$ from which we deduce that $f_i(z^{j,1}) - f_i(z^{j,-1}) = -(f_i(x^{j,1}) - f_i(x^{j,-1}))$. This implies that $J_{ij} = 0$ and $V_i = 0$ which is a contradiction. The proof is then complete.□

Proof of theorem 4.15

Throughout this proof, for any finite subsets V in $\mathbb{Z}$ and any finite measure ν on L^V, we will define $||\nu||_V = \sum_{x \in L^V} |\nu(x)|$. Furthermore, for each function $f : L^V \to \mathbb{R}$, we define $\Delta f = \sum_{t \in \mathbb{Z}} \Delta_t f$ where $\Delta_t f = \sup_{x \in L^V, l \in L} |f(x) - f(x^{t,l})|$. Now let $f : L^{\Lambda_N} \to \mathbb{R}$ be a function on L^{Λ_N} and let $Q_{T,\tau}^{\tilde{\Lambda}_N} f$ be the function on $L^{\tilde{\Lambda}_N}$ defined by $Q_{T,\tau}^{\tilde{\Lambda}_N} f(x) = \int Q_{T,\tau}^{\tilde{\Lambda}_N}(x, dy) f(y)$. A slight modification of the proof of lemma 2 in [14] gives that:

$$\Delta(Q_{T,\tau}^{\tilde{\Lambda}_N} f) \leq \delta_{T,\tau} \Delta f. \tag{45}$$

Now, let μ and ν be two probability measures on $L^{\tilde{\Lambda}_N}$, we have:

$$|\mu Q_{T,\tau}^{\tilde{\Lambda}_N} f - \nu Q_{T,\tau}^{\tilde{\Lambda}_N} f| \leq 1/2 ||\mu - \nu||_{\tilde{\Lambda}_N} \Delta(Q_{T,\tau}^{\tilde{\Lambda}_N} f). \tag{46}$$

Hence we deduce from (45) and (46) that:

$$||\mu Q_{T,\tau}^{\tilde{\Lambda}_N} f - \nu Q_{T,\tau}^{\tilde{\Lambda}_N} f||_{\Lambda_N} \leq ||\mu - \nu||_{\tilde{\Lambda}_N} \delta_{T,\tau} |\Lambda_N|. \tag{47}$$

We will now prove that:

$$||\psi_N(\mu) - \psi_N(\nu)||_{\tilde{\Lambda}_N} \leq (2r_0 + 1) ||\mu Q_{T,\tau}^{\tilde{\Lambda}_N} - \nu Q_{T,\tau}^{\tilde{\Lambda}_N}||_{\tilde{\Lambda}_N} \tag{48}$$

so that (47) and (48) will give the proposition.
The inequality (48) is a straightforward application of the following lemma.

Lemma 5.1 *Let $m, n \in \mathbb{Z}$, and $p \in \mathbb{N}^*$ such that $m \leq m + p \leq n$. Let $(\lambda^i)_{i=1,2}$ be two probability measures on $L^{\Lambda_{m,n}}$ where $\Lambda_{m,n} = \{m, m+1, \cdots, n\}$. Assuming that $\lambda^i(x) > 0$ on $L^{\Lambda_{m,n}}$, we define the two probability measures $(\nu^i)_{i=1,2}$ on $L^{\Lambda_{m-1,n}}$ by :*

$$\nu^i((x_k); m-1 \leq k \leq n) = \frac{\lambda^i((x_{j-1}); m \leq j \leq m+p)}{\lambda^i((x_{j-1}); m+1 \leq j \leq m+p)} \lambda^i((x_j); m \leq j \leq n). \tag{49}$$

Then we have,

$$||\nu^1 - \nu^2||_{\Lambda_{m-1,n}} \leq 2||(\lambda^1 - \lambda^2)_{\Lambda_{m,m+p}}||_{\Lambda_{m,m+p}} + ||\lambda^1 - \lambda^2||_{\Lambda_{m,n}}. \tag{50}$$

Proof of the lemma: Using the inequality $|ab - a'b'| \leq |(a-a')b| + |a'(b-b')|$ we compute that $||\nu^1 - \nu^2||_{\Lambda_{m-1,n}} \leq L_1 + L_2 + L_3$ with

$$L_1 = \sum_{(x_k)_{m-1\leq k\leq n}} |(\lambda^1 - \lambda^2)((x_j); m \leq j \leq n) \frac{\lambda^1((x_{j-1}); m \leq j \leq m+p)}{\lambda^1((x_{j-1}); m+1 \leq j \leq m+p)}|, \tag{51}$$

$$L_2 = \sum_{(x_k)_{m-1\leq k\leq n}} \lambda^2((x_{j-1}); m \leq j \leq n) |\frac{(\lambda^2 - \lambda^1)((x_{j-1}); m \leq j \leq m+p)}{\lambda^2((x_{j-1}); m+1 \leq j \leq m+p)}|, \tag{52}$$

and

$$L_3 = \sum_{(x_k)_{m-1\leq k\leq n}} \lambda^2((x_{j-1}); m \leq j \leq n) \lambda^1((x_{j-1}); m \leq j \leq m+p) \tag{53}$$

$$\times \quad |1/\lambda^1((x_{j-1}); m+1 \leq j \leq m+p) - 1/\lambda^2((x_{j-1}); m+1 \leq j \leq m+p)|. \tag{54}$$

Since

$$\sum_{x_{m-1}\in L} \frac{\lambda^1((x_{j-1}); m \leq j \leq m+p)}{\lambda^1((x_{j-1}); m+1 \leq j \leq m+p)} = 1, \tag{55}$$

we have

$$L_1 \leq ||\lambda^1 - \lambda^2||_{\Lambda_{m,n}}. \tag{56}$$

In the same way, one easily proves that :

$$L_2 \leq ||(\lambda^1 - \lambda^2)_{\Lambda_{m,m+p}}||_{\Lambda_{m,n}}. \tag{57}$$

Reducing to the same denominator and after a summation on $(x_k)_{m+r\leq k\leq n}$ we have :

$$L_3 \leq ||(\lambda^1 - \lambda^2)_{\Lambda_{m,m+p}}||_{\Lambda_{m,n}}. \tag{58}$$

so that we have proved the lemma. The proof of the proposition is then complete.□

References

[1] R. Azencott, Simulated annealing, Séminaire Bourbaki, 40 ième année 697, (1988).

[2] O.Catoni, Large deviations for annealing. Thesis held at university Paris-sud Orsay, March 1990.

[3] D. P. Connors and P. K. Kumar, Simulated annealing-type Markov chains and their order balance equations. SIAM Journal of Control and Optimization, Vol 27, No. 6 (1989) 1440-1461.

[4] R. L. Dobrushin, Prescribing a system of random variables by conditional distributions. Theory of probability and its applications 15 (1970) 458-486.

[5] Pablo A. Ferrari, A. Frigessi and R. H. Schonmann, Convergence of the Partially Parallel Gibbs Sampler with Annealing. (1991). Preprint.

[6] I. Gaudron and A. Trouvé, Parallelization of simulated annealing for spin glasses energies. In : Simulated annealing, Parallelization techniques, R. Azencott et al. (eds.), John Willey and Sons. To appear in 1992.

[7] S. Geman and D. Geman, Stochastic Relaxation, Gibbs Distribution and the bayesian Restauration of Images. IEEE Proc. Pattern. Anal. Mach. Intell. 6 (1984) 721-741.

[8] B. Gidas, Non stationary Markov chains and convergence of annealing algorithms, Journal of Statistical Physics 39 (1985) 73-131.

[9] B. Hajek, Cooling schedules for optimal annealing. preprints Math. Op. Reserach (1987)

[10] R. Holley and D. Stroock, Annealing via sobolev Inequalities. Communications in Mathematical Physics 115 (1988)

[11] C.R. Hwang and S.J. Sheu, Singular perturbated Markov chains and exact behaviors of simulated annealing process,preprint, Institute of mathematics, Academia Sinica, Taipei, Taiwan, 1988.

[12] S. Kirkpatrick, C. Gelatt and M. Vecchi, Optimization by simulated annealing, Science 220 (1983) 671-680.

[13] O. Kozlov and N. Vasilyev, Reversible Markov chain with local interaction. In : Multicomponent random systems, R.L. Dobrushin and Ya. G. Sinai eds. (Dekker, New- York) (1980) 451-469.

[14] J.L. Lebowitz, C. Maes and E Speer, Statistical mechanics of probabilistic cellular automata. Journal of Statistical Physics 59 (1990) 117-170.

[15] T.M. Liggett, Interacting Particle System.Berlin, Heidelberg, New York : Springer 1985.

[16] L. Miclo, Evolution de l'énergie libre. Applications à l'étude de la convergence des algorithmes du recuit simulé. Thesis held at university Paris 6, Febuary 1991.

[17] A. Trouvé, Parallel simulated annealing and effective optimization : Study of the limit measure. In : Simulated annealing, Parallelization techniques, R. Azencott et al. (eds.), John Willey and Sons. To appear in 1992.

[18] A. Trouvé, Partially synchronous kernel and parallelization of simulated annealing. Comptes Rendus de l'Académie des Sciences de Paris I 312 (1991) 155-158.

[19] A.D. Wentzell and M.I. Freidlin, Random perturbations of dynamical systems. Springer-Verlag (1984)

MARTINGALE REPRESENTATION FOR A CLASS OF PROCESSES WITH INDEPENDENT INCREMENTS AND ITS APPLICATIONS

by

Xing-Xiong Xue
Department of Statistics, Columbia University, New York, NY 10027

Introduction and summary

Every martingale with respect to the filtration generated by a Brownian motion B, can be represented as a stochastic integral with respect to B; this is a very well-known result, and goes back to K. Itô's fundamental paper "Multiple Wiener Integral" (J. Math. Soc. Japan **3** (1951), 157-169). This type of *martingale representation* result is also true when B is replaced by a multivariate compensated *Poisson process*, but not if B is replaced by an arbitrary martingale.

An important and fairly general result of Jacod ([12],[13]) states that for semi-martingales X in a wide class, and for probability measures P that are extreme points of the set of solutions to a certain "martingale problem" (X is a martingale under probability measures in this set), the above martingale representation result holds , in the following sense. Every P-martingale with respect to the filtration $\{\mathcal{F}_t^X\}$ generated by X can be represented as the sum of two stochastic integrals; one with respect to the continuous martingale part of X, and the other with respect to a *random measure* generated by X.

This result is quite general, but leaves a lot to be desired. First, it is not always straightforward to characterize the extreme points of the set of solutions to a given martingale problem. Secondly, it is usually easier – and, in certain applications, necessary – to work with stochastic integrals with respect to a *martingale*, rather than with respect to a random measure.

The purpose of this paper is to isolate a class of processes X with independent increments (PII), such that every martingale with respect to the filtration $\{\mathcal{F}_t^X\}$ generated by X can be represented as the sum of stochastic integrals with respect to the continuous martingale part and the purely discontinuous martingale part of X. This paper extends the results of [23], where the same problem was solved for a class of PII with stationary increments, i.e., Lévy processes. In particular, as was proved in [23], if X is a "mixture" of a Brownian motion and a compensated Poisson process, each $\{\mathcal{F}_t^X\}$-martingale can be written as a stochastic integral with respect to X *itself*, without any intervention of random measures. Such mixtures of Brownian and Poissonian processes are very useful in describing the dynamics of asset prices in financial market models, and the representation result alluded to in the previous sentence becomes then the key for the stochastic analysis and optimization of such models.

This paper consists of three parts. **Part I** introduces the class of PII's that form the object of this investigation (Sections 1 and 2), and establishes the martingale representation result for processes of this type (Sections 3 and 4), as well as for diffusions with jumps (Section 5). It also explores how these results are affected by an equivalent, Girsanov-type change of probability measure (Section 6); this is a consideration of particular importance in filtering and stochastic control. **Part II** discusses the nonlinear filtering problem with PII observations, thus extending results of Fujisaki-Kallianpur-Kunita [9] and Brémaud [2]. The innovation process is introduced, and the martingale representation result is proved for it (Section 2); with the help of the latter, we establish the fundamental "filtering equations" that determine the evolution of the conditional distribution of the unobserved signal, given the past of the observations (Section 3). Finally, **Part III** applies the previous results to the study of hedging and optimization problems in financial economics models with jumps in the price system (Section 2). The familiar "Equivalent Martingale Measure" is constructed, and leads to completeness of the financial market under a suitable non-degeneracy condition on the noise matrix (Sections 3 and 4). Finally, utility maximization problems are introduced, and solved explicitly with the help of this theory (Section 5); the results here generalize the work of Karatzas et. al. [16], [17] and that of Picqué & Pontier [15]. It is hoped that the same methodology will allow us to treat, in the future, models with incomplete markets and/or partial observations.

PART I

The martingale representation theorem

1. Introduction.

It is well known that every *Brownian martingale* can be represented as a stochastic integral with respect to the Brownian motion. This kind of martingale representation result, with respect to the natural filtration $\{\mathcal{F}_t^X\}$ generated by a given semimartingale X, also holds in the case where X is a *multivariate compensated Poisson process* (cf. II.T9 in [2]). However, it is not true for any arbitrary such X. A general martingale representation theorem of Jacod ([12] or Section III.4 in [13]) states that for any semimartingale X in a wide class, for which the probability measure P is an extreme point of the set of solutions to a corresponding "martingale problem", any martingale with respect to the filtration $\{\mathcal{F}_t^X\}$ can be represented as the sum of two stochastic integrals: one with respect to the continuous martingale part of X, and one with respect to a *random measure* generated by X. Especially, this is true if X is a *semimartingale and a process with independent increments (PII)* (cf. Theorem III.4.34 in [13]). Since this theorem is very general, and in some applications it is easier – in fact, necessary – to deal with stochastic integrals with respect to a *martingale* rather than to a random measure, we shall specify a class of PII X, such that each martingale of the filtration $\{\mathcal{F}_t^X\}$ can be represented as the sum of two stochastic integrals: one with respect to the *continuous martingale part*, and one to the *purely discontinuous martingale part* of X. In particular, the result holds for a corresponding class of *Lévy processes* (with stationary, independent increments). Moreover, we show that when a Lévy process X is a "mixture" of a Brownian motion and a multivariate compensated Poisson process, then each martingale of $\{\mathcal{F}_t^X\}$ can be represented as a stochastic integral with respect to X *itself*, without involving random measures. In this way, we extend the classical martingale representation theorems for Brownian motions and multivariate Poisson processes to a larger class of PII's. Since a PII can be used to describe the continuous changes and the jumps in a stochastic system, this martingale representation result is useful in stochastic control. We shall give its applications in the context of filtering theory (Part II) and in the context of optimization problems for a financial market with jumps in the price system (Part III).

2. Processes with independent increments (PII).

Let $(\Omega, \mathcal{F}, P)$ be a probability space and $X = (X_1, \cdots, X_d)$ a d-dimensional, càdlàg *process with independent increments* and a semimartingale with $X(0) = 0$. When X has also *stationary* independent increments, it is called a Lévy process. For the properties of PII's and Lévy processes, we refer the reader to Section II.4 in [13], or Section I.4 in [20]. Here we shall state some basic notations and properties. Let the filtration $\{\mathcal{F}_t\}$ be the one *generated by* X, that is,

$$\mathcal{F}_t \triangleq \cap_{s>t}\mathcal{F}_s^0; \qquad \mathcal{F}_s^0 \triangleq \sigma(X(r);\ r \le s).$$

Also, assume $\mathcal{F} = \mathcal{F}_\infty (= \vee_{t\ge 0}\mathcal{F}_t)$.

A way to characterize a PII X which is also a semimartingale, is by using the *semimartingale characteristics* $(B,\ C, \nu^X)$ associated with the truncation function $h(x) = x1_{\{\|x\|\le 1\}}$. Before we state the result (Proposition 2.1 below), we introduce some necessary notations.

Let $[M, N]$ denote the *quadratic co-variation* of two semimartingales M and N, and let $\langle M, N\rangle$ denote the *dual predictable projection* of $[M, N]$, if M and N are locally square integrable (cf. Section I.4 in [13]). Suppose that the semimartingale X is a PII, and that *canonical decomposition* of X is

$$X = M + D, \tag{2.1}$$

where M is a *martingale* and D is a *bounded variation process.* (In this paper, equality of processes is understood modulo indistinguishability.) Denote its *continuous martingale part* by X^c and its *purely discontinuous martingale part* by X^d, and introduce the random measure

$$\mu^X(\omega; dt, dx) = \Sigma_{s\ge 0} 1_{\{\Delta X(s)\neq 0\}}\varepsilon_{(s,\Delta X(s))}(dt, dx), \tag{2.2}$$

where ε_a denotes the Dirac measure at the point a. Then

$$C_{ij}(\cdot) = \langle X_i^c, X_j^c\rangle(\cdot); \qquad i, j = 1, \cdots, d, \tag{2.3}$$

the cross-variation of X_i^c and X_j^c for $1 \le i, j \le d$; ν^X is the *dual predictable projection* of the random measure μ^X; and

$$B = D - (x - h(x)) * \nu^X,$$

where $*$ means Stieltjes integration for each ω (Proposition II.2.29 in [13]). When the jumps of X are bounded by 1, we have $B = D$, and (2.1) can be written as

$$X = X^c + X^d + B, \tag{2.4}$$

The following proposition is quoted from [13], Theorem II.4.15, and can be used to provide an equivalent characterization of a PII.

2.1 Proposition: *A d-dimensional semimartingale X with $X(0) = 0$ is a PII, if and only if there is a deterministic version of its characteristics $(B,\ C,\ \nu^X)$.*

Later on, we shall assume that all the PII-semimartingales X discussed here are *quasi-left-continuous*, which implies that X has no *fixed time of discontinuity* and $\nu^X(\omega;\{t\},E) = 0,\ \forall\ E \in \mathcal{B}^d$ (Corollaries II.4.18 and II.1.19 in [13]). Moreover, we can find a *continuous*, locally integrable, increasing, real function $A(\cdot)$ such that the characteristics $(B,\ C,\ \nu^X)$ of X are of the form

$$(2.5)\qquad \begin{cases} B_i(t) = \int_0^t b_i(s)dA(s), & \\ C_{ij}(t) = \int_0^t c_{ij}(s)dA(s), & i,j = 1,\cdots,d \\ \nu^X(dt,dx) = d\,A(t)K(t,dx), & 0 \le t \le T, \end{cases}$$

where $b(\cdot)$ is a d-dimensional measurable function, $c(\cdot) = (c_{ij}(\cdot))$ is a deterministic measurable process with values in the set of all symmetric nonnegative $d \times d$ matrices, and $K(t,dx)$ is a transition kernel from $(\Re_+,\mathcal{B})$ into $(\Re^d,\mathcal{B}^d)$ which satisfies $K(t,\{0\})= 0$, and $\int_{\Re^d}(|x|^2 \wedge 1)K(t,dx) \le 1, \forall t \ge 0$. (See Proposition II.2.9 in [13]).

Let $\mathcal{P}$ be the predictable σ-field on $\Omega \times \Re_+$, and set $\widetilde{\Omega}=\Omega\times\Re_+ \times\Re^d$, $\widetilde{\mathcal{P}}=\mathcal{P}\otimes\mathcal{B}^d$; also denote by $G(\mu^X)$ the set of all real functions V on $\widetilde{\Omega}$, such that each V is $\widetilde{\mathcal{P}}$-measurable and the corresponding process $Z(t) \triangleq [\int_{[0,t]\times\Re^d} V^2(s,x)\mu^X(ds,dx)]^{1/2} = [\Sigma_{s\le t}V^2(s,\Delta X(s))1_{\{\Delta X(s)\neq 0\}}]^{\frac{1}{2}}$ is a locally integrable increasing process.

Given a function $V \in G(\mu^X)$ and a random measure γ, we define for each ω the *stochastic Stieltjes integral* $V * \gamma$ by

$$V{*}\gamma(\omega;t) \triangleq \begin{cases} \int_{[0,t]\times\Re^d} V(\omega;s,x)\gamma(\omega;ds,dx), & \text{if } \int_{[0,t]\times\Re^d}|V(\omega;s,x)|\gamma(\omega;ds,dx) < \infty, \\ \infty & , \ \text{otherwise.} \end{cases}$$

We also define the *stochastic integral of V with respect to the random measure* $\mu^X - \nu^X$, denoted by $V{*}(\mu^X - \nu^X)$ as a purely discontinuous martingale Y such that the processes ΔY and $\widetilde{V}$ are indistinguishable, where $\widetilde{V}(t,\omega) = V(\omega,t,\Delta X(t));\ t \ge 0$. It is easy to see that (cf. II.1.27 & 28 in [13]):

$$(2.6)\qquad V * (\mu^X - \nu^X) = V * \mu^X - V * \nu^X.$$

Since $V_i(\omega;t,x) \triangleq x_i$ is an element in $G(\mu^X)$, the stochastic integral $x_i * (\mu^X - \nu^X)$ exists; moreover (cf. II.2.39 in [13]), for each $i = 1,2,\cdots,d$, we have the following almost sure representation for the purely discontinuous part of X:

$$(2.7)\qquad X_i^d(t) = x_i * (\mu^X - \nu^X)(t) = \int_{[0,t]\times\Re^d} x_i\,(\mu^X(\omega;ds,dx) - \nu^X(\omega;ds,dx)),$$

$0 \le t < \infty$, where the right-hand side is the difference of two Stieltjes integrals. Rewriting (2.7) in vector form, we get

$$(2.8)\qquad X^d = x * (\mu^X - \nu^X);$$

that is, *we can regard the purely discontinuous martingale part X^d of X as a stochastic integral with respect to a random measure.*

2.2 Definition: Let $\mathcal{I}$ be the class of all PII X with $X(0) = 0$, such that each X is a d-dimensional, quasi-left continuous local martingale with jumps (if any) of each component equal to 1, and $[X_i, X_j] = 0$, $i \neq j$. ◇

A d-dimensional Brownian motion with a diagonal covariance matrix is an element of $\mathcal{I}$; so is a compensated multivariate Poisson process with a continuous, deterministic intensity process $\lambda(t)$ (for definition, cf. [2], p.20).

If $X \in \mathcal{I}$, then

(2.9) (i): For all $t \geq 0$, $i \neq j$, $\Delta X_i(t)\Delta X_j(t) = 0$ (i.e., X_i and X_j have no common jumps). This is because $[X_i^d, X_j^d](t) = \Sigma_{s\leq t}\Delta X_i(s)\Delta X_j(s)$ which implies $\Delta X_i(t)\Delta X_j(t) = \Delta[X_i^d, X_j^d](t) = 0$.

(2.9) (ii): For $i \neq j$, we have $c_{ij}(t) = 0$, $dA(t)$ $-$ $a.s.$ Actually, by (2.1), (2.3) and (i) above, and $[X_i, X_j] = 0$ by assumption.

$$C_{ij}(t) = \int_0^t c_{ij}(s)dA(s) = \langle X_i^c, X_j^c\rangle(t) = [X_i, X_j](t) - \Sigma_{s\leq t}\Delta X_i(s)\Delta X_j(s) = 0.$$

(2.9) (iii): From (2.1) and (2.3) we have $X_i^c \equiv 0$ if and only if $c_{ii}(t) = 0$, $dA(t)$ $-$ $a.s.$

(2.9) (iv): Let $I(i)$ denote the ith unit vector in $\Re^d$ (i.e. $I(1) = (1,0,\cdots,0)^{tr}, I(2) = (0,1,0,\cdots,0)^{tr}$, etc.). Then the set $Q_d \overset{\Delta}{=} \{I(i);\ i = 1,\cdots,d\}$ includes all the points of $\Delta X(t) \neq 0$, $t \geq 0$, as well as the support of $K(t,dx)$ for $dA(t)$ - almost every t. Indeed, for any Borel set $B \subset (Q_d)^c$ we have that $1_B \in G(\mu^X)$, and since $1_B * (\mu^X - \nu^X)$ is a martingale we obtain

$$\begin{aligned} 0 &= E(1_B * (\mu^X - \nu^X)(t)) \\ &= E\Sigma_{s\leq t}1_B(\Delta X(s))1_{\{\Delta X(s)\neq 0\}} - \int_0^t\int_B K(s,dx)dA(s) = -\int_0^t K(s,B)dA(s), \end{aligned}$$

for all $t \geq 0$, which implies $K(t,B) = 0$, $dA(t)$ $-$ $a.s.$

(2.9) (v): From (2.7) and the proof of (iv) above, it is also seen that $X_i^d \equiv 0$ if and only if $K(t, I(i)) = 0$, $dA(t)$ $-$ $a.s.$

(2.9) (vi): For any finite $\widetilde{\mathcal{P}}$ measurable function U and $t \geq 0$, we have

$$\Sigma_{s\leq t}U(s,I(i))1_{\{K(s,I(i))=0\}}\Delta X_i(s) = 0, \quad i = 1,\cdots,d, \quad P - a.s.$$

In fact, ν^X is the predictable projection of μ^X, therefore

$$\begin{aligned} &E|\Sigma_{s\leq t}U(s,I(i))1_{\{K(s,I(i))=0\}}\Delta X_i(s)| \leq E[|U(\cdot,I(i))|1_{\{K(\cdot,I(i))=0\}}x_i * \mu^X(t)] \\ &= E[|U(\cdot,I(i))|1_{\{K(\cdot,I(i))=0\}}x_i * \nu^X(t)] \\ &= E\int_0^t |U(s,I(i))|1_{\{K(s,I(i))=0\}}K(s,I(i))dA(s) = 0. \end{aligned}$$

2.3 Remark: For any $X \in \mathcal{I}$ we have the decomposition

(2.10) $$X_i = X_i^c + X_i^d, \quad i = 1,2,\cdots,d$$

and, since X^c and X^d are $\{\mathcal{F}_t\}$-adapted, it is easy to see that

$$\text{(2.11)} \qquad \{\mathcal{F}_t\} \text{ is generated by } X, \text{ as well as by the pair } (X^c, X^d).$$

◇

2.4 Convention: Suppose F is a d-dimensional predictable process. Then

(i): $F(\omega)^{tr}(t)x$; $\omega \in \Omega$, $t \geq 0$, $x \in \Re^d$ is a $\widetilde{\mathcal{P}}$-measurable function. We shall denote this function by $F^{tr}x$ (where tr denotes transposition):

$$\text{(2.12)} \qquad (F^{tr}x)(\omega; t, x) = F(\omega; t)^{tr}x; \qquad \omega \in \Omega, \quad t \geq 0, \quad x \in \Re^d.$$

(ii): $F \cdot M$ denotes the stochastic integral of F with respect to the local martingale M if the stochastic integral is well-defined. ◇

The following is from Proposition II.1.30b in [13]. The equation (2.14) gives us the idea of when it is possible to express a stochastic integral with respect to a random measure as one with respect to a purely discontinuous martingale.

2.5 Proposition: *Suppose that $X \in \mathcal{I}$, H is a $dP \times dA$-a.s bounded, predictable process; and $V \in G(\mu^X)$, then $HV \in G(\mu^X)$, and*

$$\text{(2.13)} \qquad H \cdot \{V * (\mu^X - \nu^X)\}(t) = (HV) * (\mu^X - \nu^X)(t), \qquad t \geq 0;$$

in particular,

$$\text{(2.14)} \qquad H \cdot X_i^d(t) = (Hx_i) * (\mu^X - \nu^X)(t), \qquad t \geq 0.$$

3. The martingale representation theorem for processes in $\mathcal{I}$.

Suppose that a semimartingale X is a PII; then each $\{\mathcal{F}_t\}$-local martingale with initial value zero can be represented as the sum of two stochastic integrals; one with respect to X^c, and another one with respect to the *random measure* $\mu^X - \nu^X$ (cf. [12], or Theorem III.4.34 in [13]). From this, we can obtain the following result to the effect that if, in addition, X belongs to the class $\mathcal{I}$ of Definition 2.2, then each $\{\mathcal{F}_t\}$-local martingale can be represented as the sum of two stochastic integrals; the first with respect to X^c, and the second with respect to the *purely discontinuous martingale* X^d. Recalling (2.11), this result can be regarded as a martingale representation theorem for the $2d$-*dimensional martingale* Y:

$$\text{(3.1)} \qquad Y \triangleq (X^c, X^d)^{tr}.$$

3.1 Theorem: *Suppose that X belongs to the class $\mathcal{I}$ of Definition 2.2; then each $\{\mathcal{F}_t\}$-local martingale M has the following form:*

$$\text{(3.2)} \qquad M(t) = M(0) + \Sigma_{i=1}^d \int_0^t H_i(s)\, dX_i^c(s) + \Sigma_{i=1}^d \int_0^t U_i(s)\, dX_i^d(s), \quad t \geq 0$$

for some predictable processes $H = (H_1, \cdots, H_d)^{tr}$ *and* $U = (U_1, \cdots, U_d)^{tr}$, *such that the stochastic integrals are well-defined. Moreover, H and U are unique in the sense that, if (3.2) holds for another pair of processes* $(\widetilde{H}, \widetilde{U})$, *then*

$$(3.3)\quad H_i(s)1_{\{c_{ii}(s)\neq 0\}} = \widetilde{H}_i(s)1_{\{c_{ii}(s)\neq 0\}}; \quad \textit{and} \quad U_i(s)1_{\{K(s,I(i))\neq 0\}} = \widetilde{U}_i(s)1_{\{K(s,I(i))\neq 0\}}$$

$dP \times dA$ *- a.s., where* $I(i)$ *is defined in (2.9) (iv).*

3.2 Lemma ([6], Remark VIII.48(b)): *Suppose that* M_i, $i = 1, \cdots, n$ *satisfy the conditions in Lemma 3.2. Then the following statements are equivalent:*

(i): For each bounded martingale N, there exists $V = (v_1, \cdots, v_n)^{tr} \in L^1(M_1, \cdots, M_n)$ *such that*

$$(3.4)\qquad N(t) = N(0) + \Sigma_{i=1}^{n} \int_0^t v_i(s)\, dM_i(s), \quad t \geq 0.$$

(ii): For each local martingale N, there exists a predictable process $V = (v_1, \cdots, v_n)^{tr}$ *such that (3.4) holds.*

Proof of Theorem 3.1: It is easy to see that $[Y_i, Y_j] = 0$, $i \neq j$, for the process Y defined by (3.1). In order to prove (3.2), by Lemma 3.2, we need only show that (3.2) holds for every *bounded* martingale M. Suppose that M is indeed a bounded martingale, namely $\sup_{\omega,t} |M(t,\omega)| \leq b$ for some real $b > 0$. A fundamental result of Jacod (Example 4.2.c in [12], or Theorem III.4.34 in [13]) implies that

$$(3.5)\qquad M(t) = M(0) + \Sigma_{i=1}^{d} \int_0^t H_i(s)\, dX_i^c(s) + V * (\mu^X - \nu^X)(t); \quad t \geq 0,$$

for some predictable process H and some $\widetilde{\mathcal{P}}$-measurable function $V \in G(\mu^X)$. Define

$$(3.6)\qquad \widetilde{U}_i(s) \triangleq V(s, I(i)) \quad \text{and} \quad U_i(s) \triangleq \widetilde{U}_i(s)1_{\{K(s,I(i))\neq 0\}}, \quad 1 \leq i \leq d;$$

then U_i, $\widetilde{U}_i$ are predictable processes, since V is $\widetilde{\mathcal{P}}$-measurable, and we have

$$\begin{aligned} V * (\mu^X - \nu^X)(t) &= \Sigma_{s\leq t} V(s, \Delta X(s))1_{\{\Delta X(s)\neq 0\}} - \int_0^t \int_{\Re^d} V(s,x)K(s,dx)dA(s) \\ &= \Sigma_{i=1}^{d} \Sigma_{s\leq t} V(s, I(i))\Delta X_i(s) - \Sigma_{i=1}^{d} \int_0^t \int_{\Re^d} \widetilde{U}_i(s) x_i K(s,dx)dA(s) \\ &= \Sigma_{i=1}^{d} \Sigma_{s\leq t} \widetilde{U}_i(s)\Delta X_i(s) - \Sigma_{i=1}^{d} \int_0^t \int_{\Re^d} U_i(s) x_i K(s,dx)dA(s) \\ &= \Sigma_{i=1}^{d} (U_i x_i) * (\mu^X - \nu^X)(t) + \Sigma_{i=1}^{d} \Sigma_{s\leq t} \widetilde{U}_i(s)1_{\{K(s,I(i))=0\}}\Delta X_i(s). \end{aligned}$$

From (2.9)(vi), the last term of the above equality is identically zero. Hence we get $V * (\mu^X - \nu^X)(t) = \Sigma_{i=1}^{d} (U_i x_i) * (\mu^X - \nu^X)(t)$. We shall prove in Lemma 3.3 below that all U_i; $i = 1, 2, \cdots, d$ are $dP \times dA$ - a.s. bounded. From the above equality, the boundedness of all U_i, and Proposition 2.5, we have (3.2).

Now we are going to prove the *uniqueness*. Because of (3.2), every $\{\mathcal{F}_t\}$-martingale M is locally square integrable; therefore, by the standard localization procedure, we may prove the uniqueness only for square integrable martingales. Suppose M is a square-integrable martingale, and that (H, U), $(\widetilde{H}, \widetilde{U})$ are two pairs for which (3.2) holds. Since $[X_i, X_j] = 0, i \neq j$, we have from (2.3)

$$E\Big(\int_0^t (H_i - \widetilde{H}_i)(s) dX_i^c(s)\Big)^2 = E\int_0^t (H_i - \widetilde{H}_i)^2(s)\, d[X_i^c, X_i^c](s)$$

$$= E\int_0^t (H_i - \widetilde{H}_i)^2\,(s) c_{ii}(s) dA(s); \quad i = 1, \cdots, d,$$

$$E\Big(\int_0^t (H_i - \widetilde{H}_i)(s)\, dX_i^c(s) \cdot \int_0^t (H_j - \widetilde{H}_j)(s)\, dX_j^c(s)\Big)$$

$$= E\int_0^t (H_i - \widetilde{H}_i)(H_j - \widetilde{H}_j)(s)\, d[X_i^c, X_j^c](s) = 0; \quad i \neq j.$$

Noting that $[X_i^c, X_j^d] = 0; \quad i, j = 1, 2, \cdots, d$, the same calculation gives

$$0 = E\Big(\Sigma_{i=1}^d \Big[\int_0^t (H_i - \widetilde{H}_i)(s) dX_i^c(s) + \int_0^t (U_i - \widetilde{U}_i)(s) dX_i^d(s)\Big]\Big)^2$$

$$= \Sigma_{i=1}^d \int_0^t [E(H_i - \widetilde{H}_i)^2(s)] c_{ii}(s) dA(s) + \Sigma_{i=1}^d \int_0^t E[(U_i - \widetilde{U}_i)^2(s)] K(s, I(i)) dA(s),$$

which implies uniqueness. ◇

3.3 Lemma: *Suppose that the martingale M is bounded by the real constant $b > 0$, uniformly in (t, ω); then the processes U_i, $i = 1, 2, \cdots, d$ in (3.10) are $dP \times dA$ - a.s. bounded.*

Proof: Let $\tau_{i,0} = 0$, $\tau_{i,k} = \inf\{t > \tau_{i,k-1} : \Delta X_i(t) > 0\}$; $k = 1, 2, \cdots$; $i = 1, 2, \cdots, d$. Then the set of stopping times $\{\tau_{i,k} : k = 1, 2, \cdots; i = 1, 2, \cdots, d\}$ exhausts the jumps of X. For each ω, the set $\cup_{i=1}^d \cup_{k=1}^\infty (\{\tau_{i,k}(\omega)\} \times \{\Delta X(\tau_{i,k}(\omega), \omega)\})$ is the support of the random measure μ^X. Since ΔM is bounded by $2b$, for each i, k we have

$$\Delta M(\tau_{i,k}) = \Delta(V * \mu^X)(\tau_{i,k}) = \int_{\{\tau_{i,k}\} \times \{\Delta X(\tau_{i,k})\}} V(s, x) \mu^X(ds, dx),$$

which implies V is bounded by $2b$ on $\cup_{i=1}^d \cup_{k=1}^\infty (\{\tau_{i,k}\} \times \{\Delta X(\tau_{i,k})\})$. Therefore, $E\int_0^\infty \int_{\Re^d} 1_{\{|V(s,x)| > 2b\}} \mu^X(ds, dx) = 0$, and hence $E\int_0^\infty \int_{\Re^d} 1_{\{|V(s,x)| > 2b\}} \nu^X(ds, dx) = 0$. From (2.5), this implies $\Sigma_{i=1}^d E\int_0^\infty 1_{\{|U_i(s)| > 2b\}} K(s, I(i))\, dA(s) = 0$, which means that $U_i(\cdot) 1_{\{K(\cdot, I(i)) \neq 0\}}$ is bounded $dP \times dA$ - a.s. ◇

4. Lévy Processes.

In this section, we review certain properties of Lévy processes, so that we can apply them later. First we state a result quoted from [13], Corollary II.4.19.

4.1 Proposition: *A d-dimensional process is a Lévy process, if and only if it is a semimartingale with characteristics (B, C, ν^X) of the form:*

(4.1) $$B(t, \omega) = bt, \quad C(t, \omega) = c \cdot t, \quad \nu^X(\omega; dt, dx) = dt K(dx),$$

where $b \in \Re^d$, c is a symmetric, nonnegative $d \times d$ matrix, and K is a positive measure on $\Re^d$ with $\int_{\Re^d}(|x|^2 \wedge 1) K(dx) < \infty$ and $K(\{0\}) = 0$. ◇

4.2 Definition: We define the class of process $\mathcal{L}$ by

$$\mathcal{L} = \{X : \ X \text{ is a Lévy process and } X \in \mathcal{I}\}.$$

◇

If $X \in \mathcal{L}$, then X has properties (2.9) as well as properties (4.2) below:

(4.2) (i): $E|X(t)|^n < \infty$, $t \geq 0$ and $n = 1, 2, \cdots$ (cf. Theorem I.34 in [20]).

(4.2) (ii): For $i \neq j$, we have $c_{ij} = 0$.

(4.2) (iii): It is straightforward from (2.9)(iii) and (4.1) that $X_i^c = 0$ if and only if $c_{ii} \equiv 0$, and from (ii) above and the P. Lévy theorem (cf. Theorem II.39 in [20]) that $\frac{X_i^c}{\sqrt{c_{ii}}}; i \in \{j : c_{jj} \neq 0\}$ is a multidimensional Brownian motion.

(4.2) (iv): As in (2.9)(iv), the set $Q_d = \{I(i);\ i = 1, \cdots, d\}$ includes the support of $K(dx)$.

(4.2) (v): From (2.9)(v) and (4.1) it is also seen that $X_i^d \equiv 0$ if and only if $K(I(i)) = 0$. Whenever $K(I(i)) \neq 0$, the point process $(x_i * \mu^X)(t) = \Sigma_{s \leq t} 1_{\{\Delta X_i(s) \neq 0\}}$, $t \geq 0$ is a Poisson process with intensity $\lambda_i = \int_{\Re^d} x_i K(dx) = K(I(i))$, from (2.7) and the Watanabe theorem (cf. II.T5 in [2]). Moreover, $x * \mu^X$ is a multivariate point process; $X_i^d, i = 1, 2, \cdots, d$ are independent processes by the Multichannel Watanabe theorem (cf. II.T6 in [2]; see also Theorem II.6.2 in [11]).

(4.2) (vi): X^c and X^d are independent; this follows from (ii), (v) and the Lévy-Hinčin theorem (cf. [11], p.65).

4.3 Definition: Suppose $X \in \mathcal{L}$. We call X a *mixture of a Brownian motion and a multivariate compensated Poisson process*, if $c_{ii} \cdot K(I(i)) = 0$, $i = 1, 2, \cdots, d$. ◇

It is easy to see that a process $X \in \mathcal{L}$ is a mixture of a Brownian motion and a multivariate compensated Poisson process, if and only if all its components are independent and each component is either a Brownian motion or a compensated Poisson process. Actually, if X is a mixture of a Brownian motion and a multivariate compensated Poisson process, then it can be written in the form $X = (W, N)^{tr}$, where W is an m-dimensional Brownian motion and N is an n-dimensional compensated Poisson process, $m + n = d$. From Theorem 3.1, the following corollary is immediate, and is a natural extension of the classical martingale representation theorem for Brownian motions and multivariate compensated Poisson processes (e.g. [11], pp. 80–84).

4.4 Corollary: *Theorem 3.1 holds for every $X \in \mathcal{L}$. Moreover, if X is a mixture of a Brownian motion and a multivariate Poisson process, then every $\{\mathcal{F}_t\}$-local martingale M can be represented as*

$$M(t) = M(0) + \Sigma_{i=1}^d \int_0^t H_i(s)\, dX_i(s); \qquad t \geq 0,$$

where H is a d-dimensional predictable process such that the stochastic integral is well-defined and H is $dP \times dt$-uniquely determined.

4.5 Remark: When $m = 1$ and $n = 1$ (hence $d = 2$), the result of Corollary 4.4 was established by Davis [5] for every locally square-integrable martingale M (also see Elliot [7] for more general cases). Here we have proved it by a different approach, which can also be used to prove a similar result for *diffusion processes with jumps*; see Theorem 5.2 below. ◇

5. Martingale representation theorem for diffusion processes with jumps.

After we have established Theorem 3.1, we can apply the method used there to show a similar martingale representation theorem for a class of *diffusions with jumps which are related to a Lévy process* $X \in \mathcal{L}$ (as introduced in Definition 4.2). We first explain its definition. Suppose $X \in \mathcal{L}$, and assume for notational simplicity that X is of the form $X = (W, N)^{tr}$, where W is an m-dimensional Brownian motion and N is an n-variate compensated Poisson process with intensity $\lambda_i(t) \equiv K(I(i)) \equiv 1$, $i = 1, \cdots, n$ (also for notational simplicity). Denote by

$$\mu^N(dt, dx) = \Sigma_{s \geq 0} 1_{\{\Delta N(s) \neq 0\}} \varepsilon_{\{s, \Delta N(s))\}}(dt, dx)$$

the counting random measure associated with N, and by $\nu^N(dt, dx) = dt\, K(dx)$ its dual predictable projection, where the positive measure K is supported on the set Q_n of (4.2)(iv). Consider the following stochastic integral equation related to X:

$$\begin{aligned}(5.1) \qquad Y(t) = Y(0) &+ \int_0^t \beta(s, Y(s))ds + \int_0^t \gamma(s, Y(s))dW(s) \\ &+ \iint_{[0,t] \times Q_n} \delta(s, Y(s-), x)(\mu^N - \nu^N)(ds, dx),\end{aligned}$$

where $\beta : \Re_+ \times \Re^r \to \Re^r$, $\gamma : \Re_+ \times \Re^r \to \Re^r \otimes \Re^m$, and $\delta : \Re_+ \times \Re^r \times Q_n \to \Re^r$ are Borel functions, and the initial condition

$$(5.2) \qquad Y(0) = Y_0$$

is an $\mathcal{F}_0$-measurable, r-dimensional random vector. Suppose that a process Y is a solution of (5.1) and (5.2); then Y is called a *diffusion with jumps* with driving term (W, N).

5.1 Assumption: The functions β, γ and δ are *locally Lipschitz continuous*; that is, for each natural number k, there exists a constant b_k such that for $s \leq k$, $\|y\| \leq k$, $\|y'\| \leq k$ and $\|z\| \in Q_n$, we have

$$(5.3) \quad \|\beta(s,y) - \beta(s,y')\| + \|\gamma(s,y) - \gamma(s,y')\| + \|\delta(s,y,z) - \delta(s,y',z)\| \leq b_k\|y - y'\|.$$

They also satisfy the *linear growth condition:*

$$(5.4) \qquad \|\beta(s,y)\| + \|\gamma(s,y)\| + \|\delta(s,y,z)\| \leq b_k(1 + \|y\|)$$

for every $0 \leq s \leq k$, $y \in \Re^r$, and $z \in Q_n$. ◇

Under Assumption 5.1, the equation (5.1), has a unique solution Y that satisfies also the initial condition (5.2) (cf. Theorem III.2.32 in [13]). Denote by $\{\mathcal{F}_t^Y\}$ the filtration generated by Y, by $\mathcal{P}^Y$ the σ-field generated by $\{\mathcal{F}_t^Y\}$-predictable processes, and let $\widetilde{\mathcal{P}}^Y = \mathcal{P}^Y \times \mathcal{B}^r$. We can also define μ^Y, ν^Y and $G(\mu^Y)$ as the counterparts of μ^X, ν^X and $G(\mu^X)$.

5.2 Theorem: *Under Assumption 5.1, each $\{\mathcal{F}_t^Y\}$-local martingale M is of the form*

$$M(t) = M(0) + \Sigma_{i=1}^m \int_0^t H_i(s)\,dW_i(s) + \Sigma_{i=1}^n \int_0^t U_i(s)\,dN_i(s); \quad t \geq 0, \tag{5.5}$$

for some H and U which are $\{\mathcal{F}_t^Y\}$-predictable processes such that the right-hand side of (5.5) is well-defined.

5.3 Remark: As in Theorem 3.1, the purpose of Theorem 5.2 is to represent a certain class of martingales, as stochastic integrals with respect to *martingales* instead of random measures (as in Jacod [12], [13]). The difference between Theorems 3.1 and 5.2 is that the processes M, H and U have different measurability properties in the two results. Theorem 5.2 implies, in particular, that each $\{\mathcal{F}_t^Y\}$-martingale is also an $\{\mathcal{F}_t\}$-martingale. In this particular case, the relationship between $\{\mathcal{F}_t^Y\}$ and $\{\mathcal{F}_t\}$ is studied in [3]. ◇

5.4 Remark: Lemma 3.2 enables us to restrict attention to bounded martingales, in the proof of the martingale representation Theorem 3.1. However, Lemma 3.2 cannot be applied directly in the proof of Theorem 5.2, since this latter has different measurability conditions and requirements. Lemma 5.5 below is a refinement of Lemma 3.2, and can be proved similar to Lemma 3.2. ◇

5.5 Lemma: *Under Assumption 5.1, the following two statements are equivalent:*
(i): Each $\{\mathcal{F}_t^Y\}$-local martingale M has the form (5.5).
(ii): Each bounded $\{\mathcal{F}_t^Y\}$-local martingale M has the form (5.5).

Proof of Theorem 5.2: By Lemma 5.5, we need only prove the theorem for each bounded $\{\mathcal{F}_t^Y\}$-martingale M with $M(0) = 0$. A result of Jacod (cf. the remark following Theorem III.4.34 in [13]) guarantees that such M is of the form

$$M(t) = \Sigma_{j=1}^r \int_0^t \tilde{H}_j(s)dY_j^c(s) + V * (\mu^Y - \nu^Y)(t); \quad t \geq 0, \tag{5.6}$$

where $\tilde{H}$ is an $\{\mathcal{F}_t^Y\}$-predictable process and $V \in G(\mu^Y)$. From (5.1),

$$1_{\{\Delta Y(s) \neq 0\}} = 1_{\cup_{j=1}^r \{\Delta Y_j(s) \neq 0\}} = \Sigma_{i=1}^n 1_{\{\delta(s,Y(s-),I(i)) \neq 0\}} 1_{\{\Delta N_i(s) \neq 0\}}.$$

Therefore,

$$\begin{aligned}\iint_{[0,t]\times\Re^r} V(s,x)\mu^Y(ds,dx) &= \Sigma_{s\leq t} V(s,\Delta Y(s)) 1_{\{\Delta Y(s)\neq 0\}} \\ &= \Sigma_{i=1}^n \Sigma_{s\leq t} V(s,\delta(s,Y(s-),I(i))) 1_{\{\delta(s,Y(s-),I(i))\neq 0\}} 1_{\{\Delta N_i(s)\neq 0\}} \\ &= \iint_{[0,t]\times\Re^n} V(s,\delta(s,Y(s-),x)) 1_{\{\delta(s,Y(s-),x)\neq 0\}} \mu^N(ds,dx),\end{aligned}$$

and

$$\iint_{[0,t]\times\Re^r} V(s,x)(\mu^Y - \nu^Y)(ds,dx) = \iint_{[0,t]\times\Re^n} \tilde{V}(t,x)(\mu^N - \nu^N)(ds,dx), \tag{5.7}$$

where $\widetilde{V}(s,x) \triangleq V(s,\delta(s,Y(s-),x))1_{\{\delta(s,Y(s-),x)\neq 0\}}$. To obtain (5.5), as in the Proof of Theorem 3.1, we need only to define the $\{\mathcal{F}_t^Y\}$-predictable processes

$$U_i(t) \triangleq \widetilde{V}(t,I(i)); \quad 0 \le t \le T, \quad i=1,\cdots,n, \tag{5.8}$$

and

$$H_i(t) \triangleq \Sigma_{j=1}^r \widetilde{H}_j(t)\gamma_{ji}(t,Y(t)); \quad t \ge 0, \quad i=1,\cdots,n. \tag{5.9}$$

◇

6. An exponential martingale.

Throughout this section, X will denote a Lévy process in the class $\mathcal{L}$ of Definition 4.2. Our aim is to study stochastic differential equations of the type (6.1) below, driven by such a process X; to introduce the equivalent probability measure $\widetilde{P}$ of (6.9), à la Girsanov; and to study the representation question for $\widetilde{P}$-martingales as stochastic integrals with respect to appropriate fundamental martingales under this measure (Theorem 6.3). This latter result will be instrumental for the developments of Part III.

6.1 Theorem: *Assume that ζ and η are two bounded, predictable, d-dimensional processes with $\eta_i > -1$, $i = 1,2,\cdots,d$, and recall (2.12) for the definition of $\eta^{tr}x$. Then the stochastic differential equation*

$$\begin{aligned} dZ(t) &= Z(t-)[\zeta(t)^{tr}\,dX^c(t) + d((\eta^{tr}x)*(\mu^X-\nu^X))(t)] \\ &= Z(t-)[\zeta(t)^{tr}\,dX^c(t) + \eta(t)^{tr}\,dX^d(t)], \qquad Z(0)=1 \end{aligned} \tag{6.1}$$

has a unique solution given by (6.4) below, which is a martingale. Moreover, Z is locally bounded and

$$E\,Z^r(t) \le e^{at}, \quad \forall\, r=1,2,\cdots \tag{6.2}$$

for some constant a which does not depend on t.

6.2 Remark: In the general case (that is, when X is not necessarily a Lévy process), the solution Z of (6.1) is only a *local* martingale. For general theorems which guarantee that Z is a uniformly integrable martingale under sufficiently strong conditions, we refer to Yan [24]. However, if X belongs to the class $\mathcal{L}$, things are easier. For example, if $\eta = 0$ in (6.1), then the Novikov theorem (cf. Corollary 3.5.13 in [18]) can be applied. If $\zeta = 0$ in (6.1), the Brémaud theorem (cf. VI.T4 in [2]) is available. Theorem 6.1 here provides us with another such result.

Proof of Theorem 6.1: The process

$$D(t) = \int_0^t \zeta(s)^{tr}\,dX^c(s) + \int_0^t \eta(s)^{tr}\,dX^d(s), \quad t \ge 0 \tag{6.3}$$

is a martingale. We can thus rewrite (6.1) as

$$dZ(t) = Z(t-)\,dD(t).$$

The unique solution of this equation (therefore, of (6.1)) is given by the Doléans-Dade exponential formula (cf. Theorem I.4.61 in [13]):

$$
\begin{aligned}
Z(t) &= \exp\Big\{\int_0^t \zeta(s)^{tr}\,dX^c(s) - \frac{1}{2}\int_0^t \zeta(s)^{tr}\cdot c\cdot \zeta(s)\,ds + (\eta^{tr}x)*(\mu^X-\nu^X)(t)\Big\} \\
&\quad \cdot \Pi_{s\le t}(1+\Delta D(s))e^{-\Delta D(s)} \qquad (6.4)\\
&= \exp\Big\{\int_0^t \zeta(s)^{tr}dX^c(s) - \frac{1}{2}\int_0^t \zeta(s)^{tr}\cdot c\cdot\zeta(s)ds - \Sigma_{i=1}^d K(I(i))\int_0^t \eta_i(s)ds\Big\} \\
&\quad \cdot \Pi_{s\le t}\Pi_{i=1}^d(1+\eta_i(s)\Delta X_i(s)),
\end{aligned}
$$

(recall Proposition 2.4). Because $\eta_i(s) > -1$, Z is positive. We have to prove that Z is a martingale. By (6.1), we need only show that

$$
\int_0^t Z(s-)\zeta_i(s)\,dX_i^c(s) \quad \text{and} \quad \int_0^t Z(s-)\eta_i(s)\,dX_i^d(s), \qquad 0 \le t < \infty
$$

are martingales for each i. For this, it is enough to prove that for any $t > 0$,

$$
E\int_0^t Z^2(s-)\zeta_i^2(s)\,ds < \infty \quad \text{and} \quad \int_0^t Z^2(-s)\eta_i^2(s)\,ds < \infty. \tag{6.5}
$$

Let $l > 0$ be an upper bound on the $\Re^d$-norms of ζ and η. In the following, we use a to denote a constant, whose value may change from statement to statement. From (6.4), the boundedness of ζ and η, it can be seen that

$$
E\,Z^r(t) \le e^{rat}E^{\frac{1}{2}}\,[(1+l)^{2r\Sigma_{i=1}^d x_i*\mu^X(t)}]E^{\frac{1}{2}}\exp\{2r\int_0^t \zeta(s)^{tr}\,dX^c(s)\}, \tag{6.6}
$$

for any given natural number r. Since $\Sigma_{i=1}^d x_i * \mu^X(t)$ has Poisson distribution with parameter $\Sigma_{i=1}^d K(I(i))t \triangleq at$, whence

$$
E\,[(1+l)^{2r\Sigma_{i=1}^d x_i*\mu^X(t)}] = e^{((1+l)^{2r}-1)at}. \tag{6.7}
$$

On the other hand,

$$
\begin{aligned}
&E\exp\{2r\int_0^t \zeta(s)^{tr}dX^c(s)\} \qquad (6.8)\\
&\quad \le e^{at}E\exp\{2r\int_0^t \zeta(s)^{tr}dX^c(s) - \frac{c_{ii}}{2}\int_0^t \|2r\zeta(s)\|^2 ds\} = e^{at},
\end{aligned}
$$

since $\exp\{2r\int_0^t \zeta(s)^{tr}\,dX_i^c(s) - \frac{c_{ii}}{2}\int_0^t \|2r\zeta(s)\|^2\,ds\}$, $t \ge 0$ is a martingale by the Novikov theorem. From (6.6)–(6.8), we have (6.2). In particular, $E\,Z^2(t) \le e^{at}$, and (6.5).

To prove the local boundedness of Z, we go back to (6.1). Since $\Delta|Z(t)| \le lZ(t-)$, the locally boundedness of Z follows from the locally boundedness of a left continuous process $Z(t-), t \ge 0$. ◇

Now let Z be the martingale of Theorem 6.1, and define the equivalent probability measure $\widetilde{P}$ by

$$\widetilde{P}(A) \triangleq \int_A Z(T)\, dP, \quad \forall\, A \in \mathcal{F}_T, \tag{6.9}$$

for an arbitrary given $T > 0$.

6.3 Theorem: *Assume $X \in \mathcal{L}$, and let Z be defined as in Theorem 6.1. Then*
(i): For each $i = 1, 2, \cdots, d$ the processes

$$\widetilde{X}_i^c(t) \triangleq X_i^c(t) - c_{ii}\int_0^t \zeta_i(s)ds \quad \text{and} \quad \widetilde{X}_i^d(t) \triangleq X_i^d(t) - K(I(i))\int_0^t \eta_i(s)ds; \quad 0 \le t \le T \tag{6.10}$$

are respectively continuous and purely discontinuous martingales under $\widetilde{P}$.
(ii): Each $\widetilde{P}$-martingale M can be represented as

$$M(t) = M(0) + \int_0^t \tilde{f}(s)^{tr}\, d\widetilde{X}^c(s) + \int_0^t \tilde{g}(s)^{tr}\, d\widetilde{X}^d(s), \quad 0 \le t \le T \tag{6.11}$$

for some predictable processes $\tilde{f}$ and $\tilde{g}$ such that the right side of (6.11) is well-defined.

Proof: (i). Notice that $[X_i, X_j] = 0$ implies $[X_i^c, X_j^c] = \langle X_i^c, X_j^c \rangle = 0$, $i \neq j$. Thus

$$X_i^c(t) - \int_0^t \frac{1}{Z(s-)} d\langle X_i^c, Z^c\rangle(s) = X_i^c(t) - \int_0^t \Sigma_{j=i}^d\, d\langle X_i^c, \zeta_j \cdot X_j^c\rangle(s) = \widetilde{X}_i^c(t)$$

is a continuous local martingale, from the Girsanov theorem. On the other hand,

$$\begin{aligned} X_i^d(t) - \int_0^t \frac{1}{Z(s-)} d\langle X_i^d, Z^d\rangle(s) &= X_i^d(t) - \int_0^t \Sigma_{j=i}^d\, d\langle X_i^d, \eta_j \cdot X_j^d\rangle(s) \\ = X_i^d(t) - \int_0^t \eta_i(s)\, d\langle X_i^d, X_i^d\rangle(s) &= X_i^d(t) - K(I(i)) \int_0^t \eta_i(s) ds = \widetilde{X}_i^d(t) \end{aligned}$$

is a local martingale by the same theorem. The local martingale $\widetilde{X}^c$ is continuous, and is actually a martingale since $\langle \widetilde{X}_i^c, \widetilde{X}_i^c\rangle(t) = c_{ii}t$. By the same theorem quoted above,

$$\langle (\widetilde{X}_i^d)^c, (\widetilde{X}_i^d)^c \rangle = \langle (X_i^d)^c, (X_i^d)^c \rangle = 0,$$

so that $(\widetilde{X}_i^d)^c = 0$, that is, $\widetilde{X}^d$ is purely discontinuous.

To prove that $\widetilde{X}_i^d$ is actually a $\widetilde{P}$-martingale, it is sufficient to show that $\widetilde{X}_i^d$ is square integrable. In fact, for any $0 \le t \le T$,

$$\begin{aligned} \widetilde{E}(X_i^d(t))^2 &= E[Z(t)(X_i^d(t))^2] \\ &\le (EZ^2(t))^{\frac{1}{2}}[2E(X_i^d(t))^2 + 2E\big(K(I(i))\int_0^t \eta_i(s)ds\big)^2]^{\frac{1}{2}} < \infty, \end{aligned}$$

by (6.2), (4.2)(i) and the boundedness of η.

(ii). From (6.10),

$$X_i(t) = \widetilde{X}_i^c(t) + \widetilde{X}_i^d(t) + (c_{ii} \int_0^t \zeta_i(s)ds + K(I(i)) \int_0^t \eta_i(s)ds); \quad 0 \le t \le T$$

is the canonical representation of X under $\widetilde{P}$, which is unique. Let (B', C', ν') be the characteristics of X under $\widetilde{P}$, then each $\widetilde{P}$-martingale M has the representation:

$$M(t) = M(0) + \int_0^t \tilde{f}(s)^{tr} d\widetilde{X}^c(s) + (g^{tr}x) * (\mu' - \nu')(t)$$

for some predictable process $\tilde{f}$ and some $g \in G(\mu')$ (Theorem III.3.11 in [13]). The rest proof is similar to the proof of Theorem 3.1. ◇

PART II
Filtering with PII observations

1. Introduction.

Suppose that we are given some filtered probability space $(\Omega, \mathcal{G}, \{\mathcal{G}_t\}, P)$ and a *signal process* $S = \{S(t):\ t \ge 0\}$ which is $\{\mathcal{G}_t\}$-adapted and cannot be observed directly. What we can observe is the process $Y = \{Y(t):\ t \ge 0\}$, which is called, for this reason, the *observation process*. We denote by $\{\mathcal{Y}_t\}$ the filtration generated by Y, that is,

$$\mathcal{Y}_t \triangleq \cap_{s>t} \mathcal{Y}_s^0; \qquad \mathcal{Y}_s^0 \triangleq \sigma(Y(r);\ r \le s).$$

The filtering problem is to find the $\{\mathcal{Y}_t\}$-optional projection of S; or even better, that of $F(S(\cdot))$, for a wide class of functions F, so that the conditional distribution of $S(t)$ given $\{\mathcal{Y}_t\}$ is completely determined. These projections are denoted by $E[S(t) \mid \mathcal{Y}_t]$ and $E[F(S(t)) \mid \mathcal{Y}_t]$ respectively, for any given $t \ge 0$. For any $\{\mathcal{G}_t\}$-measurable process H, we use the notation $\widehat{H}$:

(1.1) $\qquad \widehat{H}$ denotes its $\{\mathcal{Y}_t\}$-optional projection of H.

We shall adopt the innovation approach to the filtering problem, following the idea of Fujisaki, Kallianpur & Kunita [9]. In [9], the observation process Y is of the form

(1.2) $$Y(t) = \int_0^t h(s)\, ds + W(t); \qquad t \ge 0,$$

where W is a d-dimensional Brownian motion and h is a d-dimensional process satisfying the following condition:

(1.3) $\qquad h$ is $\{\mathcal{G}_t\}$-adapted, right continuous, and $E \int_0^t |h(s)|\, ds < \infty; \quad t \ge 0.$

The *innovation process* V is defined by

(1.4) $$V(t) = Y(t) - \int_0^t \hat{h}(s)\, ds; \qquad t \ge 0.$$

The important results in [9] are that the innovation process V is a $\{\mathcal{Y}_t\}$-Brownian motion, and that every $\{\mathcal{Y}_t\}$-martingale M with $M(0) = 0$ can be represented as a stochastic integral with respect to the innovation process. Moreover, a filtering equation is established for the recursive computation of $E[f(S(t)) \mid \mathcal{Y}_t]$. In this case, every $\{\mathcal{Y}_t\}$-martingale is continuous. When the observation process is of the form

$$(1.2)' \qquad Y(t) = M^d(t), \quad t \geq 0,$$

where M^d is a purely discontinuous martingale, there are similar results; see, for example, IV,T1, T2 as well as VIII.T9 in Brémaud [2] (and the references there).

In this part of the paper, we are going to consider the 'combination' of the two cases mentioned above; that is, we shall consider observation processes of the form

$$(1.5) \qquad Y(t) = \int_0^t c(s)h(s)\,dA(s) + X(t); \qquad t \geq 0,$$

where X is a PII in the class $\mathcal{I}$ of Definition I.2.2, and A, c and K are given by (I.2.3). Also we assume that h satisfies the condition (1.3)':

$$(1.3)' \qquad h \text{ is } \{\mathcal{G}_t\}\text{-adapted, right continuous, and } E\int_0^t |c(s)h(s)|\,dA(s) < \infty;$$

$t \geq 0$, and $\{\mathcal{Y}_t\}$ is generated by Y. Let $X = X^c + X^d$ be the orthogonal decomposition, where X^c is the continuous martingale part and X^d the purely discontinuous martingale part of X. The observation process Y reflects not only the continuous changes but also the jumps that occur in the processing of this system. In particular, when $X \in \mathcal{I}$ is a Lévy process, then if $X^d \equiv 0$, the observation Y of (1.5) goes back to the form (1.2); and if $X^c \equiv 0$, Y is of the form (1.2)'. Hence, the model of (1.5) can be regarded as the natural extension of the two filtering models mentioned above.

Let us recall the characteristics (B, C, ν^X) of X. Since $X \in \mathcal{I}$, we have

(i) $B = 0$,

(ii) $C(t) = \int_0^t c(s)dA(s)$, $\quad t \geq 0$, and $c_{ij}(\cdot) = 0$, $\quad i \neq j$, $\quad dA - a.s.$,

(iii) $\nu^X(dt, dx) = dA(t)K(t, dx)$, where $K(t, dx)$ is a positive measure with support included in the set $Q_d = \{I(i),\ i = 1, 2, \cdots, d\}$ for $dA - a.e.\ t$, is the dual predictable projection of the random measure $\mu^X(dt, dx)$ of (I.2.2).

2. Innovation process and the main results.

In this section, we shall consider the model (1.5) on the finite time-horizon $[0, T]$, where $T > 0$ is a fixed real number; it is not difficult to show that the results hold on an infinite time-horizon as well. The innovation process V is defined by

$$(2.1) \qquad V(t) = Y(t) - \int_0^t c(s)\hat{h}(s)\,dA(s); \qquad 0 \leq t \leq T.$$

The following lemma can be proved by the same way as Theorem IV.T1 in [2].

2.1 Lemma: *Suppose that u is a $\{\mathcal{G}_t\}$-measurable process with $E[\int_0^t |u(s)|dA(s)] < \infty$, $0 \le t \le T$, m is a $\{\mathcal{G}_t\}$-martingale, and define*

$$U(t) = \int_0^t u(s)\,dA(s) + m(t), \quad 0 \le t \le T.$$

Then the $\{\mathcal{Y}_t\}$-optional projection of U is of the form

$$\widehat{U}(t) = \int_0^t \hat{u}(s)\,dA(s) + \bar{m}(t); \quad 0 \le t \le T,$$

where $\bar{m}$ is a $\{\mathcal{Y}_t\}$-martingale.

2.2 Corollary: (i): *Suppose that U is a $\{\mathcal{G}_t\}$-martingale; then $\widehat{U}$ is a $\{\mathcal{Y}_t\}$-martingale.*

(ii): *Suppose that u is a $\{\mathcal{G}_t\}$-measurable process such that $E\int_0^t |u(s)|\,dA(s) < \infty$; $0 \le t \le T$; then*

$$\text{(2.2)} \qquad \widehat{\int_0^t u(s)\,dA(s)} - \int_0^t \hat{u}(s)\,dA(s); \quad 0 \le t \le T \quad \textit{is a } \{\mathcal{Y}_t\}\textit{-martingale.}$$

The following two theorems are our main results in this section.

2.3 Theorem: *Suppose that $X \in \mathcal{I}$ is a martingale. Then the innovation process V of (2.1) is a $\{\mathcal{Y}_t\}$-PII martingale with the same characteristics (B, C, ν^X) as X. Moreover, $V \in \mathcal{I}$. Let $V = V^c + V^d$ be the orthogonal decomposition of V (with V^c the continuous martingale part, and V^d the purely discontinuous martingale part); then*

$$\text{(2.3)} \qquad V_i^d(t) = X_i^d(t), \quad \textit{and} \quad V_i^c(t) = X_i^c(t) + \int_0^t c_{ii}(s)(h_i(s) - \hat{h}_i(s))dA(s);$$
$$0 \le t \le T, \quad i = 1, 2, \cdots, d.$$

2.4 Theorem: *Assume that $X \in \mathcal{I}$ is a martingale and the exponential local martingale*

$$\text{(2.4)} \qquad R(t) \triangleq \exp\{-\Sigma_{i=1}^d \int_0^t \hat{h}_i(s)dV_i^c(s) - \frac{1}{2}\Sigma_{i=1}^d \int_0^t c_{ii}(s)\|\hat{h}_i(s)\|^2\,dA(s)\}$$

is a martingale. Then for every $\{\mathcal{Y}_t\}$-martingale M, there exist two d-dimensional, $\{\mathcal{Y}_t\}$-predictable processes ϕ and ψ such that M is of the form:

$$\text{(2.5)} \quad M(t) = M(0) + \Sigma_{i=1}^d \int_0^t \phi_i(s)\,dV_i^c(s) + \Sigma_{i=1}^d \int_0^t \psi_i(s)\,dV_i^d(s); \qquad 0 \le t \le T.$$

2.5 Remark: A sufficient condition for the process R of (2.4) to be a martingale, is $E\exp\{\frac{1}{2}\Sigma_{i=1}^d \int_0^T c_{ii}|\hat{h}(s)|^2 dA(s)\} < \infty$. In particular, if h is bounded, then R is a martingale (cf. [18], Proposition 3.5.16).

Proof of Theorem 2.3: It is obvious that V is a $\{\mathcal{Y}_t\}$-optional process. From the linearity of the optional projection, we have

$$V(t)=\widehat{V}(t)=\widehat{Y}(t)-\int_0^t \hat{h}(s)dA(s)=\widehat{X}(t)+\left(\widehat{(\int_0^t c(s)h(s)dA(s))}-\int_0^t c(s)\hat{h}(s)dA(s)\right)$$

By Corollary 2.2, we see that this last expression is the sum of two $\{\mathcal{Y}_t\}$-martingales, and thus so is V. Let (B^V, C^V, ν^V) be the semimartingale characteristics of V. Since we have proved that V is a martingale, $B^V \equiv 0 \equiv B$. Since A is continuous, from (2.1), $\Delta V = \Delta Y$ which implies that ΔV is $\{\mathcal{Y}_t\}$-adapted and

$$\mu^V(dt, dx) \triangleq \Sigma_{s\geq 0} 1_{\{\Delta V(s)\neq 0\}} \varepsilon_{\{s,\Delta V(s)\}}(dt, dx) = \mu^X(dt, dx).$$

Because ν^X is deterministic, it can be seen that the dual predictable projection ν^V of μ^V is equal to ν^X. Therefore, $X^d = V^d$. On the other hand, from this , (1.5), the definition of C^V (recall (I.2.1)), and the fact that V_i, V_j have no common jumps, we have

$$\begin{aligned} C^V_{ij}(t) &= \langle V_i^c, V_j^c\rangle(t) = [V_i^c, V_j^c](t) \\ &= [Y_i - \int_0^\cdot c_{ii}(s)\hat{h}_i(s)\, dA(s) - V_i^d, Y_j - \int_0^\cdot c_{jj}(s)\hat{h}_j(s)\, dA(s) - V_j^d](t) \\ &= \langle X_i^c, X_j^c\rangle(t) = C_{ij}(t); \quad t\geq 0, \quad i\neq j, \end{aligned}$$

that is, $C^V \equiv C$. We have proved $(B^V, C^V, \nu^V) = (B, C, \nu^X)$. Therefore, by Proposition I.2.1, V is a PII. Moreover, $\Delta V_i(t) = \Delta X_i(t) = 1$ whenever $\Delta V_i(t) \neq 0$, and for $i \neq j$, noting that $c_{ij}(\cdot) = 0$, $dA - a.s.$,

$$[V_i, V_j](t)=\langle V_i^c, V_j^c\rangle(t)+\Sigma_{s\leq t}\Delta V_i(s)V_j(s)=\int_0^t c_{ij}(s)dA(s)+\Sigma_{s\leq t}\Delta X_i(s)\Delta X_j(s)=0.$$

This implies $V \in \mathcal{I}$. From (1.5) and (2.1), for each $i = 1, 2, \cdots, d$,

$$V_i^c(t) = V_i(t) - V_i^d(t) = X_i^c(t) + \int_0^t c_{ii}(s)h_i(s)\, dA(s) - \int_0^t c_{ii}(s)\hat{h}_i(s)dA(s).$$

◇

Proof of Theorem 2.4: Without loss of generality, we may assume $M(0) = 0$. First of all, we are going to define a new measure, under which Y will be a PII of class $\mathcal{I}$; thus we shall be able to apply Theorem I.3.1. From (2.4), the process R satisfies

(2.6) $$dR(t) = -R(t)\Sigma_{i=1}^d \hat{h}_i(t)\, dV_i^c(t); \quad 0 \leq t \leq T$$

by the Doléans-Dade exponential formula (cf. Theorem II.4.61 in [13]). By the assumption that R is a martingale, $E\,R(T) = 1$. Then

(2.7) $$Q(D) \triangleq \int_D R(T)\, dP; \quad D \in \mathcal{Y}_T$$

is a probability measure equivalent to P on $\mathcal{Y}_T$. By Lemma 2.6 below, Y is a Q-PII in class $\mathcal{I}$, and $Y_i^c(t) = V_i^c(t) + \int_0^t c_{ii}(s)\hat{h}_i(s)\,dA(s)$, $Y^d(t) = V^d(t)$, $0 \le t \le T$.

Suppose that M is a $(P\text{-}\{\mathcal{Y}_t\})$-martingale with $M(0) = 0$; then MR^{-1} is a $(Q\text{-}\{\mathcal{Y}_t\})$-martingale (cf. Proposition III.3.8 in [13]). From Theorem I.3.1, there exist two d-dimensional $\{\mathcal{Y}_t\}$-predictable processes $\tilde{\phi}$ and $\tilde{\psi}$ such that

$$M(t)R^{-1}(t) = \int_0^t \tilde{\phi}(s)^{tr}\,dY^c(s) + \int_0^t \tilde{\psi}(s)^{tr}\,dV^d(s); \quad 0 \le t \le T. \tag{2.8}$$

Use Itô's rule and (2.6), (2.8), to get

$$\begin{aligned} M(t) &= [M(t)R^{-1}(t)]\,R(t) \\ &= \int_0^t (M(s-)R^{-1}(s))\,dR(s) + \int_0^t R(s)\,d(M(s)R^{-1}(s)) + [MR^{-1}, R](t) \\ &= \Sigma_{j=1}^3 I_j(t), \end{aligned} \tag{2.9}$$

where

$$I_1(t) \triangleq \int_0^t (M(s-)R^{-1}(s))dR(s) = -\int_0^t M(s-)\Sigma_{i=1}^d \hat{h}_i(s)dV_i^c(s); \tag{2.10}$$

$$\begin{aligned} I_2(t) &\triangleq \int_0^t R(s)d(M(s)R^{-1}(s)) \\ &= \int_0^t R(s)\Sigma_{i=1}^d \tilde{\phi}_i(s)[dV_i^c(s) + c_{ii}(s)\hat{h}_i(s)dA(s)] + \int_0^t R(s)\tilde{\psi}(s)^{tr}dV^d(s); \end{aligned} \tag{2.11}$$

$$\begin{aligned} I_3(t) &\triangleq [MR^{-1}, R](t) \\ &= \left\langle \int_0^\cdot \Sigma_{i=1}^d \tilde{\phi}_i(s)\,dV_i^c(s), -\int_0^\cdot R(s)\Sigma_{i=1}^d \hat{h}_i(s)\,dV_i^c(s) \right\rangle (t) \\ &= -\int_0^t R(s)\Sigma_{i=1}^d \tilde{\phi}_i(s)\hat{h}_i(s)c_{ii}(s)\,dA(s). \end{aligned} \tag{2.12}$$

Define

$$\phi_i(t) \triangleq -M(t-)\hat{h}_i(t) + R(t)\tilde{\phi}_i(t);$$

and

$$\psi(t) \triangleq R(t)\tilde{\psi}(t); \quad 0 \le t \le T.$$

Then from (2.9)-(2.12), we have (2.5). ◇

2.6 Lemma: *Suppose that $X \in \mathcal{I}$ is a martingale, R in (2.4) is a martingale, and consider the probability measure Q defined by (2.7). Then Y is a Q-PII and a martingale with the same characteristics (B, C, ν^X) as the process X, and we have $Y^d = X^d$ and $Y_i^c = V_i^c + \int_0^\cdot c_{ii}(s)\hat{h}_i(s)dA(s)$. Moreover, Y belongs to the class $\mathcal{I}$ under Q.*

Proof: Let (B',C',ν') and (B^Y,C^Y,ν^Y) be the characteristics of Y under Q and P, respectively. Since A is continuous, it is obvious that $\Delta Y = \Delta V = \Delta X$, hence the random measure μ^Y generated by Y is equal to μ^X, and $\nu^Y = \nu^X$. From (2.1) and Theorem 2.3, we have $B^Y(t) = \int_0^t c(s)\hat{h}(s)dA(s)$, and $C^Y_{ij}(t) = \langle Y^c_i, Y^c_j\rangle_P(t) = \langle V^c_i, V^c_j\rangle_P(t) = C_{ij}(t)$. Since R is a continuous, $\{\mathcal{Y}_t\}$-adapted process, it is $\widetilde{\mathcal{P}}^Y$-measurable, we have $M^P_{\mu^Y}(R \mid \widetilde{\mathcal{P}}^Y) = R$, where $M^P_{\mu^Y}$ is the positive measure on $(\widetilde{\Omega}, \mathcal{Y}_T \otimes \Re_+ \otimes \Re^d)$ defined by $M^P_{\mu^Y}(W) = E(W * \mu^Y_\infty)$ for all measurable, nonnegative functions W, and

$$\langle R^c, Y^c_i\rangle(t) = \Sigma^d_{j=1}\langle \int_0^\cdot -R\hat{h}_i(s)dV^c_j(s), V^c_i\rangle(t) = (-\Sigma^d_{j=1}c_{ij}\hat{h}R)\cdot A(t).$$

From the two above observations and Theorem III.3.24 in [13], we can obtain that $C' = C^Y = C$, $\nu' = \nu^Y$ and $B'_i = B^Y_i + (-\Sigma^d_{j=1}c_{ij}\hat{h})\cdot A = 0$ immediately. Moreover, under probability Q, $Y^d = x * (\mu' - \nu') = x * (\mu^X - \nu^X) = X^d = V^d$, by Theorem 2.3. We can derive the expression of Y^c in the Lemma from (1.5) and (2.3). The remaining conclusion is easy. ◇

3. The filtering equation.

After we have established Theorem 2.4 for a martingale $X \in \mathcal{I}$, we should like to compute the integrands in Theorem 2.4 for a given signal process S. We shall consider the case X is an L^2-martingale and the time horizon $[0,T]$ is finite, and assume some structure for the signal process, as introduced in [9]; that will enable us to establish the filtering equation, as the natural extension of the Fujisaki-Kallianpur-Kunita equation. First, we shall state a result from the remark after III.4.3 in [13].

3.1 Lemma: *Suppose that M is a d-dimensional $\{\mathcal{G}_t\}_{0\le t\le T}$-martingale, bounded in L^2. Then there exist $\{\mathcal{G}_t\}_{0\le t\le T}$ - optional processes α and β, such that*

$$\langle M^c, X^c_i\rangle(t) = \int_0^t \alpha_i(s)c_{ii}(s)\,dA(s) \tag{3.1}$$

and

$$\langle M^d, X^d_i\rangle(t) = \int_0^t \beta_i(s)K(s,I(i))\,dA(s), \tag{3.2}$$

for $0 \le t \le T$, $i = 1,2,\cdots,d$. In particular,

$$\Sigma_{s\le t}\Delta M^d(s)\Delta X^d_i(s) - \int_0^t \beta_i(s)K(s,I(i))\,dA(s); \quad 0 \le t \le T \tag{3.3}$$

is a $\{\mathcal{G}_t\}$-martingale.

3.2 Theorem: *Suppose that*
(1) The signal process S is right-continuous and $\{\mathcal{G}_t\}$-adapted, with values in a Polish space S.
(2) $X \in \mathcal{I}$ is a d-dimensional L^2-martingale with $c_{ii}(t)\cdot K(t,I(i)) > 0$, $0 \le t \le T$, $i = 1,2,\cdots,d$.

(3) *A function $f \in C(S)$ is given with*

$$\sup_{0 \le t \le T} E\,|f(S(t))|^2 < \infty, \tag{3.4}$$

and suppose there exists a $\{\mathcal{G}_t\}$-optional process $(\mathcal{J}F(t))_{0 \le t \le T}$ such that

$$E \int_0^T |\mathcal{J}F(s)|^2\, ds < \infty \tag{3.5}$$

and

$$M(t) \triangleq f(S(t)) - f(S(0)) - \int_0^t \mathcal{J}F(s)\, dA(s); \quad 0 \le t \le T \text{ is a } \{\mathcal{G}_t\}\text{-martingale.} \tag{3.6}$$

Then
(i) *M is bounded in L^2, and*
(ii) *with α, β given by Lemma 3.1 for M, and $F(t) \triangleq f(S(t))$, the filtering equation is given by*

$$\begin{aligned} \hat{F}(t) &= \hat{F}(0) + \int_0^t \widehat{\mathcal{J}F}(s) dA(s) \\ &\quad + \int_0^t [\widehat{F(s-)h}(s) - \hat{F}(s-)\hat{h}(s) + \hat{\alpha}(s)]^{tr} dV^c(s) + \int_0^t \hat{\beta}(s)^{tr} dV^d(s), \end{aligned} \tag{3.7}$$

where V is the innovation process of (2.1).

Proof: Claim (i) is obvious from (3.4) and (3.5); therefore, α and β are well-defined. We can also see that $F(t)$, $0 \le t \le T$ is a $\{\mathcal{G}_t\}$-adapted, right continuous process. From (3.6) and Corollary 2.2,

$$U(t) \triangleq \hat{F}(t) - \hat{F}(0) - \int_0^t \widehat{\mathcal{J}F}(s) dA(s); \quad 0 \le t \le T \quad \text{is a } \{\mathcal{Y}_t\}\text{-martingale,} \tag{3.8}$$

which is also L^2-bounded. According to Theorem 2.4, there exist $\{\mathcal{Y}_t\}$-predictable processes ϕ and ψ, such that

$$U(t) = \int_0^t \phi(s)^{tr}\, dV^c(s) + \int_0^t \psi(s)^{tr} dV^d(s); \quad 0 \le t \le T. \tag{3.9}$$

Let

$$D = X^c + \int_0^\cdot c(s)h(s)\, dA(s) = V^c + \int_0^\cdot c(s)\hat{h}(s)\, dA(s); \tag{3.10}$$

then, D is a continuous and $\{\mathcal{Y}_t\}$-adapted process, hence it is a $\{\mathcal{Y}_t\}$-predictable process. By (3.1) and (3.6),

$$[F, D](t) = \langle M^c, X^c \rangle(t) = \int_0^t c(s)\alpha(s)\, dA(s); \quad 0 \le t \le T. \tag{3.11}$$

Now by Itô's rule, (3.10), (3.6) and (3.11),

$$
\begin{aligned}
F(t)D(t) &= \int_0^t F(s-)dX^c(s) + \int_0^t F(s-)c(s)h(s)dA(s) + \\
&\quad + \int_0^t D(s)dM(s) + \int_0^t D(s)\mathcal{J}F(s)dA(s) + \int_0^t c(s)\alpha(s)dA(s) \\
&= \int_0^t \Big(F(s-)c(s)h(s) + D(s)\mathcal{J}F(s) + c(s)\alpha(s)\Big)dA(s) \\
&\quad +(\{\mathcal{G}_t\}\text{-local martingale}).
\end{aligned}
$$

In the following, we use J to denote an $\{\mathcal{Y}_t\}$-local martingale, which may be different from statement to statement. From this equality and Corollary 2.2, we have

$$
\begin{aligned}
\text{(3.12)} \qquad \hat{F}(t)D(t) = \widehat{FD}(t) &= \int_0^t \widehat{F(s-)c(s)}h(s)dA(s) + \int_0^t D(s)\widehat{\mathcal{J}F}(s)dA(s) \\
&\quad + \int_0^t c(s)\hat{\alpha}(s)dA(s) + J(t).
\end{aligned}
$$

On the other hand, by Itô's rule, (3.10), (3.7), (3.8) and (3.9), we have, for $i = 1, 2, \cdots, d$,

$$
\begin{aligned}
\hat{F}(t)D(t) &= \int_0^t \hat{F}(s-)[dV^c(s) + c(s)\hat{h}(s)dA(s)] \\
\text{(3.13)} \qquad &\quad + \int_0^t D(s)\widehat{\mathcal{J}F}(s)dA(s) + \int_0^t D(s)dU(s) + [\hat{F}, D](t) \\
&= \int_0^t [\hat{F}(s-)c(s)\hat{h}(s) + D(s)\widehat{\mathcal{J}F}(s)]dA(s) + \langle U^c, V^c\rangle(t) + J(t) \\
&= \int_0^t [\hat{F}(s-)c(s)\hat{h}(s) + D(s)\widehat{\mathcal{J}F}(s) + c(s)\phi(s)]dA(s) + J(t)
\end{aligned}
$$

From (3.12) and (3.13), we can see that

$$
\int_0^t c(s)[\widehat{F(s-)}h(s) + \hat{\alpha}(s) - \hat{F}(s-)\hat{h}(s) - \phi(s)]dA(s); \qquad 0 \le t \le T
$$

is a continuous $\{\mathcal{Y}_t\}$-local martingale of bounded variation, and therefore evanescent, by Corollary I.3.16 in [13]. Hence we have

$$
c(s)\phi(t) = c(s)[\widehat{F(s-)}h(s) + \hat{\alpha}(s) - \hat{F}(s-)\hat{h}(s)]; \qquad dA - a.s.
$$

By assumption that c is of full rank, we have

$$
\text{(3.14)} \qquad \phi(t) = [\widehat{F(s-)}h(s) + \hat{\alpha}(s) - \hat{F}(s-)\hat{h}(s)]; \qquad P \times dA - a.s.
$$

We can calculate ψ of (3.9) by the same method, noting that $X^d = V^d$. From (3.6), it is easy to see that

$$
\begin{aligned}
(3.15) \qquad [F, X^d](t) &= \Sigma_{s\leq t}\Delta F(s)\Delta X^d(s) \\
&= \Sigma_{s\leq t}\Delta M^d(s)\Delta X^d(s) = [M^d, X^d](t),
\end{aligned}
$$

and by Itô's rule, (3.6), (3.15) and (3.3),

$$
\begin{aligned}
F(t)X_i^d(t) &= \int_0^t F(s-)dX_i^d(s) + \int_0^t X_i^d(s-)[dM(s) + \mathcal{J}F(s)dA(s)] \\
&\quad + [M^d, X_i^d](t) - \int_0^t \beta_i(s)K(s, I(i))dA(s) + \int_0^t \beta_i(s)K(s, I(i))\, dA(s) \\
&= \int_0^t [X_i^d(s-)\mathcal{J}F(s) + \beta_i(s)K(s, I(i))]dA(s) + (\{\mathcal{G}_t\}\text{-local martingale}).
\end{aligned}
$$

Using Lemma 2.1, we obtain

$$
(3.16) \qquad \hat{F}(t)X_i^d(t) = \widehat{FX_i^d}(t) = \int_0^t [X_i^d(s-)\widehat{\mathcal{J}F}(s) + \hat{\beta}_i(s)K(s, I(i))]dA(s) + J(t).
$$

On the other hand, using Itô's rule, (3.8) and (3.9), we have, for $i = 1, 2, \cdots, d$,

$$
\begin{aligned}
\hat{F}(t)X_i^d(t) &= \int_0^t \hat{F}(s-)dX_i^d(s) + \int_0^t X_i^d(s-)dU(s) \\
(3.17) \qquad &\quad + \int_0^t X_i^d(s-)\widehat{\mathcal{J}F}(s)dA(s) + [U^d, X_i^d](t) - \langle U^d, X_i^d\rangle(t) + \langle U^d, X_i^d\rangle(t) \\
&= \int_0^t X_i^d(s-)\widehat{\mathcal{J}F}(s)dA(s) + \langle U^d, X_i^d\rangle(t) + J(t) \\
&= \int_0^t X_i^d(s-)\widehat{\mathcal{J}F}(s)dA(s) + \int_0^t \psi_i(s)K(s, I(i))dA(s) + J(t).
\end{aligned}
$$

From (3.16), (3.17), and following the argument that led to (3.14), we have $\psi = \hat{\beta}$, $P \times dA - a.s.$ Finally, (3.7) follows from (3.8), (3.9), (3.14) and (3.17). ◇

PART III

Optimization in a market with jumps in the price system

1. Introduction.

In this part, we study a financial market which consists of a bond and several stocks. The bond is a risk-free asset and the stocks are risky. The fluctuations in the stock prices are caused by numerous factors of the real world. If all of these factors are relatively insignificant, the Invariance Principle of Donsker (cf. Theorem 2.4.20 in [18]) suggests that their effects can be thought of as being caused by a Brownian motion, and therefore, the price system is then governed by a number of stochastic differential equations with respect to a Brownian motion. This kind of financial market, in which the price process is continuous and the market is complete, is studied thoroughly, for example, in [17] or the survey paper [16] by Karatzas. In reality, however, some abnormal events do happen in the process of trading, and their effects cannot be neglected. In this case, jumps will occur in the price process.

Many papers investigate financial markets with jumps in the price system; for example, [1], [4], [19] and recently [15] and [8]. In [1], the investigation of the market is based on a random measure, which is introduced in order to characterize the price jumps. For the purpose of obtaining more explicit results, the authors of [15] introduce a Brownian motion and an independent Poisson process as the driving processes, and use stochastic integrals with respect to martingales rather than random measures, to describe the market with price jumps. In this way, the portfolio process can be regarded as part of the integrands, and one can expect to obtain results which parallel those of [16], [17]. The key point here is the martingale representation property. The purpose of this part of our work is to use the results in the first part of this paper to generalize the model of [15] to the d-dimensional case. After our paper had been completed, we became aware of the existence of paper [21], in which the author deals with a model similar to ours. The martingale representation property established here can be used in [21]. We believe that our approach provides the necessary martingale representation results for this kind of models.

We approach this market by introducing a Lévy process $X \in \mathcal{L}$ (cf. Definition I.4.2) as the driving process, and use the natural filtration generated by X itself. If $X \in \mathcal{L}$, then the magnitude of all the jumps of X is 1. Moreover, if $i \neq j$, the quadratic co-variation $[X_i, X_j]$ is equal to zero, i.e. $X_i X_j$ is a martingale, and X_i and X_j have no common jumps. Essentially, each component of such a Lévy process is then the sum of a Brownian motion and a compensated Poisson process. In Part I, we proved the martingale representation theorem for this kind of Lévy processes, so that we can use it here to study the market on a solid basis. In this part of our paper, we prefer to introduce a *mixture of a Brownian motion and a multivariate compensated Poisson process* (cf. Definition I.4.3) as our driving process because of its simplicity. In fact, the method used here can also be applied in a financial market in which the price system is driven by a PII X, but we shall not go through the details. Foldes [8] studies an optimization problem for a special utility function in such a market. In the model formulated below, we shall see that the continuous martingale part and the purely discontinuous martingale part of each component of the Lévy process, are regarded as different sources of uncertainty in the market.

Our focus here is to establish the market model, to construct the equivalent martingale measure, and to show how the method used in [16] in studying the market without jumps in the price system can be applied here smoothly.

2. The model.

Suppose $(\Omega, \mathcal{F}, P)$ is a probability space, M is a d-dimensional mixture of a Brownian motion and a multivariate compensated Poisson process, i.e., $M = (W, N)^{tr}$ where W is an m-dimensional Brownian motion, and N is an n-dimensional compensated multivariate Poisson process with intensity $\lambda_i(t)$ (for notational simplicity, we shall assume that $\lambda_i(t) \equiv K(I(i)) \equiv 1)$, $i = 1, \cdots, n$; that is, $\tilde{N}(t) \triangleq \Sigma_{s \le t} \Delta N(s), t \ge 0$ is an n-variate Poisson Process and $N_i(t) = \tilde{N}_i(t) - t$, $t \ge 0$ is a martingale for $i = 1, \cdots, n$. Note that

$$d = m + n$$

and

$$M_i = \begin{cases} W_i, & \text{if } i = 1, \cdots, m; \\ N_{i-m}, & \text{if } i = m+1, \cdots, d. \end{cases} \tag{2.1}$$

Recall the random measure

$$\mu^N(dt, dx) = \Sigma_{s \ge 0} 1_{\{\Delta N(s) \ne 0\}} \varepsilon_{\{s, \Delta N(s)\}}(dt, dx); \quad \nu^N(dt, dx) = K(dx)dt$$

is the dual predictable projection of μ^N. The filtration $\{\mathcal{F}_t\}$ is generated by M, that is,

$$\mathcal{F}_t \triangleq \cap_{s > t} \mathcal{F}_s^0; \quad \mathcal{F}_s^0 \triangleq \sigma(M(r); \ r \le s).$$

Note that N is a purely discontinuous martingale, and we know from (I.2.8)

$$N(t) = x * (\mu^N - \nu^N)(t), \tag{2.2}$$

and

$$\int_0^t g(s)^{tr} \, dN(s) = (g^{tr} x) * (\mu^N - \nu^N)(t) \tag{2.3}$$

for an n-dimensional locally bounded predictable process g.

There are $d+1$ assets in the market. One of them is a bond, and its price evolves according to the differential equation

$$dp_0(t) = p_0(t) r(t) \, dt, \quad 0 \le t \le T; \quad p_0(0) = p_0, \tag{2.4}$$

where $T > 0$ is a fixed real number. The remaining assets are risky stocks; their price processes are given by the following linear differential equations:

$$dp_i(t) = p_i(t-)[b_i(t) \, dt + \Sigma_{j=1}^m \psi_{ij}(t) \, dW_j(t) + \Sigma_{j=1}^n \phi_{ij}(t) \, dN_j(t)]; \quad 0 \le t \le T \tag{2.5}$$

with $p_i(0) = p_i$, $i = 1, 2, \cdots, d$. We assume that the *interest rate process* r *of the bond*, the *appreciation rate processes* b_i *of the stocks*, and the *stock volatilities* ψ_{ij}, ϕ_{ij} are bounded, predictable processes for each i, j. Denote

$$\psi(t) = (\psi_{ij}(t))_{d \times m}; \quad \phi(t) = (\phi_{ij}(t))_{d \times n}$$

and

$$\sigma(t) = (\psi(t) \vdots \phi(t))_{d \times d}. \tag{2.6}$$

2.1 Lemma: *(i) For each $i = 1,2,\cdots,d$, the equation (2.5) has a unique strong solution given by*

$$(2.7)\quad p_i(t)=p_i(0)\exp\Big\{\int_0^t b_i(s)\,ds + \Sigma_{j=1}^m \int_0^t \psi_{ij}(s)dW_j(s) - \frac{1}{2}\Sigma_{j=1}^m\int_0^t \psi_{ij}^2(s)ds - \Sigma_{j=1}^n\int_0^t \phi_{ij}(s)ds\Big\}\cdot\Pi_{s\le t}\Pi_{j=1}^n[1+\phi_{ij}(s)\Delta N_j(s)].$$

(ii) For every $i = 1,2,\cdots,d$ the process p_i is locally bounded, and for any given $n > 0$,

$$(2.8)\qquad E\,|p_i^n(t)| \le e^{at},\quad 0\le t\le T,\quad \textit{for some } a>0.$$

Proof: Consider the semimartingale

$$Y_i(t) = \int_0^t b_i(s)\,ds + \Sigma_{j=1}^m \int_0^t \psi_{ij}(s)\,dW_j(s) + \Sigma_{j=1}^n \int_0^t \phi_{ij}(s)\,dN_j(s).$$

Then the equation (2.5) can be written as $dp_i(t) = p_i(t-)dY_i(t)$, and the proof of the Lemma is similar to Theorem I.6.1. ◇

From (2.7), we can see the one of the main differences between price systems with and without jumps is that, in a price system without jumps, the price of the stocks will not be negative, but in the price system with jumps, the price may be negative, i.e. the stocks may crash down. In order to avoid this possibility, we need

2.2 Assumption: $\phi_i(t) > -1,\quad 0\le t\le T;\quad i=1,\cdots n.$

3. An equivalent martingale measure.

We shall impose the conditions, that the matrix $\sigma(t)\sigma^{tr}(t)$ be strongly nondegenerate (Assumption 3.1). This will imply the existence of a unique probability measure $\widetilde{P}$, equivalent to P, under which discounted stock prices are martingales; cf. (3.5) and Lemma 3.2.

The existence of such an "equivalent martingale measure" guarantees the absence of arbitrage opportunities in models of this sort (cf. [10], [16], [22]); its uniqueness amounts to the property of "completeness" of the model, which is taken up in Section 4.

3.1 Assumption: With the notation $\sigma(t) = (\psi(t)\vdots\phi(t))_{d\times d}$ of (2.6), the matrix $\sigma(t)\sigma^{tr}(t)$ is strongly non-degenerate; i.e., there exists a number $\epsilon > 0$ such that

$$(3.1)\qquad \xi^{tr}\sigma(t)\sigma(t)^{tr}\xi \ge \epsilon\|\xi\|^2,\quad \forall\,\xi\in\mathcal{R}^d,\quad 0\le t\le T$$

holds almost surely. ◇

This implies

$$\|\sigma^{-1}(t)\xi\| \le \frac{1}{\sqrt{\epsilon}}\|\xi\|$$

and

$$\|\sigma^{-1}(t)^{tr}\xi\| \leq \frac{1}{\sqrt{\epsilon}}\|\xi\|$$

for every $\xi \in \mathcal{R}^d$ and every $(t,\omega) \in \mathcal{R}^+ \times \Omega$. Define

$$\zeta_i(t) = -\big(\sigma^{-1}(t)(b(t) - r(t)\underset{\sim}{1})\big)_i, \quad i = 1,\cdots,m \tag{3.2}$$

and

$$\eta_i(t) = -\big(\sigma^{-1}(t)(b(t) - r(t)\underset{\sim}{1})\big)_{i+m}, \quad i = 1,\cdots,n, \tag{3.3}$$

where $\underset{\sim}{1} = (1,\cdots,1)^{tr}$. From Assumption 3.1 and the boundedness of b and r, σ, the processes ζ and η are bounded. Define Z by the following stochastic differential equation:

$$dZ(t) = Z(t-)[\Sigma_{i=1}^m \zeta_i(t)\,dW_i(t) + \Sigma_{i=1}^n \eta_i(t)\,dN_i(t)], \quad Z(0) = 1. \tag{3.4}$$

Then by Theorem I.6.1, Z is a martingale. Introduce the probability measure

$$\widetilde{P}(A) = \int_A Z(T)\,dP, \qquad A \in \mathcal{F}_T. \tag{3.5}$$

From Theorem I.6.3, the processes

$$\widetilde{M}_i(t) \triangleq W_i(t) - \int_0^t \zeta_i(s)ds = W_i(t) + \int_0^t \big(\sigma^{-1}(s)[b(s) - r(s)\underset{\sim}{1}]\big)_i ds, \quad i = 1,\cdots,m; \tag{3.6}$$

and

$$\widetilde{M}_{m+i}(t) \triangleq N_i(t) - \int_0^t \eta_i(s)ds = N_i(t) + \int_0^t \big(\sigma^{-1}(s)[b(s) - r(s)\underset{\sim}{1}]\big)_{m+i} ds, \quad i = 1,\cdots,n, \tag{3.7}$$

are martingales on $[0,T]$, under $\widetilde{P}$. The equations (3.6) and (3.7) can be written in vector form:

$$\widetilde{M}(t) = M(t) + \int_0^t \sigma^{-1}(s)[b(s) - r(s)\underset{\sim}{1}]\,ds, \tag{3.8}$$

and with

$$D(x) \triangleq \begin{pmatrix} x_1 & 0 & \dots & 0 \\ 0 & x_2 & \dots & 0 \\ \vdots & \vdots & \ddots & \vdots \\ 0 & 0 & \dots & x_d \end{pmatrix}, \quad \forall\, x \in \mathcal{R}^d, \tag{3.9}$$

the equation (2.5) is rewritten in vector form as

$$\begin{aligned} dp(t) &= D(p(t-))[b(t)\,dt + \sigma(t)\,dM(t)] \\ &= D(p(t-))[r(t)\underset{\sim}{1}\,dt + \sigma(t)\,d\widetilde{M}(t)]. \end{aligned} \tag{3.10}$$

It can be seen that

$$d(\beta(t)p_i(t)) = \beta(t)p_i(t-)\sigma(t)\, d\widetilde{M}(t), \tag{3.11}$$

where

$$\beta(t) \triangleq \exp\{-\int_0^t r(s)\, ds\} \tag{3.12}$$

Therefore, the discounted price process βp_i is a $\widetilde{P}$-local martingale. Actually we have the following stronger result.

3.2 Lemma: *The discounted price process* $\{\beta(t)p_i(t)\}_{0\le t\le T}$ *is a* $\widetilde{P}$*-martingale, for every* $i = 1, 2, \cdots, d$.

Proof: From (3.11), βp_i is a $\widetilde{P}$-local martingale. By (2.8), (I.6.2) and the boundedness of ϕ and ψ we can prove that $\widetilde{E}(\beta(t)p_i(t))^2 < \infty$ as in the proof of Theorem I.6.3 (ii), and thus the $\widetilde{P}$-local martingale βp_i is a $\widetilde{P}$-martingale. ◇

3.3 Remark: Lemma 3.2 justifies the terminology *equivalent martingale measure* for $\widetilde{P}$. ◇

4. The completeness of the market.

Let us consider now an economic agent, who invests in these assets of (2.4), (2.5), and whose decisions cannot affect the prices in the market. We shall denote by $\pi_i(t)$ the amount that he invests in the ith stock ($i = 1, 2, \cdots, d$) at time t, by $c(t)$ the rate at which he withdraws money for consumption, and by $X^{\pi,c}(t)$ the *wealth* corresponding to his strategy pair (π, c) at that time. We assume that the *portfolio* process π and the *consumption rate* process c are predictable, take values in $\Re^d$ and $[0, \infty)$, respectively, and satisfy $\int_0^T [\|\pi(s)\|^2 + c(s)]\, ds < \infty$ almost surely. Then $X^{\pi,c}$ is governed by the differential equation

$$\begin{aligned} dX^{\pi,c}(t) &= [X^{\pi,c}(t-) - \Sigma_{i=1}^d \pi_i(t)]r(t)\, dt - c(t)\, dt + \Sigma_{i=1}^d \pi_i(t)\, dp_i(t)/p_i(t-) \\ &= [r(t)X^{\pi,c}(t-) - c(t)]dt + \pi(t)^{tr}[b(t) - r(t)\underline{1}]dt + \pi(t)^{tr}\sigma(t)dM(t) \\ &= [r(t)X^{\pi,c}(t-) - c(t)]\, dt + \pi(t)^{tr}\sigma(t)\, d\widetilde{M}(t). \end{aligned} \tag{4.1}$$

The equation (4.1) has a unique strong solution (cf. Theorem V.7 in [20]), and it is easy to check that this solution is given by

$$\beta(t)X^{\pi,c}(t) = x - \int_0^t \beta(s)c(s)\, ds + \int_0^t \beta(s)\pi(s)^{tr}\sigma(s)\, d\widetilde{M}(s). \tag{4.2}$$

It is thus clear that the sum of the 'discounted' wealth and the updated 'discounted' consumption $\beta(\cdot)X^{\pi,c}(\cdot) + \int_0^{\cdot} \beta(s)c(s)\, ds$ is a $\widetilde{P}$-local martingale.

4.1 Definition: Suppose (π, c) is given and $X^{\pi,c}(t) \ge 0,\ \ 0 \le t \le T$, *a.s.* Then this strategy (π, c) is called *admissible.* ◇

For fixed $x > 0$, denote

$$\mathcal{A}(x) \triangleq \{(\pi, c) : \ (\pi, c) \text{ is admisssible for the initial wealth } x\}.$$

Suppose that $(\pi, c) \in \mathcal{A}(x)$; then

$$\beta(t)X^{\pi,c}(t) + \int_0^t \beta(s)c(s)\,ds = x + \int_0^t \beta(s)\pi(s)^{tr}\sigma(s)\,d\widetilde{M}(s), \qquad 0 \le t \le T$$

is a nonnegative $\widetilde{P}$-local martingale, therefore a $\widetilde{P}$ supermartingale, and

$$\widetilde{E}[\beta(T)X^{\pi,c}(T) + \int_0^T \beta(s)c(s)\,ds] \le x. \tag{4.3}$$

To study the completeness of the market, let us define

(i) $\mathcal{C}(x) \triangleq \{c:\ c$ is a consumption rate process with $\widetilde{E}\int_0^T \beta(s)c(s)ds \le x\}$;

$\mathcal{D}(x) \triangleq \{c:\ c \in \mathcal{A}(x)$ and $\widetilde{E}\int_0^T \beta(s)c(s)\,ds = x\}$.

(ii) $\mathcal{L}(x) \triangleq \{B:\ B \ge 0$ and is $\mathcal{F}_T$ – measurable with $\widetilde{E}\,(\beta(T)B) \le x\}$;

$\mathcal{M}(x) \triangleq \{B:\ B \in \mathcal{L}$ and $\widetilde{E}\,(\beta(T)B) = x\}$.

4.2 Theorem: *(i) For any $c \in \mathcal{C}(x)$, there exists a portfolio process π, such that $(\pi, c) \in \mathcal{A}(x)$. Moreover, if $c \in \mathcal{D}(x)$, then π is unique and $X^{\pi,c}(T) = 0$, a.s.*
(ii) For any $B \in \mathcal{L}(x)$, there exists some $(\pi, c) \in \mathcal{A}(x)$, such that $X^{\pi,c}(T) = B$, a.s. Moreover, if $B \in \mathcal{M}(x)$, then (π, c) is unique and c is identically zero.

Proof: (i). Suppose that $c \in \mathcal{C}(x)$, and introduce the $(\widetilde{P}, \{\mathcal{F}_t\})$-martingale

$$u(t) = \widetilde{E}\left[\int_0^T \beta(s)c(s)\,ds \,\middle|\, \mathcal{F}_t\right] - \widetilde{E}\left[\int_0^T \beta(s)c(s)\,ds\right].$$

The martingale representation result (Theorem I.6.3) shows that u can be represented as a stochastic integral with respect to the process $\widetilde{M}$ of (3.8):

$$u(t) = \int_0^t f(s)^{tr}\,d\widetilde{M}(s) = \int_0^t \beta(s)\pi(s)^{tr}\sigma(s)\,d\widetilde{M}(s)$$

for some predictable process f and an associated portfolio process π, which is defined by $\pi(t) \triangleq \beta^{-1}(s)\sigma^{-1}(s)^{tr}f(s)$. According to (4.2),

$$\begin{aligned}\beta(t)X^{\pi,c}(t) &= \int_0^t \beta(s)c(s)\,ds + \int_0^t \beta(s)\pi(s)^{tr}\sigma(s)\,d\widetilde{M}(s) \\ &= x - \widetilde{E}\left[\int_0^T \beta(s)c(s)\,ds\right] + \widetilde{E}\left[\int_t^T \beta(s)c(s)\,ds \,\middle|\, \mathcal{F}_t\right] \ge 0,\end{aligned} \tag{4.4}$$

since $c \in \mathcal{C}(x)$. Therefore, $(\pi, c) \in \mathcal{A}(x)$. Suppose, furthermore, that $c \in \mathcal{D}(x)$; we have then from (4.4) that $\widetilde{E}\,\beta(T)X^{\pi,c}(T) = 0$, and this implies $X^{\pi,c}(T) = 0$, a.s.

Now, we are going to prove the uniqueness of the strategy pair. Suppose $(\pi^1, c) \in \mathcal{A}(x)$ is another strategy, then, from (4.3), we also have $X^{\pi^1,c}(T) = 0$, a.s. Moreover, $\beta(t)X^{\pi,c}(t) + \int_0^t \beta(s)c(s)\,ds$ and $\beta(t)X^{\pi^1,c}(t) + \int_0^t \beta(s)c(s)\,ds$ are not only

$\widetilde{P}$-supermartingales, but also $\widetilde{P}$-martingales because their expectations are constant (equal to x). Hence

$$v(t) \triangleq \int_0^t \beta(s)(\pi(s) - \pi^1(s))^{tr}\sigma(s)\, d\widetilde{M}(s) \tag{4.5}$$

is a $\widetilde{P}$-martingale, and it is easy to see that $v(0) = v(T) = 0$ *a.s.*, which implies that v is identically equal to zero. Thus, both

$$v^c(t) = \int_0^t \beta(s)\Sigma_{i=1}^d(\pi_i(s) - \pi_i^1(s))\Sigma_{j=1}^m \sigma_{ij}(s)\, dW_j(s)$$

and

$$v^d(t) = \int_0^t \beta(s)\Sigma_{i=1}^d(\pi_i(s) - \pi_i^1(s))\Sigma_{j=1}^n \sigma_{i,m+j}(s)\, dN_j(s)$$

are identically equal to zero, whence Thus we have that

$$0 = \langle v^c, v^c\rangle(t) = \Sigma_{i=1}^d\Sigma_{j=1}^m\Sigma_{k=1}^d \int_0^t \beta^2(s)(\pi_i(s) - \pi_i^1(s))\sigma_{ij}(s)\sigma_{kj}(s)(\pi_k(s) - \pi_k^1(s))ds,$$

and

$$\begin{aligned} 0 = \langle v^d, v^d\rangle(t) &= \Sigma_{i=1}^d\Sigma_{j=m+1}^d\Sigma_{k=1}^d \int_0^t \beta^2(s)(\pi_i(s) - \pi_i^1(s))\sigma_{ij}(s)\sigma_{kj}(s)\times \\ &\quad \times (\pi_k(s) - \pi_k^1(s))\, d\langle N_{j-m}, N_{j-m}\rangle(s) \\ &= \Sigma_{i=1}^d\Sigma_{j=m+1}^d\Sigma_{k=1}^d \int_0^t \beta^2(s)(\pi_i(s) - \pi_i^1(s))\sigma_{ij}(s)\sigma_{kj}(s) \times (\pi_k(s) - \pi_k^1(s))\, ds. \end{aligned}$$

Summing the two equalities together,

$$\begin{aligned} 0 &= \Sigma_{i=1}^d\Sigma_{j=1}^d\Sigma_{k=1}^d \int_0^t \beta^2(s)(\pi_i(s) - \pi_i^1(s))\sigma_{ij}(s)\sigma_{kj}(s)(\pi_k(s) - \pi_k^1(s))\, ds \\ &= \int_0^t \beta^2(s)\|(\pi - \pi^1)(s)\sigma(s)\|^2\, ds, \end{aligned}$$

and consequently $\pi(t) = \pi^1(t)$, *a.e.* $dt \times dP$. We have uniqueness in this sense; (ii) can be proved in much the same way. ◇

5. Optimization problems.

The optimization problems of maximizing expected utility from terminal wealth, or from consumption, or from both, are studied thoroughly in [17], [16], in the context of a complete market without price jumps. The core of the successful method employed there, is the completeness of the market and the martingale representation theorem for the martingales under the new equivalent martingale measure. In our model, the prices exhibit jumps, which complicate the situation. Although we have the martingale representation result of Jacod [12], this involves stochastic integrals with respect to a random measure, which are not easy to handle. In this paper, we have proved the martingale representation theorem (without involving stochastic integrals with respect to a random measure) for any Lévy process belonging to the class $\mathcal{L}$ of Part I, and have shown that the market model with price jumps introduced in Section 2 is complete (Theorem 4.2). In this way we can apply the method of [16] to deal the optimization problem without difficulties. As an example, we shall solve the problem of maximizing expected utility from terminal wealth; other such problems are handled similarly.

5.1 Definition: A function $U : (0,\infty) \to \mathcal{R}$ is called a *utility function,* if it is strictly increasing, strictly concave and continuously differentiable, with $U'(\infty) \triangleq \lim_{c\to\infty} U'(c) = 0$, $U'(0+) \triangleq \lim_{c\downarrow 0} U'(c) \leq \infty$ and $U(0+) \triangleq \lim_{c\downarrow 0} U(c) \geq -\infty$.

The inverse $I : [0, U'(0)] \overset{\text{onto}}{\to} [0,\infty]$ of the strictly decreasing function $U' : [0,\infty] \overset{\text{onto}}{\to} [0, U'(0)]$ will be extended by continuity on all of $[0,\infty]$: $I(y) = 0$, $\forall\ y \in [U'(0), \infty]$. It is then easily checked that

$$U(I(y)) \geq U(c) + y[I(y) - c]; \ \forall\ 0 \leq c < \infty,\ 0 < y < \infty. \tag{5.1}$$

We want to solve the following optimization problem:

For a given initial capital $x > 0$ and a given utility function U, to maximize the expected utility from terminal wealth

$$EU(X^{\pi,0}(T)) \tag{5.2}$$

over the class $\mathcal{A}_1(x) = \{(\pi, 0) \in \mathcal{A}(x);\ EU^-(X^{\pi,0}(T)) < \infty\}$.

A pair $(\hat{\pi}, 0) \in \mathcal{A}_1(x)$ which achieves the supremum of the expression in (5.2) over $(\pi, 0) \in \mathcal{A}_1(x)$ is called *optimal.* The supremum is denoted by $V(x)$ and is called the *value function* of the problem.

Define

$$\mathcal{X}(y) \triangleq \widetilde{E}\,[\beta(T) I(y \cdot \beta(T) Z(T))], \quad 0 < y < \infty. \tag{5.3}$$

5.2 Theorem: *Fix $x > 0$. Suppose that there exists some $y_0 \in (0,\infty)$, such that $\mathcal{X}(y_0) < \infty$. Then there exists a unique $\hat{\pi}$ such that $(\hat{\pi}, 0) \in \mathcal{A}_1(x)$ is optimal for maximization of utility from terminal wealth. In particular,*

$$\widehat{X}(T) = I(y^* \cdot \beta(T) Z(T)) \tag{5.4}$$

is the unique optimal level of terminal wealth, where y^* is determined by (5.5) below. Moreover, $\widehat{X}(T) \in \mathcal{M}(x)$ and $\hat{\pi}$ is given by Theorem 4.2.

Proof: Define

$$\bar{y}_0 = \inf\{y :\ \mathcal{X}(y) < \infty\} (\geq 0)$$

and

$$\bar{y}_1 = \sup\{y :\ \mathcal{X}(y) \text{ is strictly decreasing on } (0, y)\}.$$

Notice that $\bar{y}_1 > y_0$. Then $\mathcal{X}(\bar{y}_0) \triangleq \lim_{y\to y_0} \mathcal{X}(y) = \infty$, and $\mathcal{X}(\bar{y}_1) = 0$. Let $\mathcal{Y} : [0,\infty] \to [\bar{y}_0, \bar{y}_1]$ be the inverse of $\mathcal{X}$. Since $\mathcal{Y}$ is strictly decreasing on $[\bar{y}_0, \bar{y}_1]$ and continuous on $(\bar{y}_0, \bar{y}_1]$, there exists y^*, such that

$$y^* = \mathcal{Y}(x). \tag{5.5}$$

Choosing $\widehat{X}(T)$ as in (5.4), we have

$$\widetilde{E}[\widehat{X}(T)\beta(T)] = \widetilde{E}\,\beta(T)I(y^*\beta(T)Z(T))] = x, \tag{5.6}$$

hence $\widehat{X}(T) \in \mathcal{M}(x)$. Now for any $\pi \in \mathcal{A}_1(x)$ such that $X^{\pi,0}(T) \neq \widehat{X}(T)$, *a.s.*, by the property (5.1) of a strictly concave function,

$$\begin{aligned} E\,U(\widehat{X}(T) &> E\,[U(X^{\pi,0}(T) + y^*\beta(T)Z(T)I(y^*\beta(T)Z(T)) \\ &\quad - y^*\beta(T)Z(T)X^{\pi,0}(T)] \\ &= E\,U(X^{\pi,0}(T)) + y^*x - y^*\widetilde{E}\,[X^{\pi,0}(T)\beta(T)] \\ &\geq E\,U(X^{\pi,0}(T)). \end{aligned}$$

Therefore, $E\,U(\widehat{X}(T)) \geq V(x)$. Since $\widehat{X}(T) \in \mathcal{M}(x)$, there exists a unique $\hat{\pi}$ such that $(\hat{\pi}, 0) \in \mathcal{A}_1(x)$ and $X^{\pi,0}(T) = \widehat{X}(T)$, *a.s.* by Theorem 4.2. This $(\pi, 0)$ is optimal. ◇

Concluding remarks

In this paper, we have established a martingale representation theorem for a class of PII $\mathcal{I}$ and in particular for a class of Lévy processes $\mathcal{L}$ and applied it successfully in filtering theory and in trading. It is still unknown whether it would be possible to extend this result to an even larger class of processes.

In Part III, we have solved the optimization problem of maximizing expected utility in a complete financial market. If a financial market is incomplete, or we have only partial observations of the market available, this optimization problem is still open.

Acknowledgements

The author is very grateful to Professor Ioannis Karatzas for his constant encouragement and numerous helpful suggestions and comments on this paper. He is also grateful to Professor Daniel Ocone for his helpful remarks of the original version of this paper and an anonymous referee for many useful remarks. Thanks are also due to Professors Monique Jeanblanc-Picqué and Monique Pontier, and to Professor Hiroshi Shirakawa, for advancing preprint versions of their papers [15] and [21], respectively. This research was supported in part by the National Science Foundation, under grants DMS-87-23078 and DMS-90-22188.

References

[1]. AASE, K. K. & ØKSENDAL, J. Admissible investment strategies in continuous trading. *Stochastic Processes and Their Applications* **30**, 291–301. 1988.

[2]. BRÉMAUD, P. *Point Processes and Queues: Martingale Dynamics.* Springer-Verlag, New York, 1981.

[3]. BRÉMAUD, P. & YOR, M. Changes of filtrations and of probability measures. *Z. Wahr. verw. Gebiete.* **45**, 269–295. 1978.

[4]. COX, J. & ROSS, S. The valuation of options for alternative stochastic processes. *J. Financial Economics* **3**, 145–166. 1976.

[5]. DAVIS, M. H. A. Martingales of Wiener and Poisson processes. *J. London Math. Soc. (2)* **13**, 336–338. 1976.

[6]. DELLACHERIE, C. & MEYER, P. *Probabilities and Potential B: Theory of Martingales.* North-Holland, New York. 1982.

[7]. ELLIOTT, R. Double martingales. *Z. Wahr. verw. Gebiete.* **34**, 17–28. 1976.

[8]. FOLDES, L. Optimal sure portfolio plans. *Mathematical Finance* Vol. **1** No. **2**, 15–55. 1991.

[9]. FUJISAKI, M. KALLIANPUR, G. & KUNITA, H. Stochastic differential equations for the nonlinear filtering problem. *Osaka J. Math.* **9**, 19–40. 1972.

[10]. HARRISON, J. H. & PLISKA, S. R. Martingales and stochastic integrals in the theory of continuous trading. *Stochastic Processes and Their Applications* **11**, 215–260. 1981.

[11]. IKEDA, N. & WATANABE, S. *Stochastic Differential Equations and Diffusion Processes.* North-Holland, New York. 1981.

[12]. JACOD, J. A general theorem of representation for martingales. *Proc. Symp. Pure Math.* **31**, 37–53. 1977.

[13]. JACOD, J. & SHIRYAEV, A. N. *Limit Theorems for Stochastic Processes* Springer-Verlag, New York. 1987.

[14]. JACOD, J. & YOR, M. Étude des solutions extrémales et représentation intégrale des solutions pour certains problèmes de martingales. *Z. Wahr. verw. Gebiete.* **38**, 83–125. 1977.

[15]. JEANBLANC-PICQUÉ, M. & PONTIER, M. Optimal portfolio for a small investor in a market with discontinuous prices. *Appl. Math. & Optimization* **22**, 287–310. 1990.

[16]. KARATZAS, I. Optimization problems in the theory of continuous trading. *SIAM J. Control & Optimization* **27**, 1221–1259. 1989.

[17]. KARATZAS, I., LEHOCZKY, J. & SHREVE, S. Optimal portfolio and consumption decisions for a "small investor" on a finite horizon. *SIAM J. Control & Optimization* **25**, 1557–1586. 1987.

[18]. KARATZAS, I. & SHREVE, S. *Brownian Motion and Stochastic Calculus.* Springer-Verlag, New York. 1988.

[19]. MERTON, R. C. Option pricing when underlying stock returns are discontinuous. *J. Financial Economics* **3**, 125–144. 1976.

[20]. PROTTER, P. *Stochastic Integration and Differential Equations: A New Approach.* Springer-Verlag, New York. 1990.

[21]. SHIRAKAWA, H. Optimal dividend and portfolio decisions with Poisson and diffusion type return processes. Preprint. 1990.

[22]. STRICKER, C. Arbitrage et lois de martingale. *Ann. Inst. Henri Poincaré* **26**, 451-460. 1990.

[23]. XUE, X. X. *The martingale representation theorem for a class of Lévy processes, and its applications.* Doctoral Dissertation, Columbia University, May 1991.

[24]. YAN, J. A. A propos de l'intégrabilité uniforme des martingales exponentielles. *Lecture Notes in Mathematics* **920**, 338–347. Springer-Verlag, New York. 1982.

Lecture Notes in Control and Information Sciences

Edited by M. Thoma and A. Wyner

Vol. 137: S. L. Shah, G. Dumont (Eds.)
Adaptive Control Strategies for Industrial Use
Proceedings of a Workshop
Kananaskis, Canada, 1988
VI, 360 pages. 1989

Vol. 138: D. C. McFarlane, K. Glover
Robust Controller Design Using Normalized Coprime Factor Plant Descriptions
X, 206 pages. 1990

Vol. 139: V. Hayward, O. Khatib (Eds.)
Experimental Robotics I
The First International Symposium
Montreal, June 19-21, 1989
XVII, 613 pages. 1990

Vol. 140: Z. Gajic, D. Petkovski, X. Shen
Singularly Perturbed and Weakly Coupled Linear Control Systems
A Recursive Approach
VII, 202 pages. 1990

Vol. 141: S. Gutman
Root Clustering in Parameter Space
VIII, 153 pages. 1990

Vol. 142: A. N. Gündeş, C. A. Desoer
Algebraic Theory of Linear Feedback Systems with Full and Decentralized Compensators
V, 176 pages. 1990

Vol. 143: H.-J. Sebastian, K. Tammer (Eds.)
System Modelling and Optimization
Proceedings of the 14th IFIP Conference
Leipzig, GDR, July 3-7, 1989
X, 960 pages. 1990

Vol. 144: A. Bensoussan, J. L. Lions (Eds.)
Analysis and Optimization of Systems
Proceedings of the 9th International Conference, Antibes, June 12-15, 1990
XII, 992 pages. 1990

Vol. 145: M. B. Subrahmanyam
Optimal Control with a Worst-Case Performance Criterion and Applications
X, 133 pages. 1990

Vol. 146: D. Mustafa, K. Glover
Minimum Entropy H_∞ Control
IX, 144 pages. 1990

Vol. 147: J. P. Zolésio (Ed.)
Stabilization of Flexible Structures
Third Working Conference,
Montpellier, France, January 1989
V, 327 pages, 1990

Vol. 148: In preparation

Vol. 149: K. H. Hoffmann, W. Krabs (Eds.)
Optimal Control of Partial Differential Equations
Proceedings of the IFIP WG 7.2 International Conference, Irsee, April 9-12, 1990
VI, 245 pages. 1991

Vol. 150: L. C. G. J. M. Habets
Robust Stabilization in the Gap-topology
IX, 126 pages. 1991

Vol. 151: J. M. Skowronski, H. Flashner, R. S. Guttalu (Eds.)
Mechanics and Control
Proceedings of the 3rd Workshop on Control Mechanics in Honor of the 65th Birthday of George Leitmann
January 22-24, 1990, University of Southern California
IV, 497 pages. 1991

Vol. 152: J. D. Aplevich
Implicit Linear Systems
XI, 176 pages. 1991

Vol. 153: O. Hájek
Control Theory in the Plane
X, 272 pages. 1991 .

Vol. 154: A. Kurzhanski, I. Lasiecka (Eds.)
Modelling and Inverse Problems of Control for Distributed Parameter Systems
Proceedings of IFIP (W.G. 7.2)-IIASA Conference
Laxenburg, Austria, July 24-28, 1989
VII, 179 pages. 1991

Vol. 155: M. Bouvet, G. Bienvenu (Eds.)
High-Resolution Methods in Underwater Acoustics
V, 249 pages. 1991

Lecture Notes in Control and Information Sciences

Edited by M. Thoma and A. Wyner

Vol. 156: R. P. Hämäläinen, H. K. Ehtamo (Eds.)
Differential Games –
Developments in Modelling and Computation
Proceedings of the Fourth International Symposium on Differential Games and Applications
August 9-10, 1990, Helsinki University of Technology, Finland
XIII, 292 pages. 1991

Vol. 157: R. P. Hämäläinen, H. K. Ehtamo (Eds.)
Dynamic Games in Economic Analysis
Proceedings of the Fourth International Symposium on Differential Games and Applications
August 9-10, 1990, Helsinki University of Technology, Finland
XIII, 311 pages. 1991

Vol. 158: K. Warwick, M. Kárný, A. Halousková (Eds.)
Advanced Methods in Adaptive Control for Industrial Applications
X, 331 pages. 1991

Vol. 159: X. Li, J. Yong (Eds.)
Control Theory of Distributed Parameter Systems and Applications
Proceedings of the IFIP WG 7.2 Working Conference, Shanghai, China, May 6–9, 1990
VIII, 219 pages. 1991

Vol. 160: P. V. Kokotović (Ed.)
Foundations of Adaptive Control
IX, 525 pages. 1991

Vol. 161: L. Gerencsér, P. E. Caines (Eds.)
Topics in Stochastic Systems:
Modelling, Estimation and Adaptive Control
IV, 401 pages, 1991

Vol. 162: C. Canudas de Wit (Ed.)
Advanced Robot Control
IX, 314 pages, 1991

Vol. 163: V. L. Mehrmann
The Autonomous Linear Quadratic Control Problem
Theory and Numerical Solution
VI, 177 pages, 1991

Vol. 164: I. Lasiecka, R. Triggiani
Differential and Algebraic Riccati Equations with Application to Boundary/Point Control Problems: Continuous Theory and Approximation Theory
XI, 160 pages, 1991

Vol. 165: G. Jacob, F. Lamnabhi-Lagarrigue (Eds.)
Algebraic Computing in Control
Proceedings of the First European Conference
Paris, March 13–15, 1991
IX, 385 pages. 1991

Vol. 166: L. L. M. van der Wegen
Local Disturbance Decoupling with Stability for Nonlinear Systems
V, 135 pages, 1991

Vol. 167: M. Rao
Integrated System for Intelligent Control
VIII, 133 pages. 1992

Vol. 168: P. Dorato, L. Fortuna, G. Muscato
Robust Control for Unstructured Perturbations – An Introduction
VI, 117 pages. 1992

Vol. 169: V. M. Kuntzevich, M. Lychak
Guaranteed Estimates, Adaptation and Robustness in Control Systems
IV, 209 pages. 1992

Vol. 170: J. M. Skowronski, H. Flashner, R. S. Guttalu (Eds.)
Mechanics and Control
Proceedings of the 4th Workshop on Control Mechanics, January 21–23, 1991
University of Southern California, USA
IV, 301 pages. 1992

Vol. 171: P. Stefanidis, A. P. Papliński, M. J. Gibbard
Numerical Operations with Polynomial Matrices
Application to Multi-Variable Dynamic Compensator Design
VIII, 205 pages. 1992

Vol. 172: H. Tolle, E. Ersü
Neurocontrol
Learning Control Systems
Inspired by Neuronal Architectures and Human Problem Solving Strategies
X, 211 pages. 1992

Vol. 173: W. Krabs
On Moment Theory and Controllability of One-Dimensional Vibrating Systems and Heating Processes
VII, 174 pages. 1992

Lecture Notes in Control and Information Sciences

Edited by M. Thoma and A. Wyner

Vol. 174: A.J.M. Beulens, H.-J. Sebastian (Eds.)
Optimization-Based Computer-Aided
Modelling and Design
Proceedings of the First Working Conference
of the IFIP TC 7.6 Working Group,
The Hague, The Netherlands, 1991
VIII, 270 pages, 1992

Vol. 175: E. Rogers, D.H. Owens
Stability Analysis for Linear Repetitive Processes
VII, 197 pages, 1992

Vol. 176: B.L. Rozovskii, R.B. Sowers (Eds.)
Stochastic Partial Differential Equations
and Their Applications
Proceedings of IFIP WG 7/1 International Conference
University of North Carolina at Charlotte, NC
June 6 - 8, 1991
VIII, 251 pages, 1992

Vol. 177: I. Karatzas, D. Ocone (Eds.)
Applied Stochastic Analysis
Proceedings of a US-French Workshop,
Rutgers University, New Brunswick, N.J.
April 29 - May 2, 1991
X, 311 pages, 1992